Science Dept

£8.75

A Laboratory Manual for Schools and Colleges

John Creedy, B.Sc., M.I.Biol.

Heinemann Educational Books
London

Heinemann Educational Books Ltd
LONDON EDINBURGH MELBOURNE AUCKLAND TORONTO
HONG KONG SINGAPORE KUALA LUMPUR NEW DELHI
NAIROBI JOHANNESBURG LUSAKA IBADAN
KINGSTON

ISBN 0 435 57130 3

Published by Heinemann Educational Books Limited
48 Charles Street, London W1X 8AH

Filmset by Keyspools Limited, Golborne, Lancs.
Printed in Great Britain by Morrison and Gibb Limited,
London and Edinburgh

Preface

The management of a good laboratory in a modern educational system, with the varying requirements for project work, places heavy demands on the reference material available in most school and college laboratories, and there is an increasing need for a manual which places between two covers much of the information which is needed to run a good laboratory.

My objective has not been to replace the many specialized monographs already available, but rather to produce a reference work which is very broadly based, yet which has depth and detail in those areas where I have felt this to be necessary. To keep such a book to a reasonable length the author must make some arbitrary decisions, and a number of sections (such as glassworking) have been kept to a basic level which will cover the work likely to be carried out in all but a handful of school laboratories. Where the scope has been restricted in this way however, I have tried to suggest suitable references for those who may seek more information.

Another consideration when compiling this manual has been the widely differing backgrounds of its potential users, which will range from the untrained assistant to the highly-skilled technician and scientist. I would ask the indulgence of highly-skilled technicians for those sections which are necessarily written at an introductory level; I trust that they will find much of value elsewhere in the book. The very basic section on hand tools should enable even the least experienced technician to select and use the correct tool for each job, and thereby to grow in confidence.

Safety in laboratories has long been a concern of mine, and I make no apology for chapters 1 and 2 being devoted to this essential aspect of laboratory work. With the implementation of the 1973 Health and Safety at Work Act, teachers and technicians are deeply concerned about safety matters.

My thanks are due to those organizations and persons who have kindly consented to the reproduction of material in this book. I am particularly grateful to Graham Taylor of Heinemann Educational Books for the help which he has given me on the organization and presentation of the book. Above all, I acknowledge my debt to the highly-skilled technicians with whom I have had the good fortune to work over a period of many years. I would thank particularly Ted Gaite of Haberdashers' Aske's School, Elstree, and Frank Taylor and Christopher Walker of the Department of Education and Science Laboratory, Knockholt, from whose skills I have learned so much about the 'laboratory arts'.

1977   J.C.

Contents

List of Tables

SI Units

The Système International d'Unités, or SI system, was agreed in 1960 as a coherent and rational system of units for all scientific work; it replaces the previous M.K.S. system (Meter, Kilogram, Second). *Coherent* means that the product or quotient of any two units is another unit in the system, e.g. unit of length × unit of area = unit of volume ($1 m \times 1 m^2 = 1 m^3$). Such a system precludes the use of incoherent units such as acres or gallons. *Rational* means that equations in electricity and magnetism are brought within the single system of units.

The SI system is based on seven units.

metre The length equal to 1 650 763.73 wavelengths in vacuum of the radiation corresponding to the transition between two chosen energy levels in the krypton-86 atom.

kilogram Equal to the mass of the international prototype of the kilogram. The kilogram is the only remaining unit based on a prototype and it is expected that this will be replaced by a definition based on the inertia of the proton.

second The duration of 9 192 631 770 periods of the radiation corresponding to the transition between two chosen energy levels of the ground state of the caesium-33 atom.

ampere The intensity of a constant current which, when maintained in two parallel rectilinear conductors of infinite length and negligible cross section, and placed 1 metre apart in vacuum, would produce between these conductors a force equal to 2×10^{-7} newton per metre of length.

kelvin The fraction 1/273.16 of the thermodynamic temperature of the triple point of water.

candela The luminous intensity, in a perpendicular direction, of a surface of 1/600 000 square metre of a black body at the temperature of freezing platinum under a pressure of 101 325 newton per square metre.

mole The amount of substance of a system which contains as many elementary entities as there are atoms in 0.012 kilogram of carbon-12. When the mole is used, the elementary entities must be specified, e.g. atoms, molecules, ions, electrons.

The base units and their symbols are:

Physical quantity	SI base unit	Symbol
length	metre	m
mass	kilogram	kg
time	second	s
electric current	ampere	A
thermodynamic temperature	kelvin	K
luminous intensity	candela	cd
amount of substance	mole	mol

In addition to these base units there are two supplementary (dimensionless) units in the system:

| plane angle | radian | rad |
| solid angle | steridian | sr |

By division and multiplication of these nine units, derived units can be obtained for all other physical quantities, for example:

area	square metre	m^2
volume	cubic metre	m^3
velocity	metre per second	$m s^{-1}$
acceleration	metre per second per second	$m s^{-2}$
density	kilogram per cubic metre	$kg m^{-3}$
pressure	newton per square metre	$N m^{-2}$
luminance	candela per square metre	$cd m^{-2}$

For a comprehensive list of derived units for physical quantities, consult *SI Units, Signs, Symbols and Abbreviations* published by the Association for Science Education, College Lane, Hatfield, Herts.

Some derived SI units have special names and symbols; a list of those likely to be encountered in schools is on page xii.

Conventions when using SI units

1. *Names* of units are always written with a small letter, e.g. watt, farad.
2. *Symbols* are always written with small letters, unless they are named after a person, e.g. watt W, luminous flux lm.
3. Symbols do not have a plural form – they name the unit in which the preceding magnitude is measured, e.g. 3.6 kg **not** 3.6 kgs.
4. When unit symbols are multiplied together, the product may be represented: mV, or m.V, or m·V, or m × V, but mV is preferred.

Physical quantity	Derived unit	Symbol	Unit expressed in terms of base or derived unit
frequency	hertz	Hz	s^{-1}
force	newton	N	$kg\,m\,s^{-2}$
pressure and stress	pascal	Pa	$N\,m^{-2}$
work, energy, quantity of heat	joule	J	$N\,m$
power	watt	W	$J\,s^{-1}$
quantity of electricity	coulomb	C	$A\,s$
electric potential, potential difference, electromotive force	volt	V	$J\,C^{-1}$
electric capacitance	farad	F	$C\,V^{-1}$
electric resistance	ohm	Ω	$V\,A^{-1}$
flux of magnetic induction, magnetic flux	weber	Wb	$V\,s$
magnetic flux density (magnetic induction)	tesla	T	$Wb\,m^{-2}$
inductance	henry	H	$Wb\,A^{-1}$
luminous flux	lumen	lm	$cd\,sr$
illumination	lux	lx	$lm\,m^{-2}$

5. A combination of units as a quotient may be expressed as: $m\,s^{-1}$, or m/s. The index notation is preferred to the solidus, but for complex expressions the use of not more than one solidus may improve the clarity, e.g. $cm\,s^{-1}/V\,cm^{-1}$ instead of $cm^2\,V^{-1}\,s^{-1}$, but *never* cm/s/V/cm.

6. The decimal sign between the digits in a number should be a point on the line, or a comma, but never a centralised dot, e.g. 33.3 or 33,3, but never 33·3.

7. Numbers of more than four digits on either side of the decimal sign should be arranged in groups of three digits for easy reading, but never separated by a dot or a comma unless this indicates the decimal place, e.g. 1 743 348.13 not 1,743,348.13; 64.452 815 98 not 64.452,815,98.

8. The prefixes in the table opposite may be used to indicate decimal sub-multiples or multiples of both base and derived units.

9. It is usual to choose a unit for a physical quantity such that its numerical value is between 0.1 and 1000. This preferred value range may be exceeded in drawing up tables, to preserve a constant unit.

10. Only base units or their derived units should be used in calculations; in only a few cases is it permissible to use multiples or sub-multiples.

Multiple or sub-multiple	Prefix	Symbol
10^{12}	tera	T
10^9	giga	G
10^8	mega	M
10^3	kilo	k
10^2	hecto	h
10	deca	da
10^{-1}	deci	d
10^{-2}	centi	c
10^{-3}	milli	m
10^{-6}	micro	μ
10^{-9}	nano	n
10^{-12}	pico	p
10^{-15}	femto	f
10^{-18}	atto	a

Conversion into SI units

Some common conversion factors are given on page 215 to enable existing data in Imperial units to be converted into data in SI units. Much time in calculation can often be saved by the use of conversion tables such as 'Imperial/Metric Conversion Tables' published by Heinemann Educational Books.

1. Laboratory Management and Safety

These two subjects are considered together in this chapter because they are so interlinked. A well-managed laboratory is a safe place in which to work but there can be few places more dangerous than a badly organized and ill-managed laboratory.

Whenever a laboratory or workshop accident is investigated, a predisposing factor is usually found; at the time of the accident human error has been added to this predisposing factor. In the teaching situation, the teacher and technician have control over only two of the sources of human error in the laboratory, and only partial control over the other thirty. In such circumstances, the best hope of maintaining safe laboratories lies in meticulous attention to the *prevention* of hazardous situations; the essential ingredients of accident prevention are:

1. **Good housekeeping**
2. **Knowledge**
3. **Good storage**
4. **Safe working procedures**

In this chapter, no pretence is made at an exhaustive account of safety and laboratory management matters, but a perusal of the points described below will perhaps indicate to the reader some areas where more thought is required in their own laboratory.

FIRE FIGHTING

The fire fighting equipment provided in schools is for *first aid use only* and can only be expected to tackle minor fires. Partly-used extinguishers must be recharged after use. **It is essential to understand that the portable extinguishers provided will only cover at the most about two square metres of fire** – in the event of a larger outbreak caused, say, by dropping a winchester of flammable liquid, the building should be evacuated **immediately**, and the fire brigade summoned. Firemen should be informed immediately on arrival if gas cylinders, particularly oxygen, hydrogen, or acetylene, are stored in the fire area. People are rarely killed by fire, **they die first from asphyxiation by smoke and fumes**. Always evacuate a building as soon as the fire gets out of control or threatens the escape route. If a running solvent fire threatens other solvent bottles or gas cylinders it **must** be left to the professional fireman.

All laboratories should be provided with:

A carbon dioxide extinguisher
A fire blanket
A sandbucket, with a long handled scoop, in all areas where sodium, potassium, or phosphorus are used. The function of *dry* sand is to smother sodium and phosphorus fires to allow time to evacuate the laboratory. When fires such as these occur, the fire brigade **must** be called. The only type of extinguisher effective on strongly burning sodium is a low pressure graphite powder type which is not supplied to schools – burning sodium will split CO_2 into oxygen and soot!

The following types of extinguisher may be found in school laboratories:

Foam—good for small fires of immiscible solvents such as petrol, benzene, and oil.

Tetrachloromethane (carbon tetrachloride)— **unsuitable** because poisonous fumes, including carbonyl chloride (phosgene), can be evolved. Tetrachloromethane will detonate sodium and potassium.

Soda acid—cannot be used on electrical fires, or near electrical supplies, or on sodium, potassium, magnesium, or calcium fires. No effect on solvent fires other than causing them to spread. **Unsuitable for laboratories.**

Carbon dioxide—the best general-purpose type of extinguisher for school laboratories since it is effective on most types of small fire. It has no cooling effect however, and re-ignition may occur when the gas disperses.

Dry powder—good for electrical fires.

BCF—excellent for all-round use, but many fire officers regard them as unsuitable for use by amateurs as the vapour has some degree of toxicity.

STORAGE AND STOCK CONTROL SYSTEMS

Most laboratories seem to suffer from poorly-designed storage systems, often because the architect has spent the available money on providing unnecessary floor space rather than an adequate storage capacity. If it is necessary to store materials and equipment on bench tops or the floor, this is an architectural fault – often due to the architect being inadequately briefed. Preparation rooms are usually dual purpose storage and preparation areas, yet rarely does the storage capacity of the room exceed 10 per cent of its total volume. By the use of mobile shelving and industrial storage systems it is possible to increase the storage capacity to around 50 per cent without loss of preparation area. Small equipment and components should always be stored in efficient, stackable, plastic drawers or bins.

In a large school there is much to be said for keeping a card index of all equipment and materials. A stock book which gives a shelf or cupboard number will prevent confusion when the technician is absent.

Chemicals and materials should be purchased in quantities which will be turned over in one to two years. The storage of thirty-year stocks of such substances as phosphorus and bromine is not only uneconomic, but also thoroughly irresponsible. For most reagents a six-month to one-year supply should be stocked and the technician should be given a reordering level so that replacements can be ordered in good time. The storage and handling of chemicals, *as well as disposal techniques*, are fully described in a publication of the Institute of Science Technology entitled *The Care, Handling, and Disposal of Dangerous Chemicals* by P. J. Gaston (obtainable from Northern Publishers (Aberdeen) Ltd., 11 Albyn Terrace, Aberdeen). No school should be without a copy of this excellent handbook. An excellent introduction to school laboratory safety is to be found in a pamphlet entitled *Safety in the Science Laboratory* (Department of Education and Science, 1973).

Storage of chemicals

1 Store containers in *single rows*, **never** banked.
2 Do **not** store on *high shelves*; keep corrosives below shoulder height.
3 Do **not** use *glass stoppers for alkalis* – polythene stoppers are best.
4 *Bottles* should never be completely filled. (Except for substances which form explosive peroxides during storage – as they are used up they should be transferred to smaller containers, or glass beads should be added to displace the air.)
5 *Chemicals known to deteriorate on storage* should be purchased in small quantities and the *date of purchase should be marked on the container*.
6 All *containers* should bear the *full name* of the substance they contain (see Appendix on Organic Nomenclature), its concentration, and any hazards that may be involved. Labels should be suitably protected by varnish or 'lasso tape'.

7 *Shelving* should be strong. If traffic vibration is a problem, shelves should be lipped, although this in itself can be a hazard when picking up bottles.
8 Store chemicals *systematically* – alphabetical order is probably best, with scheduled poisons stored in a separate locked cupboard.
9 Any *chemicals which react violently* should not be stored close together. For example, oxidizing agents such as nitric acid, chromic acid, potassium permanganate, hypochlorites, and peroxides should be separated both vertically and horizontally from organic substances such as alcohols, glycerol, sugar, cotton waste, wood, and paper.
10 *Floor-level (ideally, lead-lined) troughs* help to reduce the risk of running fires if large bottles of flammable solvents are kept.
11 Chemical storage areas should be supplied with *permanent, cool ventilation*, and fitted with external switchgear.
12 All chemical stocks should be *examined regularly*; any which show signs of deterioration should be disposed of with great care. Chemicals which have no further use, or for which the use is not known, should be disposed of by the recognised methods.
13 *Yellow phosphorus* should be stored under water.
14 *Sodium, lithium, and potassium* should be stored under a hydrocarbon oil, such as naphtha oil.
15 *Bromine* is best purchased in ampules of a size which will be used up in individual experiments.
16 A number of *organic solvents* and *carbon disulphide* have *very low flashpoints*, boiling points, and auto-ignition points. These are often combined with a wide range of explosive mixtures in air (see Table 15.6, page 226). Ethanol (acetaldehyde), propanone (acetone), ethanoic acid (acetic acid), benzene, petroleum ether, and ethoxyethane (diethyl ether) all need careful storage. Carbon disulphide, which is used for a couple of trivial experiments in many schools, is perhaps the most hazardous – it has a flashpoint of $-27°C$, boils just above room temperature, and has an auto-ignition temperature of $100°C$ i.e. the vapour can ignite on contact with the surface of an electric light bulb, or even with steam. Carbon disulphide forms an explosive mixture in air if the concentration is between 2 per cent and 50 per cent. *The dropping of a bottle in a laboratory with naked lights would result in an instantaneous explosion.*
17 Chemicals which deteriorate through oxidation (such as photographic chemicals) may be stored in Air Evac bottles (available in various sizes

from Pelling and Cross Ltd, 104 Baker Street, London W1) which can 'concertina' to expel air as the level of chemical falls, so that the liquid level is always up to the neck of the bottle. The bottles are brown to prevent further deterioration by exposure to light.

Organisation of the technicians' work to service teaching requirements

As far as is possible, the work of technicians should be planned over a twelve-month period so that major equipment servicing, storage reorganisation, and routine requisitioning of materials and reagents can be carried out during the low pressure periods of the school holidays. If several technicians are employed, a holiday rota will mean that there is always someone present to receive incoming goods and to feed and clean livestock.

In very small establishments it may not be essential to have formal systems for teaching staff to order their requirements, but it is usually desirable to set up a system which will enable the laboratory technician to plan ahead and to spot overlapping requirements. Any such system imposes upon the teaching staff the duty to give due warning to technicians of their requirements – if possible planning a week ahead.

A simple system which is operated in many schools is for staff to enter their requirements in a *requisition book* kept in the preparation room. In all but the smallest establishments this system suffers from the drawback that several staff and the technician may all need to use the book at the same time.

A more sophisticated system, such as the one used at Thomas Bennett School, Crawley, depends on a visual display board with a space for each period of the week. The staff complete request slips which are placed in a box. The slips are then checked by a technician and attached to the appropriate part of the visual board. Such a system helps to prevent double booking and gives the technicians a quick visual check on the demands to be made on them. Slips can be completed anywhere and sent to the preparation room – a boon in split premises. Notes on coloured paper can also be pinned to the board to indicate disruptions to normal routine, such as speech days or examinations. Clothes pegs or spring clips hung on a nail in each space are a convenient way of holding the requests. The sketch of the board (Figure 1.1) and sample request slip (Figure 1.2) will clearly indicate how the system operates.

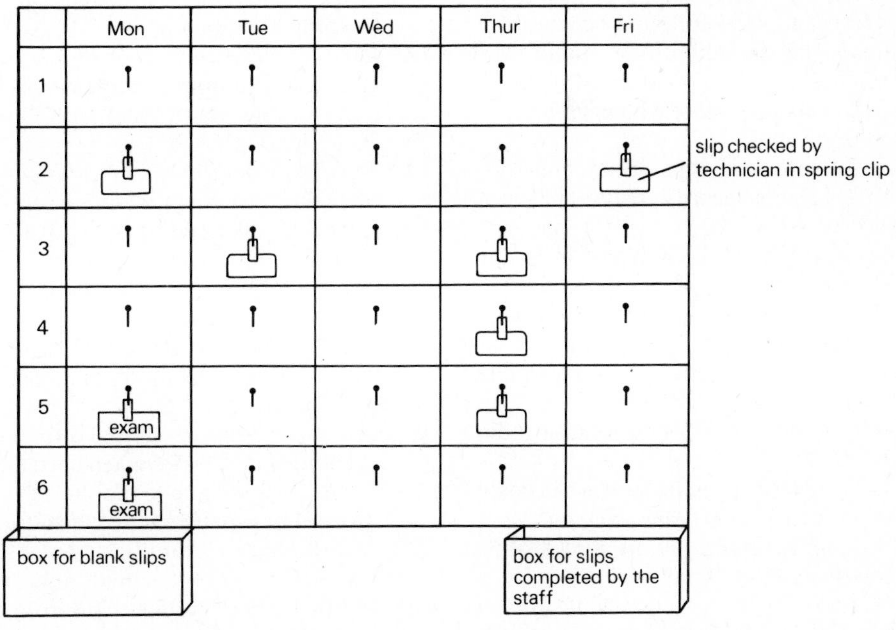

Figure 1.1 A visual requirements board

```
Name _____          Mon. Tues. Wed. Thurs. Fri.          Date _____
                            (please ring)

Lab. _____ Class ____    Expt. No. [      ]          Periods 1 2 3 4 5 6
                                                        (please ring)
                            No. of sets [      ]
                            required

                                         Other requirements
    audio visual equipment
    required. Please mark ✓  Equipment        Reagents        Biological material

    lecture theatre
    loop projector Std. 8 mm
    "        "      super 8
    dual guage 8 (not loop)
    16 mm projector
    slide projector
    overhead projector
    microprojector
```

Figure 1.2 A sample teaching requirements slip

DAY-TO-DAY LABORATORY MANAGEMENT

1 All adults using the laboratory should know the *position of gas and water main stopcocks* and of the *main electrical switch* for pupil benches.

2 All persons using the laboratory should know the *whereabouts of the fire-fighting appliances* and of the *first aid kit* – all of which should be in prominent positions.

3 Unless a member of staff is present, laboratories where any forseeable hazards are present should be kept *locked*.

4 *Toxic chemicals* should be handled in a fume cupboard, with protective gloves if necessary.

5 Use a *safety screen* for all experiments where there is the slightest risk of explosion. Use two screens if the demonstrator works behind the bench.

6 Wear a *simple dust mask* when hand-grinding chemicals.

7 Always carry *lengths of glass tubing* vertically. Handle them with a cloth when working.

8 Keep *sinks and waste traps* clean, and disinfect them regularly.

9 Always carry *large bottles* such as winchesters in a proper carrier.

10 Bring only small quantities of *highly flammable solvents* (not more than 500 cm^3) at a time into teaching laboratories. The main stocks should not leave the reserve chemical store.

11 Apparatus not in immediate use should be *put away – untidyness* is often the predisposing factor in accidents.

12 Do not place *flammable material* near exits.

13 If an accident does occur, *clear up thoroughly at once*. Fragments of glass can be collected with plasticine.

14 Always dispose of *broken glass and swarf* from workshop machines in special waste bins, **never** in the waste paperbasket, to avoid injury to unsuspecting colleagues.

15 Always *label bottles* with the full name and the strength of their contents (see Appendix on Organic Nomenclature). Do not use abbreviations. Hazards should be clearly marked on all containers.

16 *Corrosive chemicals* should only be brought into teaching laboratories when needed. They should never be stored above the shoulder height of the shortest person using the laboratory.

17 Always make certain that *carboys and winchesters* are properly vented and/or have loose stoppers when moving them from cool stores to warm preparation rooms. The thought of forty litres of concentrated acid releasing itself on the preparation room floor is not amusing.

4

18 *Winchesters of 0.88 s.g. aqueous ammonia* must be kept cool at all times. When opening a new bottle, do so in a fume cupboard with the glass drawn down far enough to protect the eyes, or wear a safety visor. There is almost always a slight spray as the bottle is opened.

19 Return any *empty bottles* or dispose of them safely. They are hazardous if they clutter the store, and even though apparently empty, they may contain a small amount of vapour in an explosive mixture with air.

20 Do not store *reagents* in direct sunlight.

21 Use a cork borer to insert *glass tubing* into a rubber bung.

22 Always wear *protective clothing* when applicable – but remember that good practice is the surest defence. Technique should never be 'sloppy' because a labcoat is being worn. Wear a well-fitting coat, correctly buttoned up, and if you work with highly flammable organic solvents, choose a cotton one. The static electricity from a nylon coat can ignite flammable atmospheres. *Cotton labcoats absorb far more spillage than nylon ones before liquid reaches the garment below.*

23 *Machines in the preparation room* must have the appropriate guards which are required by common sense, and by the Factory Acts. Machines should be sited so that the operator is not working in a 'traffic' area. Loose clothing, ties, and long hair can be lethal near machines. Never use a grindstone without eye protection – grit and metal particles may leave the periphery of the wheel at a velocity of 100 km h^{-1}.

24 The peripheral speed of a *centrifuge* with 10 cm rotors running at 5000 r.p.m. is 180 km h^{-1}. Do **not** try to slow it by hand!

25 Do not charge *accumulators* near naked flames or cigarettes.

26 When *diluting strong acids* add very small quantities of acid to water, **never water to acid**. Wear protective goggles, or better still a visor.

27 *Eating, drinking,* and *smoking* in the laboratory are hazardous. If you cannot restrain yourself, choose a less hazardous job. Do not drink from laboratory taps.

28 Always *wash your hands* before leaving the laboratory, and leave your labcoat behind. Other people using the staffroom and dining accommodation do not wish to share your chemical contaminants.

29 *Mouth pipettes* should never be used. A range of mechanical fillers and alternative measuring techniques exist. Mouth pipettes are, at best, unhygienic and fatal accidents occur every year with toxic reagents.

30 Care should be taken when setting up *vacuum or pressure equipment* – the dangers of implosion and explosion are very real. Remember that if a glass vessel implodes at high vacuum the fragments move inwards at high velocity, but if they do not meet fragments of equal mass travelling in the opposite direction, they come out the other side, and the effect is the same as if an explosion had occurred. Safety screens and visors are sensible precautions when working with pressurised or vacuum glassware.

31 *All organic chemicals* other than foodstuffs should be treated as toxic. Many solvents dissolve the protective layers of the body easily and are either rapidly absorbed, or cause dermatitis, or both

Water emergencies

It is most important when setting up apparatus which uses running water, particularly if it is to run unattended or overnight, to fasten all hoses and rubber tubing with wire or screw-clips. Remember that most apparatus is set up at times of peak water consumption, and in many areas water pressure can double in a matter of minutes as factories close for the day. Also make quite certain that the drainage system is foolproof – flooding of a floor can lead to fires though shorting of electrical conductors. Some tap washers expand and reduce the flow of water when left for long periods.

Equipment left overnight

Never set up equipment to run overnight without considering the consequences of any possible supply failure. Such equipment **must** be designed on the 'fail safe' principle.

Gas cylinders

For identification colours of gas cylinders see Table 15.8, page 228.

1 *Stiff valves* should be treated with caution; **never** use wrenches or hammers. Cylinders with stiff valves which cannot be turned by hand should be returned unused to the supplier with a covering note.

2 Cylinders, valves, pressure reducers, gauges, etc. for *combustible gases* have left-hand thread fittings; those for *non-combustible gases* have a right hand thread outlet.

3 Cylinders must not be used without *pressure regulators* – this is very important if the gas is being passed into glass vessels.

4 Do not store cylinders where there are *flammable solvents*. On one occasion when a fire enveloped propane cylinders they burst with such force that the 'shrapnel' travelled half a mile.

5 Remember that 50 per cent of all accidents involving gas cylinders are caused by *oxygen*. Oxygen-enriched atmospheres enormously increase fire hazards – to give but one example, clothing which smoulders in an atmosphere containing 21 per cent oxygen will burn fiercely in an atmosphere of 24 per cent oxygen.

6 *Handle cylinders with care* – a heavy blow on an ethyne (acetylene) cylinder can cause it to explode, and the valve gear of any cylinder is easy to damage. Leakage of hydrogen or oxygen could be disastrous.

7 *Valves of hydrogen cylinders* should be opened slowly. Too rapid opening may cause an explosion due to static electricity.

8 *Valve gear* must never be greased or oiled.

9 Care should be taken with *hydrogen sulphide cylinders*; hydrogen sulphide is more toxic than hydrogen cyanide gas.

10 Cylinders must be *firmly secured* at all times, either in properly designed stands or trolleys, or by chaining or clamping to the wall or bench.

ELECTRICAL SAFETY

1 The *resistance of the human body* to the passage of an electrical current can vary between a few hundred ohms and 10 000 ohms depending largely on the moisture content of the skin. At 15 mA and 50 cycles per second a.c., or 70 mA d.c., the casualty is likely to lose muscular control and be unable to remove himself from the supply. At 20 mA and 50 cycles per second a.c., or 80 mA d.c., danger to life occurs. 100 mA a.c. or d.c. is almost certain to be fatal. Even presuming a resistance of 10 000 ohms, a 240 V a.c. supply will provide a current greater than 20 mA, which could be fatal. Personal variation is enormous and the lowest voltage recorded as causing death is as low as 60 V. It will be understood that these figures apply to current passing through the body, say from hand to hand. If a finger shorts two contacts there may only be contact burns.

2 The above information makes it quite clear that *all mains electrical apparatus is potentially lethal* and that it is of paramount importance that the equipment and cables are maintained in good condition. Equipment which is intended to be earthed should **never be used without an earth connection**. In many new laboratories voltage or current earth trip devices are provided as an additional protection, but these must not be relied upon as a first line of defence against electrocution since they may well develop faults in the corrosive atmosphere of some laboratories.

3 *Large capacitors* are a source of danger even when not connected to the mains – a point worth remembering if dismantling old television sets in the laboratory. Large capacitors will recover about 10 per cent of their initial charge if left on open circuit after discharge by a phenomenum known as 'dielectric hysteresis'. As an example, a 20 kV capacitor will build up a charge of 2 kV after discharge. The discharge of 10 joules of electrical energy into the body is likely to be fatal, and as little as 0.25 joules will give a heavy shock. A shock of 10 joules would be delivered by the discharge of a capacitor charged to the voltage shown in the following table:

Capacity (μF)	Charged to (kV)
0.2	10
20	1
80	0.5
320	0.25
2000	0.1

Very large capacitors rarely, if ever, have a place in school laboratories, but if they *are* used they *must* be kept shorted when not in use.

4 *High voltages* in excess of 5 kV can produce *radiation hazards* in some experiments, and for this reason the possession of equipment capable of developing a potential difference of more than 5 kV has been made **illegal in UK schools**. Some of the large induction coils of Victorian origin still found in schools fall into this category.

5 *Rubber-insulated, cotton-covered flex* is unsuitable for school use and should be replaced.

6 The old UK, continental, and American *colour codes for cable* are all different. When fitting plugs on new equipment, check which system is in use, and make quite sure that the correct connections have been made by using a test meter before plugging into the mains.

7 *Trailing cables* and additions to the mains are dangerous practice.

8 The great majority of electrical accidents are attributable to *faulty workmanship* or *worn-out equipment* – both are avoidable.

BIOLOGICAL HAZARDS

1   Students should be warned *not to eat parts of plants* provided in the laboratory, as some may be poisonous naturally, and seeds may have been treated by the seed merchant with organic fungicides.

2   *Wild mammals* must on no account be used in the laboratory as they can transmit disease to man. Wild rodents are particularly dangerous as they carry Weal's disease, which can be **fatal**.

3   *Wild birds*, recaptured budgerigars, and pigeons are potential carriers of psittacosis which can also be **fatal** to man.

4   All *dissection instruments* should be washed in disinfectant after use.

5   *Plastic syringes* have a fascination for children. They should be broken before disposal and should never be obtained second hand.

6   When using a syringe to inject *methanal (formaldehyde)*, a safety visor should be worn in case the needle blows off.

7   *Bacteriological and fungal cultures* should be soaked in disinfectant before disposal.

8   *Primates* (monkeys) must not be kept in schools. They are dangerous carriers of disease, including rabies, hookworm, and B virus infection. The latter is normally **fatal** to man.

9   *Terrapins* are known to carry salmonella food poisoning.

10 *Preserved animal skins* should not be handled unnecessarily as they may have been preserved with arsenical compounds and mercuric chloride.

11 *Benches* in biological laboratories should be wiped down occasionally with disinfectant solutions.

12 No one should leave a biological laboratory without *washing their hands*.

Refrigerators

It is not clever to store the milk for staff tea in the biological refrigerator alongside microbiological cultures and dissection material.

The refrigerators used in school laboratories often have a micro-switch for the internal light which is not spark-proof. About three to five cm^3 of ether, propanone (acetone), or petroleum ether evaporated in a refrigerator of three cubic feet capacity will produce an explosive mixture (see Table 15.6, page 226).

DANGEROUS EXPERIMENTS

All experiments carry at least a minor risk. In the majority of cases such slight risks are eliminated if the experiment is planned and carried out in a sensible way. Some experiments however are known to be particularly dangerous, and the Department of Education and Science gives a clear indication in its publication *'Safety in the Science Laboratory'* that such experiments should not be carried out in schools:

'The making of explosive, and in particular the mixing of sulphur or phosphorus with chlorates, is forbidden by law, and for reasons of safety the making of other chlorate mixtures is strongly deprecated. Some schools have experimented with rockets and rocket fuels but because of the ease with which such materials can frequently be obtained outside the school, such experiments are also strongly deprecated. Such experiments are quite unnecessary in a school course, and others would include the preparations of carbonyl chloride (phosgene), hydrocyanic acid, cyanogen, nitrogen triiodide, the oxides of chlorine, the explosions of hydrogen and chlorine, and of ethyne (acetylene) and oxygen. Nor are sealed-tube combustions necessary in organic chemistry in schools. The preparation of oxygen from an oxygen mixture is unnecessary.'

The D.E.S. also outlaws all experiments using carcinogens, including asbestos wool, and lays down clear codes of practice for schools using radioactive materials and lasers. Any teacher carrying out these experiments would have a very weak case if involved in subsequent litigation. The D.E.S. publication also lists the following as *'requiring special care'*:

Action of water on sodium peroxide
Action of sodium on water
Action of potassium on water
Action of chlorine on ammonia solution
Burning of hydrogen
Explosion of hydrogen and oxygen
Any use of hydrofluoric acid
Qualitative organic tests involving fusions with sodium

Radioactivity

Young persons are particularly susceptible to genetic damage by ionising radiations. For this reason the Department of Education and Science has issued *Administrative Memorandum 2/76* 'The use of Ionising Radiations in Educational Establishments' which sets out precisely what is legal in schools and technical colleges, as well as establishing a code of practice.

It is important to check from time to time that the storage arrangements conform to the regulations.

Careful records of the purchase and disposal of all open sources is essential, and when an individual teacher licensed to use radioactive isotopes moves away, the attention of the LEA science adviser should be drawn to the fact. The school can only store open sources legally with the consent of the Department of Education and Science, and this consent is only given with respect to individual teachers who have attended approved training courses.

Ultraviolet radiation

Ultraviolet radiation in the wavelength band between 4^{-6} and 1^{-7} m (4000 and 100 Å) is absorbed in the outer layers of the eye, the cornea, and conjunctiva. Conjunctivitis occurs four to eight days after exposure, and last several days; the damage is not permanent, but is very painful. Ultraviolet goggles should always be worn and the light source shielded from direct view. Skin should not be exposed to ultraviolet radiation.

Lasers

Lasers of low power are permitted in schools. The Department of Education and Science has published a code of practice set out in *Adminstrative Memorandum 7/70* which should be followed implicitly.

CARCINOGENIC SUBSTANCES

A number of chemicals are known to induce cancer. Cases are recorded where exposure had only been for a few weeks. It is wise to handle all organic substances with extreme caution, avoiding skin contact and inhalation.

It is **illegal** to keep, use, or manufacture the following substances in school laboratories:

Alpha- or beta-naphthylamine

Diphenyl substituted by:
  (1) at least one nitro- or primary amino-group, or by at least one nitro- and primary amino-group.
  (2) in addition to substitution as above, further substitution by halogens, methyl or methoxy-groups, but not by other groups.
  Note – The most frequently used substituted diphenyls in the laboratory are the following (and their salts):

benzidine	4-nitrodiphenyl
4-aminodiphenyl	o-toluidine
(xenylamine)	o-dianisidine

The nitrosamines

The nitrosophenols (para-nitrosophenol is regarded as safe and may be used in schools)

The nitronaphthalenes

Preparations likely to contain any of the above as impurities should be avoided. For example, alpha-naphthyl-thiorea may contain alpha and beta-naphthylamine as impurities.

Some of these substances may be present as impurities in non-Analytical grade reagents.

Certain forms of asbestos are known to cause two forms of cancer and they should be avoided in school laboratories. (Mineral wool can be used as a substitute for asbestos wool.)

DANGEROUS MIXTURES

Chemicals which by themselves present no great hazard may become extremely hazardous when mixed with other reagents. Table 1.1, although not exhaustive, should warn against many potential hazards in school chemistry departments.

CHEMICALS WITH SPECIFIC HAZARDS

The vast majority of reagents have some degree of hazard associated with them. Where they are toxic, flammable, or have an irritant vapour, this information is usually given on the manufacturer's label. Whenever an unfamiliar reagent is used the label should be read. When placing new stock of any chemicals, including ones with which you are familiar, on the storage shelves it is sound practice to check the label as new hazards become known from time to time, and manufacturers depend to a large degree upon safety warnings printed on the label to inform their customers of hazards. **It would be impossible in a manual of this size to list every hazard. In Table 1.2 many substances which are flammable or toxic are ommitted as this will be clearly marked on the bottle label.** When handling any substance which is not on the list, and about which little is known, it should be checked in one of the standard references. Many dangerous substances are not listed because they are unlikely to be found in schools.

Table 1.1. Dangerous mixtures

Substance	Hazardous when mixed with:
Water	Alkali metals and their oxides; oleum and sulphuric acid; sulphur trioxide; phosphorus(V) chloride and oxide; calcium carbide (acetylide); chlorosulphonic acid
Aluminium (particularly powdered)	Chlorates; nitrates; ammonium persulphate and water; organic compounds in solutions of nitrates and nitrites
Chlorates	Sulphuric acid; sulphur or sulphides; ammonium salts; phosphorus; picric acid and picrates; metal powders; gallic acid; easily oxidized materials such as sugar, sawdust, etc.
Chromium trioxide	Alcohols; glycerol; glacial ethanoic acid; ethanoic anhydride; other oxidizable organic materials; sulphur
Ethanoyl (acetyl) chloride	Lower aliphatic alcohols; water
Lithium aluminium hydride	Ethyl ethanoate (which has been recommended in the past for destroying lithium aluminium hydride – now known to be responsible for some explosions); water
Magnesium (particularly powdered)	Silver nitrate and water; peroxides; perchlorates; phosphates; heavy metal oxalates
Mercury(II) oxide	Sulphur
Nitric acid	Hydriodic acid; ethanol; methanol; easily oxidizable organic substances (e.g. paper, wood, sawdust, and sugar)
Nitrates	Zinc powder and water; esters; phosphorus; sodium ethanoate; tin(II) chloride
Nitrites	Sodium cyanide, potassium cyanide
Oxygen	Metallic dusts; many organic compounds and powders such as flour and starch
Peroxides (including hydrogen peroxide)	Aluminium, magnesium, or zinc powders
Perchloric acid	Very dangerous with organic compounds, particularly alcohols – **do not heat**
Phosphorus	Chlorates; nitrates; nitric acids; perchlorates
Picric acid (2,4,6-trinitrophenol)	Explosive when heated with salts of heavy metals or ammonia, metallic oxides, or by friction with oxidizing agents. Has been known to detonate when dry on contact with a metallic spatula
Potassium ferricyanide Potassium mercuricyanide	Halogens in the presence of ammonia

Table 1.2 Dangerous chemicals found in school laboratories

Reagent	Hazard
Acetaldehyde	See *ethanal*
Acrylic acid (propenoic acid)	Highly irritant
Acrylonitrile (propenonitrile)	The vapour and liquid are very toxic and may give rise to skin burns. A 3–17 per cent mixture with air is explosive
Allyl alcohol (prop-2-en-1-ol)	Toxic irritant; skin absorption; fumes may cause eye damage
Aluminium arsenide	Very toxic. Evolves arsine gas in a moist atmosphere
Aluminium carbide	Evolves highly explosive methane on contact with water
Aluminium nitrate	A very powerful oxidizing agent
Aluminium trialkyls	Instantaneous explosion on contact with water
Ammonium nitrate	Can be explosive. Fire hazard
Ammonium perchlorate	Heat or shock may cause detonation
Ammonium permanganate	A strong oxidizing agent
Amyl chloride (pentyl chloride)	Flash point 2 °C
Amyl nitrate (pentyl nitrate)	Affects vision. Explodes if heated
Aniline (phenylamine)	Highly toxic; skin absorption
Antimony and its salts	All are toxic, and are skin irritants which may cause dermatitis or an allergy
Aqua regia	Gives off suffocating, toxic nitrosyl chloride fumes. A powerful oxidizing agent; highly corrosive
Asbestos wool and powder	A cancer-inducing agent. Its use is illegal in British schools
Benzidrene (amphetamine)	Toxic
Benzene	Very toxic; skin absorption is very rapid, causing liver and kidney damage. Destruction of bone marrow may cause severe anaemia. Often contains carcinogenic substances as impurities
Benzoyl chloride (benzenecarbonyl chloride)	Damages skin and eyes
Benzoyl peroxide	A very strong oxidizing agent. May explode on touch if allowed to dry (lauryl peroxide is a safer alternative for plastic polymerisation experiments)
Benzyl halides	Highly corrosive, affecting skin, eyes, and mucous membranes
Bromine	Both liquid and vapour are very dangerous to the skin and eyes, as well as to the respiratory system. Can ignite organic waste or sawdust. **Purchase in 1 or 2.5 cm³ phials** to minimize danger when handling
Bromoform	See *Tribromomethane*
Butyl ether	Forms very explosive peroxides. Skin contact may lead to dermatitis
Butyl methyl ketone	The vapour causes depression and narcosis
Calcium	Reacts with water to release hydrogen – fire risk
Calcium carbide	Reacts with water to release ethyne (acetylene). Carbide dust is a respiratory irritant
Calcium chloride	The dust can give rise to conjunctivitis

Calcium hydride	A fire risk, as it evolves hydrogen in a moist atmosphere. Can cause dermatitis
Carbon disulphide	Vapourises freely at room temperature. A 2–50 per cent mixture with air is explosive and the self ignition temperature is 100 °C. Can be detonated by steam, the hot surface of an electric light bulb, or even by sparks from nylon clothing charged with static electricity
Catechol (benzene-1,2-diol)	Causes dermatitis and corneal burns
Chlorates of calcium, potassium, and sodium	Dangerous oxidizing agents. Shock or friction may detonate them when mixed with certain other reagents
Chlorinated compounds of benzenes, diphenyls, hydrocarbons, naphthalenes, triphenyls	Cause dermatitis, acne, ulcers, liver and kidney damage, and central nervous system damage
Chlorine	Very toxic – can cause blindness
Chlorobenzene	Flash point 29 °C
Chloroform	See *trichloromethane*
Chlorosulphonic acid	Ignites most combustible materials, and corrodes most metals with the evolution of hydrogen. Has a violent reaction with water, releasing a large amount of hydrogen chloride gas. Very corrosive on skin
Chromic(VI) acid (chromium trioxide)	Dangerous to eyes. A powerful oxidizing agent
Cresols	See *methyl phenols*
Diazonium salts	Highly explosive when dry, detonated by shock
Dimethylbenzene	See *xylene*
Ethanal (acetaldehyde)	The vapour is dangerous to the eyes and mucous membranes
Ethanoic acid, ethanoic anhydride	Cause conjunctivitis, skin burns, and dermatitis, as well as skin ulcers. The anhydride reacts dangerously with water and with sulphuric acid
Ethanoyl bromide, ethanoyl chloride	Fuming liquids, giving off toxic vapours which can cause corneal burns. Violent reactions with water or alcohol
Ether	Forms explosive peroxides on storage
Ethyl ethanoate	Toxic, affecting eyes, liver, and kidneys
Ethene	Explodes with chlorine in sunlight
Formaldehyde	See *methanal*
Formic acid	See *methanoic acid*
Furfural	The fumes attack eyes and mucous membranes
Furfuryl alcohol	Has explosive reactions with strong organic acids and dilute mineral acids
Hydrazine and its salts	Anhydrous hydrazine is especially unstable. It is very caustic, attacking skin and mucous membranes. Used as rocket fuels
Hydrochloric acid	Prolonged exposure to the vapour can lead to intestinal ulcers and dental decay
Hydrofluoric acid	**Very corrosive** – the vapour may cause blindness. Dangerous on contact with any part of the body. Dissolves glass – keep in plastic or gutta-percha bottles
Hydrogen peroxide (100 vols upwards)	A powerful oxidizing agent which may react violently with combustible materials. It may decompose with explosive violence on contact with metallic dusts, or with catalytic metals such as iron, copper, chromium, and their salts; causes skin burns.

Hypochlorites	Strongly alkaline oxidizing agents which may cause dermatitis
Lysol	A poisonous disinfectant of the phenol group. Skin absorption (best replaced by Cetavalon)
Magnesium	A serious fire hazard – water or tetrachloromethane will cause an explosion with burning magnesium – dry sand is the only safe extinguisher
Maleic anhydride (*cis*-butanedioic anhydride)	Eye damage from the vapour. Explodes on contact with even minute traces of alkali
Mercury	The vapour is particularly toxic, causing kidney damage and brain cell destruction, leading to 'mad-hatter's disease'. At room temperature mercury vaporises until equilibrium is reached at 200 times the concentration permitted in industry
Mercury fulminate	When dry, it may be detonated by friction or shock. Causes dermatitis
Methanal (formaldehyde)	Used as aqueous solution, which gives off an irritant vapour which can cause catarrh. Skin contact may lead to dermatitis
Methanoic acid	Causes acid burns, and the fumes attack eyes and mucous membranes
Methanol	Very toxic, attacking the central nervous system causing blindness
Methylbenzene (toluene)	Highly toxic: skin absorption
Methyl phenols (cresols)	The liquid or vapour is very dangerous to eyes and may cause skin burns and ulcers
α and β naphthols	Skin irritants; skin absorption can cause kidney damage
α and β naphthylamines	Carcinogenic, inducing tumours and cancer of the bladder. Use illegal in schools
Nitrobenzene	Skin absorption; inhalation very toxic
Osmium tetroxide (osmic acid)	A very dangerous vapour, absorbed by skin and by inhalation. Causes derangement of vision by depositing particles on the cornea
Oxygen (from cylinders)	Greatly increases fire hazards
Phenol (carbolic acid)	The solid, liquid, and vapour are all highly corrosive. Skin absorption is rapid; the central nervous system and all internal organs are affected. Severe skin burns may appear some time after contact – all cases of skin contact should be referred for hospital treatment
Phenolphthalein	Some people develop an allergy
Phenylamine	See *aniline*
Phosphoric(V) acid	Corrosive to eyes and skin
Phosphorus(V) oxide	Contact may cause skin burns. Has a violent reaction with water
Phosphorus (red)	A fire hazard
Phosphorus (white or yellow)	Burns spontaneously in air. Always cut under water in the preparation room, and only take the quantity to be used into the teaching laboratory Small amounts of phosphorus absorbed over long periods may cause bone damage

Picric acid (2,4,6-trinitrophenol)	Causes dermatitis, and is explosive in contact with ammonia and metals
Potassium	Has violent reactions with water, acids, and lower aliphatic alcohols. A dangerous fire hazard
Potassium chlorate	A strong oxidizing agent — fire and explosive hazard
Potassium dichromate	Known to cause dermatitis
Potassium permanganate	A strong oxidizing agent — will ignite organic substances, and explode with glycerol
Pyridine	Very toxic liquid and vapour. Absorption can lead to male sterility
Resorcinol (benzene-1,3-diol)	A systemic poison, affecting blood and nervous systems. A skin irritant, causing dermatitis
Silver nitrate	An oxidizing agent. Explosive with magnesium and water
Sodium	Has a violent reaction with water; fire hazard — cannot be extinguished with normal laboratory extinguishers — use dry sand and evacuate the building
Sodium dichromate	Harmful to eyes; the dust causes skin irritation and ulcers
Sodium perchlorate	Very unstable — fire and explosion danger
Sodium peroxide	Has a violent reaction with water, releasing oxygen — therefore a fire hazard
Sulphur dioxide	Attacks lungs and eyes
Tetrachlorethene	Causes liver damage and dermatitis
Tetrachloromethane (carbon tetrachloride)	Continuous exposure to very low concentrations can cause irreversible liver damage. Some people are allergic to it, and the vapour interferes with the functioning of the cardiac nerve, leading to heart failure
Tetrahydrofuran	An irritant, causing severe kidney damage
Thiourea	Induces carcinoma in mice
Thorium	Causes dermatitis; radioactive
Toluene	See *methylbenzene*
o-, m-, p-toluidine (methylphenylamines)	Highly toxic; skin absorption. o-toluidine is carcinogenic, and its use is illegal in schools
Tribromomethane (bromoform)	A metabolic poison causing liver damage
Trichlorobenzene	Causes liver damage
Trichloroethane	May cause corneal burns and liver damage
Trichlorethene	Causes dermatitis and liver damage. Has a violent reaction with strong alkalis
Trichloromethane (chloroform)	An anaesthetic; causes nervous aberration, and can be addictive
Trinitrophenol	See *picric acid*
Uric acid	Gives off hydrogen cyanide gas when heated
Xylene (1,2-dimethylbenzene)	Skin absorption; toxicity only slightly less than that of benzene

2. Safety in the Laboratory Workshop

Accidents in the workshop result most commonly from the following:

1 *Loose clothing or long hair* becoming caught in the revolving parts of machines, a particular hazard being the vertical drill.
2 *Failure to secure work properly* before commencing machining.
3 *Lack of care* when handling hot materials.
4 *Careless and heavy movements* in the workshop which may cause collisions or the knocking over of tools or materials. Particular care should be taken when moving near machinery.
5 *Bad housekeeping*; an untidy workshop with badly stored tools and cluttered work surfaces and floor surfaces is a serious hazard.
6 *Poorly maintained tools and equipment*. Blunt tools and machine parts which have become worn. Use of tools for purposes for which they were not intended.

PORTABLE POWER TOOLS

Portable power tools are being increasingly used. They should always be employed for their primary purpose; if attachments are used special care is necessary. Many modern drills have powerful motors, some developing up to $\frac{1}{2}$ hp, and there is a great risk that attachments, particularly if started or restarted in contact with work, can lead to an accident.

Another important aspect of safety when using portable electric drills is to remember that the tool is carrying mains electricity. Most modern portable drills are of a double insulated type and often have an external casing made of insulating material. Older patterns are not always double insulated and may well have a metal exterior casing. If these are used when the operator is standing on a concrete floor, or near a radiator water-pipe, or an earth connected machine, a considerable risk is involved. If a breakdown of the drill insulation occurs a heavy current can flow to earth through the body of the operator. It is worth noting that 50 per cent of all fatal electrical accidents which occur in industry involve the use of portable equipment.

MACHINES

Before using a machine make sure that all *loose clothing*, particularly cuffs, belts, ties, and aprons are *firmly secured*.

Always switch off machines before clearing swarf away from the work or from the cutting tool. Swarf should always be removed with a brush, never with the hands. Never clean, or attempt to adjust, moving machinery; never apply a rag or cotton waste to revolving work.

Work being machined should never be held in the fingers, but must always be secured in a chuck or vice, or, if appropriate, bolted to the drill table or the face or angle-plate of the machine in use. A block of hardwood beneath sheet material which is being drilled reduces the risk of sudden breakthrough and 'snatching'.

Never switch on a machine until all guards are securely in place.

Always check that the *chuck key* has been removed from the chuck.

When work has been secured in a machine, before switching on the machine should be slowly rotated by hand to ensure that nothing will foul the moving parts.

Only one person should be involved in the operation of a machine. When a machine is in use other people using the workshop should *keep at a safe distance*, and in particular should ensure that they are not in line with the path of any workpiece which may leave the machine by accident.

A machine should never be situated in an area where there is a *great deal of traffic*. Not only is there a danger of collision between somebody coming into the room and the machine operator, but there is also considerable risk if the operator is distracted.

Good lighting is essential. This is ideally provided by a low voltage light attachment on the lathe. If modern fluorescent lighting is used this should be satisfactory, although hazards can arise from some older fluorescent fitments because they may produce a stroboscopic effect with machine tools. When a lathe is being sited in the workshop, avoid placing it so that the operator faces a large window; such a position means that the operator may be distracted and also that the work is being lit from behind.

Goggles or safety spectacles should always be worn when using machines. This is particularly important when using lathes and grinding wheels.

A swarf guard and/or safety spectacles should always be used when a *pillar drill* is being used. If the pillar drill is fitted with a foot-operated safety switch it is essential to check regularly that swarf and woodshavings have not accumulated under the switch rendering it inoperative.

All machine switches, guards, and brakes should be checked regularly.

When any machine has been *set up by a pupil* or an inexperienced technician it is desirable that it should be *checked* by an experienced technician or teacher before the machine is switched on. Inexperienced workers should use low machine speeds until they have gained some knowledge of the work.

When material is fed through the hollow mandrel of the lathe headstock it should be *adequately guarded* at the free end; under no circumstances should excessive lengths of metal be handled in this way.

Well designed guards are essential on grinding machines and on the spindles of buffing machines. When work is being buffed a cloth or, worse still, an apron should *never* be used to hold the material. If material is too small to be held safely in the hand, it should be gripped in a hand-vice or pliers.

Wood-turning may introduce some hazards. It is inadvisable to turn blocks of wood which have been made up by gluing together smaller pieces. They are prone to failure of the adhesive and are very likely to split apart and fly across the room. Wood-turning tools should be maintained in sharp condition and stored in racks in front of or underneath the lathe. It should *never* be necessary to lean across a machine to pick up a tool.

It is unlikely that a laboratory workshop will contain a *circular saw*, but if one is fitted it is most important that the correct design of riving knife is always used. Care must always be taken with circular saws to keep the hands well away from the blade. This is particularly important when machining small pieces of material, for which appropriate push-sticks and guide-sticks are essential. If there is any doubt as to whether a small piece of work can be controlled *safely* it must *not* be machined.

Metal shears and guillotines should always have their blades locked, or the handle locked or chained, when not in use. They should always be securely bolted to a strong table or fixed bench at a height convenient to give good control.

HAND TOOLS

Blunt cutting tools are a common source of danger. Considerable pressure has to be applied to a blunt chisel before it will move through a piece of wood; if it slips from the wood, it will do so at tremendous speed, and the risk of personal injury is great.

Whenever hand tools are being used the wood or metal should be *gripped firmly* in a vice or by means of a bench hook. Particular care should be paid to the *positioning of both hands*.

Hand files should never be used without a handle.

Hammers should be checked occasionally to ensure that the head is not loose.

HOT-WORKING AREAS IN THE LABORATORY WORKSHOP

The main uses of heat in the laboratory workshop will be for glass-blowing, hard soldering, and soft soldering. *Careful siting* of the hot-working area is essential. Ideally this should be well away from the window area so that flames show up clearly and should never be underneath wall cupboards; it is desirable that the area be backed by a screen of good quality, heavy density, asbestos board and that the bench be topped with this material, or with stainless steel. (Ordinary asbestos cement sheet must be avoided in this situation as it may shatter dangerously when exposed to a flame.)

When *soft soldering*, if corrosive fluxes are used care must be taken to avoid injury to hands or to eyes.

It is essential wherever *brazing torches* or *soldering irons* are used that correctly designed stands are available to hold the tools when they are not in use.

PLANNING

When new laboratory workshop accommodation is being planned careful consideration should be given to the following points:
1 *Lighting.*
2 *Workshops should be warm*; if working conditions are not comfortable this is an increased danger when handling tools and machinery.
3 *Good ventilation is essential.*
4 *Storage of materials* requires special consideration, particularly for long materials such as metal bars, timber, and glass tubing. All these materials should be so stored that sharp and uneven edges are not sited where they may cause personal injury.

5 A workshop should always be planned to have *definite working zones*. If each zone is designed so that its tools and materials can be kept within the zone, this will reduce movement and consequently the risk of accident when people rush to fetch materials or tools which are not to hand.

6 *Fire precautions:*
Every workshop should have a *fire blanket*, *sand bucket*, and *an appropriate extinguisher*. Foam extinguishers are best avoided as they are unsuitable for use on metal fires or electrical fires.

MISCELLANEOUS CONSIDERATIONS

1 *Modern sandals* which have no strap at the heel can be regarded as dangerous in the workshop or preparation room.

2 If the workshop is used by pupils as well as by staff, it is desirable that any machine can have its isolating or on/off switch *looked in the 'off' position*.

3 It is essential that any *'V' belts and pulleys* are properly guarded.

4 Certain *imported hardwood timbers*, e.g. mansonia, dahoma, teak, and iroko, produce skin irritation in some individuals, and their use should be avoided particularly in operations which produce fine dust. This may seriously irritate the upper respiratory passages.

5 When *adhesives based on synthetic resins* are used, skin contact should be minimised and any contamination removed as soon as possible.

6 *Glass fibre particles* are very irritating to the skin, and are best removed by rinsing in a stream of cold water.

7 Many modern *finishes and adhesives* are based on highly flammable organic solvents. They should never be used in a poorly ventilated workshop, and care should be taken to ensure that *no naked flames* are present.

8 *The guard covering the gearbox of a lathe* should be fitted with a micro-switch.

9 The *casting of metals*, particularly alloys containing lead and zinc, should not be undertaken in a science workshop.

3. Educational Technology

The term *Educational Technology* is used to embrace all that was at one time covered by the phrase *Audio and Visual Aids*, but the term also implies that visual and audio materials are employed, not just as aids, but as an essential and highly planned part of the communication process. It is not possible in this manual to discuss all the professional aspects of educational technology, nor is it possible to describe in detail general photographic techniques; both of these topics are well covered in the literature already available. Rather, the aim of this chapter is to outline a few points which may be of specific interest to the technician who from time to time will be called upon to set up equipment and produce material.

SERVICING EQUIPMENT

The amount of servicing which can be undertaken will be related to the training of the technician. At the lowest level, servicing will require that routine cleaning and lubrication (only if specified by the equipment manufacturer) and the changing of spent bulbs be carried out. The competent technician will ensure that:

1 *Instruction manuals* for all equipment are filed where they can be easily found. (Duplicates can usually be purchased from the manufacturer for both current and most older models.)

2 At least one *spare bulb* is kept available for all equipment. (Do not overlook the exciter bulb for sound cine projectors which have optical sound-tracks – the type normally used.)

3 *Spare fuses* are available for all equipment – not only the two or five amp fuses for the ring-mains plug, but also the internal fuses which protect tape recorders and cine projectors.

4 All *projection lenses* are covered with a lens cap when not in use – this reduces the frequency with which the tedious task of cleaning lenses comes around.

5 A *blower brush* (available from photographic shops) and *lens cleaning tissues* are always available, and stored in a dust-free place or in a plastic bag. Never use 'clean' handkerchiefs or clothing to clean lenses as they often carry grit which will damage the blooming on a modern lens, and possibly the lens itself. The correct method of cleaning the lens is to remove as much dust as possible with the blower brush and then polish gently with the lens tissue which should be folded several times and handled in such a way that the natural grease from the fingers does not reach the lens. If microscope slide mountants such as Canada balsam or 'Micrex' contaminate microprojector lenses, they should be cleaned off immediately with a suitable solvent such as xylol.

Note that many older microscope objectives have the front lens element cemented in place with Canada balsam, so care must be taken.

6 *Spare drive belts* and *focusing belts* (if applicable) are available for ciné and loop film projectors.

7 That the *gate* of ciné and loop film projectors is always clean. Dust trapped here can do enormous damage to a film, and the cost of a colour 16 mm film often exceeds £100 per copy. A special gate cleaning brush is often provided with the projector or one can be purchased from photographic shops.

8 *Screens*, which are very easily damaged and almost impossible to clean, are rolled and put away after each use.

9 *Supplies of materials* such as acetate sheet and rolls, overlay films, and suitable pens are always available for use with the overhead projector.

10 Equipment is set up for use without trailing leads in areas where class movement occurs.

11 *Equipment leads* should be checked formally at least once per year for faulty insulation, earth wires, and anchorage. This check is a 'longstop' measure and the worthwhile technician will check cable condition almost automatically whenever equipment passes through his or her hands.

12 *Teaching notes* for films and film loops should be suitably filed.

13 All equipment should be stored in as *dust-free* a place as possible. Use cans and dustcovers when available.

SETTING-UP PROJECTION EQUIPMENT

The choice of equipment and its setting-up in school laboratories inevitably involves some degree of compromise. Often the major problem is that stray light finds its way onto the screen and degrades the projected image – this may be of little importance where line diagrams are concerned but can destroy the value of many biological photographs. As far as possible:

1 Avoid stray light falling directly onto the screen.
2 Avoid trailing cables.
3 Site the projector and screen to give good viewing to all the class, and ensure that the screen is at right angles to the axis of projection to avoid 'keystoning' (Figure 3.1).
4 Do not site overhead projectors so that the lecturer blocks the view of most of the class (Figure 3.2).
5 Complete the setting-up of equipment before the class arrives to avoid both time wastage and distraction.
6 Do not move a projector whilst the bulb is hot since hot bulbs tend to blow more readily than cold bulbs.

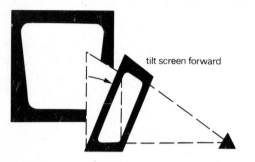

Figure 3.1 Tilt the screen to prevent keystoning with the overhead projector

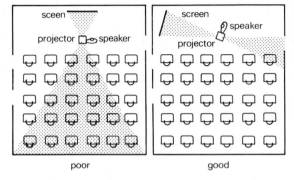

Figure 3.2 Siting the overhead projector

Table 3.1 Choice of projection lenses for cine films to produce a suitable picture size

Focal length of lens for 16 mm film in inches	Focal length of lens for 8 mm film in inches	Distance to screen from front of projector lens in feet					
1 (25.4 mm)	$\frac{1}{2}$ (13 mm)	6	9	11	14	18	21
$1\frac{1}{2}$ (38 mm)	$\frac{3}{4}$ (19 mm)	9	13.5	16.5	21	27	32
2 (51 mm)	1 (25.4 mm)	12	18	22	28	36	42
3 (76 mm)	$1\frac{1}{2}$ (38 mm)	18	27	33	42	54	63
4 (102 mm)	2 (51 mm)	24	36	44	56	72	84
Picture size in inches		27×20	40×30	52×40	63×47	80×60	96×72

Table 3.2 Choice of projection lenses for 35 mm slides to produce a suitable picture size (36×24 mm format, with the long side horizontal)

Focal length of lens in inches	Distance to screen from front of projector lens in feet							
2 (51 mm)	2.5	3	4	4.5	5	6	8	9	10.5
3 (76 mm)	3.5	5	6	7	8	9	11.5	14	16
4 (102 mm)	5	6	8	9	10	12	15	18	21
5 (127 mm)	6	8	10	12	13	15	19	23	26
6 (152 mm)	7	9.5	12	14	15	18	23	27	32
$7\frac{1}{2}$ (191 mm)	9	12	15	18	20	23	29	35	40
Projected picture width in inches	18	24	30	36	40	48	60	72	84

PREPARING VISUAL MATERIAL

The quality, and hence the impact, of all visual material stems from good preparation and an understanding of the medium used. When preparing visuals, either for the overhead projector or as 35 mm transparencies, the following points should be born in mind:

1 Keep diagrams simple and uncluttered. Prepare originals on paper and copy with transparency makers such as the 'Fordigraph' or 3Ms copier.
2 Ten lines of copy, with five to six words per line, is an absolute maximum for readability.
3 Define subjects clearly with headings.
4 A good visual communicates an idea – it is usually best to introduce one idea per visual.
5 Use large, clear lettering (see *Lettering* below).
6 Colour can be used for both emphasis and variety. In topics where standard colour codes are accepted, stick to the convention to avoid confusion – for purely visual effect, use a colour outside the code.
7 Where there is a progressive development of an idea, arrange transparencies so that they can either be progressively unmasked or added to with overlays, so that unwanted information is not present on the screen as a distraction.
8 Labelling lines should never cross each other.

Overhead projectors have one or two specialised pieces of equipment associated with them which, on occasion, may be useful in the laboratory:

Polarizing spinners enable motion to be introduced in circuit and fluid flow diagrams.

Polarizing filters, in conjunction with perspex models of bones or mechanical structures, can be used to show internal stressing.

Transparent models and equipment are available including slide rules, verniers, ripple tanks, and avometers. (Always bear in mind that, in the school situation, some of the specially produced apparatus may be of lesser value than direct demonstration.)

Fliptrans books of ready prepared diagrams in transparency form are available (Figure 3.3). The same firm provides 'Fliptran' book spines for the formation of books from specially prepared personal material.

LETTERING

Stencils

Brush stencils of the 'Econasign' type (Figure 3.4) are easy to use, and, with care, very good work can be produced. The styles are clear, and are ideal for photographic work.

For the best results with brush stencils, the brush should be well charged with *almost dry* pigment and the paper surface should be firm.

Pen stencils of the 'Uno' type produce clear accurate lettering if skillfully used (Figure 3.5). They tend to be more expensive than brush stencils. Pen

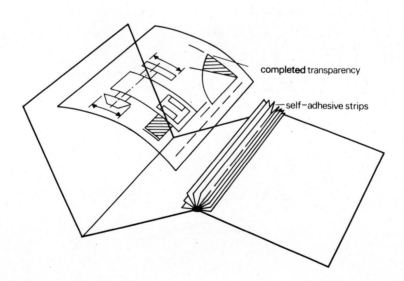

completed transparency

self–adhesive strips

Figure 3.3 'Fliptran' strip books

stencils are available in a wide range of sizes and styles, but the small sizes are particularly tricky to use as alignment is difficult and discrepancies are greatly magnified when projected. A major disadvantage with pens is their tendency to run or clog unless the operator has considerable experience in their use.

When choosing stencils, select a make in which all stencils have a common base line. This enables the user to change styles and sizes in a single line without alignment problems.

Stencils are rather easier to store than some other means of lettering.

Freehand

This is undoubtedly the cheapest method and can be very effective if the writer is at all accomplished. If the writer is not skilled in lettering, very good results which reproduce well on film may be obtained by drawing faint pencil guide-lines to regulate height, and then lettering in a bold roundhand style.

The most suitable pens are the Pelican Graphos (available in 58 nib sizes and styles), Rapidograph, felt-tip pens, or a broad-nibbed drawing pen. They can be obtained from any good stationers.

Figure 3.4 'Econasign' brush stencils

Figure 3.6 An effective roundhand style

Figure 3.5 'Uno' stencils

Instant 'Letraset'

Although more expensive than the preceding methods, 'Letraset' gives the most perfect results with rapidity and extreme ease. The finished product can be indistinguishable from good quality printing. A slide prepared by this method will show no loss of quality when projected on a large screen.

'Letraset' instant lettering is very easy to use (see Figure 3.7). The transfer is laid over any smooth surface and rubbed with a pencil or ball point pen, the letter adheres to the surface, and the backing sheet can be peeled off. Should a mistake be made, it is possible to scrape off the offending letter with a sharp knife or to lift it off with adhesive tape. This method is particularly valuable as a means of temporarily lettering black and white photographs which are to be rephotographed for slides or filmstrips. Similarly the method can be used to label museum specimens prior to photographing.

White 'Letraset' mounted on glass and photographed against a black background can be used to caption any colour transparency by double exposure (only the white letters will register on the film).

Figure 3.7 'Letrasetting'

Other methods

Charts and diagrams can be labelled with self adhesive plastic letters such as those sold in many ciné title outfits. A wide range of plastic film letters is available commercially under the name of 'Zell em' self-adhesive P.V.C. letters. (These will only adhere to a smooth surface such as plastic, glass, or gloss paint.) They are ideal for use in conjunction with a plastigraph.

Magnetic letters are now available and enable charts to be photographed with varied labelling for different age groups.

PREPARATION OF DIAGRAMS FOR PHOTOGRAPHIC REPRODUCTION

Any well-drawn diagram of sufficient contrast can be reproduced as a transparency for projection. Although wall charts can be photographed, it is preferable to use specially prepared diagrams on a standard-sized white card. If a 10 in. × 8 in. card is used and the picture area is limited to 9 in. × 6 in., the white margin will allow some latitude when setting up the camera.

Diagrams prepared on a standard sized card have two big advantages:

   a) They may be stored and rephotographed if the original slide becomes damaged or lost.
   b) A number of diagrams may be prepared and then photographed very rapidly in a ciné titler (or a simple, home-made stand) without adjustment of lighting, focus, or exposure. Once the diagrams are prepared and the camera set up a full roll of thirty six transparencies can be taken in less than half an hour.

When preparing diagrammatic material, draw a rough sketch to help eliminate all extraneous detail and to choose colours and labelling. Once the rough sketch is complete, draw the actual diagram lightly in pencil on a white card. Take care that detail is drawn sufficiently large to prevent close lines from appearing to merge when viewed at a distance – the effect must be bold if the slide is to project well. Now add the colour – a colour wash of water paint or poster paint is suitable. When the colour is completely dry, draw the outline in indian ink and add the labelling. Often indian ink labelling lines will cross dark parts of the drawing and will only be clear in the mind of the observer if they are systematically arranged, i.e. radial or parallel. Remember that direct labelling has much greater visual impact than a key system.

It will be found that certain styles of lettering are more legible, size for size, than others. This can be important if labelling space is at a premium, or if 'Letraset' is used; e.g. 'Letraset' Grotesque 216 (see Figure 3.8) in the 10 pt. size is as clear as many styles in the 24 point size. As there are more than twice as many letters on a 10 point sheet of transfers as on the larger size for the same price, it is obviously important to choose the type style and size carefully.

If pen stencils are used, the smallest sizes should be avoided as the characters do not photograph too well and a slight variation in the angle at which the pen is held will affect the alignment adversely.

If water-colour washes are found difficult, diagrams may be coloured by using self adhesive colour sheets cut to shape with a sharp blade and pressed into position. 'Plastitone' and 'Raster Silhouette Shading' films, sold by office equipment shops, provide a wide range of cross hatching and screen dots in a number of colours.

Comparative graphs and histograms may be similarly prepared, or by using a 'Plastograph' method on the 'Vitras Com-Pare Board'. The 'Com-Pare Board' is particularly useful if a series of slides are required, with additional steps or factors shown on successive slides.

PHOTOGRAPHING DIAGRAMS

If diagrams are prepared on standard 10 in. × 8 in. cards a permanent device to hold the camera, lights, and cards at the correct distances can be used. If the money is available, a Gnome Cine Titler is ideal.

For a small outlay, a similar device can be constructed from wood or metal. The essential require-

10 pt . . . **10 pt Grotesque 216 is as clear as many 24 pt typefaces**

18 pt . . . **10 pt Grotesque 216 is as clear as**

24 pt . . . **10 pt Grotesque 216 is as**

Figure 3.8   Grotesque 216

ments are careful alignment of the picture holder in relation to the camera, and a firm support for the camera. If a reflex camera is used there will be no alignment problems, and it is easy to check that the diagram almost fills the frame. If the camera is not of the reflex type, it may be accurately centred by moving the camera, on its support, up to the card holder in which has been placed a special card with a number of concentric circles at its centre; the card holder is then adjusted so that the lens mount coincides with a circle. Once alignment has been obtained by this method the camera can be slid back to a position where the viewfinder indicates a sufficiently wide field of view to cover the diagram. Few cameras have a parallax correction which is accurate at short distances, but this difficulty is negligible if a reasonable margin of white card surrounds the diagram. With non-reflex cameras, focusing can be accurately carried out by measurement. Any difficulties with precise focusing can be overcome by the use of a slow shutter speed or time exposure in conjunction with a small aperture, thus increasing the depth of field. To obtain a first class slide the whole of the frame should be filled by the diagram. This can be done on most of the cheaper cameras by means of a one or two dioptre close-up lens or by the use of extension rings on cameras with interchangeable lenses.

If only a few diagrams are required, and the acquisition of close-up devices is not considered to be worthwhile, the following method may be used instead:

1) Find out, by use of the camera viewfinder and a ruler, the exact area covered by the camera at its minimum focusing distance. (If the camera has a standard lens, focal length 50 mm, it will cover a rectangle approximately 56 cm × 36 cm when focused at 76 cm.)

2) Then prepare the diagram as a wall chart on a card slightly larger than the area covered by the lens, leaving a fairly wide margin of white card so that the diagram will fill most of the transparency without being cut off by the edge of the slide through parallax. If a short exposure is used, a tripod will not be necessary.

Lighting

The lighting must be completely even for copy work, and is best provided by the use of two number one photoflood bulbs placed one either side of the camera. As the photofloods produce a standard amount of light, the exposure will depend on the distance from the bulbs to the object. Film manufac-

turers always include an exposure table for number one photofloods on the information sheet packed with each film, or an exposure meter reading can be taken.

If colour film is used with photofloods, check that it is of the artificial light type if accurate colour rendering is important. If a colour filmstrip is made with a mixture of outdoor shots and diagrams, a daylight type film must be used, and the diagrams photographed in good sunlight. Beware of diffuse shadows from window bars, or dirty marks on the window. If a conversion filter is available, it may be used by making the exposure changes recommended by the manufacturer.

When photographing relief maps, models, and objects such as fossils, very good effects can be obtained by using strong lighting from one side of the camera only. The shadows will give 'depth' to the slide, but considerable care is necessary if colour film is used.

COPYING PHOTOGRAPHS

It is often desirable to copy black and white photographs. This provides no special problems provided that the prints have good contrast and are printed on glossy paper. Illumination must be even, and the lights adjusted so that no reflections are visible from the camera position. The ciné titler again provides an ideal means of holding lights, camera, and photograph.

By using 'Letraset' lettering, there is the possibility of adding labelling and arrows as well as a caption; in most photographs there is a predominance of dark tones, and white 'Letraset' stands out clearly. This method is particularly useful for biological and geographical aids such as soil profile photographs and for photographs of archaeological excavations.

Colour photographs and illustrations from books and magazines can be treated in the same way, but care should be taken that copyrights are not infringed. The photographic work involved is the same as for diagrams.

Note that whenever published material is copied, the publisher's consent must be obtained. This is usually freely granted for slides to be used in schools.

PHOTOGRAPHING SMALL OBJECTS

One of the most important uses of slides and film strips is to show the audience objects which are too small or too delicate for ordinary class examination. The chief methods are:

1 *Close-up photography* – the image on the transparency is the same size as, or smaller than, the original object.
2 *Macrophotography* – the image on the transparency is larger than the original object.
3 *Microphotography* – the object is so small that it is photographed with the aid of a microscope.

Close-up photographs may be taken with three types of accessory: extension rings, extension bellows, and close-up lenses. The main problems encountered are camera shake, small depth of field, and exposure. The first two can be overcome by the use of electronic flash, if it is available, or by the use of a tripod and an aperture not larger than f/8.

Exposure presents a special problem in close-up photography. When an object is brought inside the normal focusing range of a lens by the use of extension tubes or bellows, the image on the film is often larger than the original. As an object reflects a fixed quantity of light the exposure will vary according to the area of film over which the light is spread. Thus with a magnification of ×2 on the film, the available light will be spread over an area of film four times that of the object, and the exposure must be increased accordingly. If extension rings are purchased, the factor by which the exposure must be increased will be given in a table supplied with the rings.

With macrophotography, small objects such as insect mouthparts are photographed without the aid of a microscope, the camera lens on long extension tubes being used to obtain some degree of magnification on the film. Lighting and exposure are tricky with macrophotography, whether direct or transmitted light is used. The beginner would be well advised to make two or three trial exposures at the end of a film, recording details of the exposure and lighting set-up. This will enable further shots to be taken using exposures derived from experience.

With photomicroscopy, a microscope is used in the place of the camera lens. There is an elaborate method of calculating the correct exposure, but again the beginner is best advised to take a few shots at the end of a film – with so many variables involved, calibration is often easier than calculation. If the camera has a fixed lens and cannot be used with a micro-adaptor, it is possible to take photomicrographs by setting the camera at infinity and placing it over the microscope, but by this method the field of view will not fill the slide.

Accurate colour rendering is not easily achieved and lies beyond the scope of this book. Black and white photomicrographs can be conveniently labelled with 'Letraset' for re-photographing and display.

GENERAL PHOTOGRAPHY

Although many good visual aid pictures are taken by chance, it will usually be found that the best are taken with a specific purpose in mind. Visual aids are not a substitute for teaching, and their value usually falls into either of the following categories:

1) *General background material* in which the detail is of secondary importance to the atmosphere. Such slides often have a high pictorial value and their role is to bring the class into temporary contact with the subject.

2) *Record material* which usually conveys specific detail. In this type of shot the principal object should occupy most of the picture area, and all detail should be sharply focused. If an obtrusive background has to be included, its effect should be minimised by use of differential focus – careful focusing on the object and the use of full aperture, in order to throw the background out of focus. Alternatively, if the object is small, a card or even a plain raincoat may be placed behind it to form a neutral background. Do everything possible to eliminate superfluous detail; for example, when photographing a commemorative plaque on a building of historical interest, take care that no other point of interest, such as a 'no parking' sign, appears on the edge of the picture.

Lighting

For *colour* photography, the best results are usually obtained with the source of light more or less directly behind the camera, thus avoiding excessive contrast and shadows which cannot be recorded well in colour. If *black and white* film is used, lighting from the side is often the most effective as it will give 'depth', texture, and atmosphere to the picture.

Correct exposure is of the utmost importance, and the use of an exposure meter or calculator is highly desirable. Unfortunately many amateurs believe that the purchase of an exposure meter will put an end to exposure problems and are often disappointed with their results. A meter cannot take the place of the photographer's brain; the meter gives accurate information which makes decision easy, but it cannot make the decision. For example, if the meter is pointed towards a scene there will be a great difference between readings for the sky or distant landscape, and for people or buildings in the foreground. This is because of the tremendous amount of reflected light in the sky. If the foreground is unimportant, the correct exposure will be that shown for the sky. If on the other hand the foreground subject is to be the main part of the picture, a longer exposure will be needed and will result in over-exposure of the sky, so much so on very bright days that the sky may not register at all.

FILMSTRIPS

Material for filmstrips should be very carefully considered before photography begins as the shots should form a planned sequence. General shots should be followed by close-ups, not vice versa. Careful scripting is essential as the sequence and exposure time of pictures on a home-produced strip cannot be altered. If a *colour* filmstrip is to be made, make sure that the processing labroatory is willing to return the film uncut, and be sure that clear instructions are attached to the film. Some laboratories will produce a filmstrip from your own transparencies, but this service is expensive compared to taking the material direct onto film.

If *black and white* filmstrips are required, they can be processed at home by the reversal method. Most makes of film are suitable, and instructions are obtainable from the manufacturers. The process is similar in all cases; detailed information for Ilford materials is set out below.

Reversal of Ilford 35 mm film

Reversal processing enables black and white transparencies (*positives*) to be obtained directly from ordinary negative material exposed in the camera in the usual way. This technique is of value when making filmstrips or slides of which only one copy is required. Transparencies produced by the process are of a pleasing warm tone – particularly warm in the case of Pan F – and have an extremely fine grain.

Films recommended

For general photography – Ilford Selochrome Pan, FP3, HP3, HPS, and Ilford Pan F.

For copying prints and drawings – Pan F or, when greater contrast is desired and a slower, non colour-sensitive film is acceptable, Fine Grain Safety Positive Film.

Exposure

When exposing films in the ordinary way for the production of negatives, considerable variation in exposure is permissible. This is because variations in negative density can be largely compensated for at the printing stage. When a film is reversed this opportunity for compensation no longer exists, and

consequently very little variation of the camera exposure can be permitted if transparencies of consistent quality are to be obtained.

For general outdoor and indoor exposures the use of an exposure meter is recommended. The published meter setting of the film should be used as a guide, but the setting that will give the best results for any one user may be slightly different and should be determined by practical tests.

When copying prints or drawings on Pan F or Fine Grain Safety Positive film using two series one photofloods in reflectors 60 cm from the centre of the copy, the exposure required will be approximately:

Pan F	$\frac{1}{2}$ second at f/22
Fine Grain Safety Positive	$\frac{1}{2}$ second at f/8

Alternatively, exposures for copying may be determined with an exposure meter by the 'artificial highlight' method. For this a sheet of clean white paper about 30 cm square should be placed over the evenly-illuminated original. The meter should be held about 15 cm away, facing the paper, and the maximum reading obtained as the meter is moved slowly from side to side should be noted. The settings when using a Weston meter by this method are:

Pan F	25
Fine Grain Safety Positive	4

When copying, the exact exposure for a given set of conditions must be found by trial and error, whether or not a meter is used. It is suggested that a series of trial exposures are made at half, equal to, and double the exposure indicated. Over-exposure will lead to a final positive that is too light, under exposure to one which is too dark.

Safelight

This will vary with the film used. The following are recommended:

Fine Grain Safety Positive	S, No. 902 (light brown)
Pan F, FP3, Selochrome Pan, HP3	GB, No. 908 (very dark green)
HPS	Total Darkness

All films can of course be processed in total darkness if a suitable safelight is not available.

Solutions required

Reversal developer The recommended developer is Ilford PFP developer or ID-36, made up to contact paper strength with the addition of the following quantity of hypo crystals:

Pan F, Selochrome Pan, and HP3 — 8 grams per litre of working strength developer

FP3 and HPS — 12 grams per litre of working solution

Bleaching solution

Solution A: 2 grams of potassium permanganate in water, made up to 500 cm^3

Solution B: 10 cm^3 of sulphuric acid (conc.) in water, made up to 500 cm^3

These stock solutions will keep for a long time. For use, mix equal parts of A and B; mix a fresh working solution for each film.

Clearing solution 25 grams of sodium or potassium metabisulphite in water, made up to one litre.

Fixing bath A hardening fixing bath should be used as the bleaching operation renders the emulsion tender. Ilford Hypam Fixer – with hardener – is recommended.

Processing

A negative image is first obtained by developing of the original latent image. This negative image is then dissolved away in a bleaching bath; the silver halide remaining is exposed and developed giving a positive image. The second development is followed by fixing in an acid hardening-fixing bath. The time required to carry out all the operations from the beginning of the first development to the beginning of the final wash is about thirty minutes. The details of the process are as follows:

(1) First development

An ordinary spiral tank is filled with the recommended quantity of developer at 20 °C (68 °F). Place the tank in a water bath to maintain this temperature. Working in total darkness, load the film into the spiral and gently lower this into the developer, lifting it out again to remove air-bells before placing the cover on the tank. Agitate during development by giving a five second rotation to the spiral every fifteen seconds.

A development time of twelve minutes at 20 °C (68 °F) is recommended for all the Ilford films listed above. It is important that this time and temperature be closely adhered to if consistent results are to be obtained. *Careful control of the time and temperature of the first development is as important as the control of camera exposure.* It is best to remove the cover from the tank before the end of development so that the spiral containing the film may be removed quickly from the developer and plunged into water for washing, thus stopping development immediately at the predetermined time. The developer which will be required later for the second development should be poured into a convenient vessel until needed. The tank itself should be washed and then filled with the bleaching solution (mixed at working strength) in readiness for the bleaching operation.

(2) First washing

Wash for three minutes in running water.

(3) Bleaching

Treat the film for five minutes in the bleaching solution at 18 °C–21 °C (65 °F–70 °F) with continuous strong agitation, lifting the spiral from the solution rapidly and repeatedly. This vigorous agitation is necessary to ensure complete bleaching. White light may be switched on after the film has been in the solution for thirty seconds, and all subsequent operations may be carried out in the light.

(4) Second washing

Wash for two minutes in running water.

(5) Clearing

Soak the film for two minutes in the clearing solution.

(6) Third washing

Wash for two minutes in running water.

(7) Second exposure

Remove the film from the spiral and expose it for the equivalent of thirty seconds at 45 cm distance from a 100 watt tungsten-filament lamp. Insufficient second exposure will result in a reduction in density when the film is fixed. Two to four times the specified exposure may safely be given, but overexposure beyond this may lead to foggy highlights.

(8) Second development

If the spiral tank permits reinsertion of the film this should now be done. It may help if the operation is carried out under water. If, however, reinsertion of the film is not possible, the second development and subsequent operations may be carried out in a dish by the 'see-saw' method.

Develop for six minutes at 20 °C (68 °F) in the

developer left from the first development. The presence of hypo in this developer will help in giving clear highlights. The time of development is not critical; continue until maximum density results with no appreciable loss in density when the film is fixed.

(9) Fixation

Fix for ten minutes in the acid hardening-fixing bath. The purpose of this operation is to remove any insensitized silver halide in the highlights and to harden the emulsion layer.

(10) Fourth washing

Wash in running water for thirty minutes. After washing, surplus moisture may be removed by gently wiping with a clean, wrung-out chamois leather, or by giving the film a final rinse in water containing Ilford Wetting Agent (1 part in 500). The film should then be hung up to dry.

When dry the film may be projected as it stands, or it may be cut and mounted as transparencies.

Short filmstrips

Some subjects are suitable for filmstrips of fewer frames than the usual commercial products and strips can be designed to suit short periods. When producing short strips it is possible to make two or three on the same 36 exposure film. Each strip needs a short 'head' and 'tail', the minimum for most projectors being two to three frames. When winding on blank frames remember to cover the lens when releasing the shutter. This will ensure a continuous black strip when reversal processed. If two or more short strips are being made on the same film leave four to six blank frames *between* the strips to allow for both 'head' and 'tail'.

PHOTOGRAPHIC EQUIPMENT

Camera

The choice of camera will influence the type of work which can be undertaken. If no close-up work is contemplated, and any diagrams which may be required can be drawn wall-chart size, then almost any 35 mm camera with a good lens is suitable. The ideal camera is a single lens reflex with interchangeable lenses, which can be used with extension tubes for extreme close-ups. With this type of camera parallax and focusing problems are resolved. The most refined instruments in this class will cost well over £100, but in the medium price range some versions of the Edixa Reflex will be found to be extremely versatile. A recent introduction of great value to schools is the Exa Reflex which is very robust with a simple shutter (1/25, 1/50, 1/125, B). The Exa Reflex is very suitable for school journeys, and can be obtained with a good lens for a modest outlay. Its major advantage is that it is fully interchangeable, and will accept most lenses and extension tubes designed for the far more elaborate Kine-Exacta. Although a wide range of shutter speeds can be very convenient, the extremes are not often used and if a camera with only a few shutter speeds is chosen, much can be achieved, both in poor light and with action photographs, by using flash. When an inexpensive camera is chosen it is important that most of the cost is reflected by the quality of the lens rather than by the body work.

Tripod

At shutter speeds slower than 1/200 second, a stable tripod will often result in a considerable improvement in image sharpness. This is particularly noticeable with transparencies which are to be greatly enlarged when projected.

Exposure meter

This is a very useful accessory if used with skill (see page 24). For reliability and robustness over a long period of time, the Weston Master is a very good investment – it is the choice of almost all professionals.

Built-in exposure meters are very useful, but when damaged will necessitate sending the whole camera for repair. In extreme lighting conditions a coupled meter is often unreliable as judgment and experience are the only guides.

Close-up equipment

If the camera is of the interchangeable-lens type, extension tubes will be available and, probably at a much higher price, extension bellows.

With fixed lenses, close-up work is done by using a supplementary lens in front of the camera lens (held in the filter holder). There are two common supplementary lenses available – 1 Dioptre, focusing from 1 m to 0.5 m, and 2 Dioptre, focusing from 0.5 m to 0.25 m.

Flash

Occasionally useful, particularly the more expensive electronic flash which has a flash duration so short that it can be used to 'freeze' movement in close-up photographs of insects.

Filters

If colour material is used for outdoor shots, the rendering of distant landscapes is often improved by the use of an ultra violet (U.V.) filter.

Conversion filters are available which enable daylight shots to be taken on artificial light type films and vice versa. They are not to be recommended, as they greatly affect exposure times and colour balance is not as good as with the correct type of film. In spite of this drawback however, they can be useful in the production of amateur film strips.

Colour correction filters are available which can be used to prevent the warm or cool 'colour cast' associated with early morning or evening sunlight and with very cloudy conditions. Although very useful on occasion, they must be used with discretion as they can destroy 'atmosphere'.

4. Care of Animals

CULTURE OF LABORATORY ANIMALS

Experiments on vertebrates are rigorously controlled by The Cruelty to Animals Act (1876). Every care should be taken not to infringe this rather complicated piece of legislation. Any experimental work about which the teacher has doubt should be checked by seeking the advice of the Home Office Inspector for his area, whose address is obtainable from the Under Secretary of State, Home Office, London S.W.1.

A licence is required for any experiment which causes harm to a vertebrate, or for an experiment in which the results are unknown. Injections for experimental purposes are legal only if the results are known and the animal suffers no harm, e.g. the injection of Pregnyl in reasonable doses, to induce spawning in Xenopus, is legal.

The pithing of frogs is regarded as a means of killing and a pithed frog is regarded as a dead animal within the meaning of the Act. Hence kymograph experiments using pithed frogs are legal. **The author feels that such experiments should be kept to an absolute minimum and that pithing should never be carried out in the presence of a class. The teacher has an obligation not to act in any way which might encourage a callous attitude on the part of the student.**

GENERAL NOTES ON THE CARE OF LABORATORY ANIMALS

Cleanliness

If laboratory animals are to be kept in a healthy condition, it is essential to keep the animal room scrupulously clean and free from unnecessary equipment – it should not be used as a store. Cages should be cleaned weekly and all litter swept up. It is advisable to spray or wash all floors and benches with a suitable disinfectant such as Tego MGH at regular intervals. This precaution will very considerably reduce the incidence of respiratory infections, intestinal infections, and skin conditions. A very useful leaflet, *Animal House Disinfection*, is available from the distributors of Tego, Messrs. Hough, Hoseason and Co. Ltd., Chapel Street, Manchester 19.

Ventilation, humidity, and light

Fortunately small mammals are tolerant of considerable differences in their physical environment and this makes it possible to house a number of species in the same room. The ventilation system should allow 6–10 changes of air per hour without exposing the animals to draughts. Good ventilation plays an important role in reducing respiratory infection. Optimum humidity varies from 45 per cent in the case of rabbits, to 65 per cent for mice. For a school animal house any figure between these points will be found to be suitable for a mixed stock.

Lighting conditions are not critical but should be diurnal. Seasonal variation in litter frequency is related to day length.

Cages

Make sure that the cages are appropriate to the size of the animal. Freedom of movement and room to stretch out fully are essential for health.

Wire cages are easy to sterilise with Tego but are much more expensive than the wooden boxes which are preferred by many large scale users. For school use polypropylene mouse and rat cages with wire tops incorporating a food hopper and water bottle are ideal.

Whatever type of cage is chosen, the floor should be lined with at least one centimetre of sawdust which should be changed each week. If possible use sterile sawdust obtained from animal foodstuff suppliers as the untreated sawdust obtained from a local sawmill is often contaminated by the faeces of cats, dogs, mice, and rats and can be an unexpected source of parasitic and bacterial infections. When wire cages are used it is most important to see that the floor is covered by sawdust to prevent the development of sores on the feet of the occupants. Pregnant animals should be given a good supply of wood shavings for nesting. If rabbits are kept, hay is the most suitable nesting material.

Cages should be clearly labelled or numbered in such a way that the animals are unable to chew the cards.

Food and water

A balanced diet in the form of pellets is available to suit the needs of most mammals and can be ordered through a local corn-merchant. Mashes are satisfactory but are much more labour-consuming than pellets. When working on a tight budget, daily feeding is more economical than weekly filling of a hopper. If animals are kept on a minimal diet, litter frequency and size may be reduced.

Water should always be available, independently of the food supply. If there is a considerable fluctuation in the air temperature, difficulty may be experienced with the inverted type of water bottle as expansion of an air bubble may force water into the cage. Water bottles should be changed and disinfected regularly.

Disease

The diagnosis and treatment of disease is a job for the expert and in a school animal house should not normally be attempted. Ailing animals and those with lesions should be humanely killed and the cage sterilised. If this precaution is taken and animals are only obtained from accredited breeders, epidemics should not occur. Animals should never be introduced from pet shops.

Noise

Mice and rats are seriously affected by high-frequency noise such as the emptying of dustbins and hammering.

Killing

The following methods are humane and are suitable for use in the school:
1. Breaking the spinal cord. The animal should be held by the hind legs and its cervical region cracked sharply across the edge of a bench.
2. An overdose of chloroform, nitrogen, or coal gas may be administered. Ether should not be used on young mice or rats as they may recover from its affects some time after apparent death.
   Chloroform should never be administered in the animal house as male mice are very susceptible to the vapour and may die from degeneration of the liver two to three days after exposure.
3. Place the animal in a strong polythene bag and fill this with CO_2 from a cylinder. It may be convenient to place the cage and the animal in the bag. This method is recommended by the Universities Federation of Animal Welfare.

As far as possible, animals should not be killed in the presence of a class.

SOME NOTES ON SPECIFIC ANIMALS

Mongolian gerbil

Oestrous cycle	Has not yet been precisely defined, but would appear to be about 4–7 days in length
Gestation	24–28 days
Weaning age	24 days
Mating age	10–12 weeks
Litter frequency	Post-partum mating may take place and hence litters may be produced monthly
Litter size (average)	5–7

Breeding

They are best kept in monogamous pairs. The male can be left with the female throughout the breeding cycle.

There is a tendency for some pairs to cannibalise their young. Several factors, such as dietary deficiency and inbred stock, seem to cause this. It is best to leave the litter completely undisturbed for the first few days.

Bedding should not be changed for about two weeks.

Juvenile animals may be kept together in large groups of the same sex, and it is possible to continue keeping them in this manner when the animals are adult. If two strange adults are placed together they will fight and severely injure each other.

Mice

Oestrous cycle	4–5 days
Gestation	19–21 days
Weaning age	21 days
Mating age	6–8 weeks
Litter frequency	8–12 yearly
Litter size (average)	7–8
Optimum temperature	21 °C
Humidity	50–65 per cent

Breeding

Although mice may mate at 6 weeks, the usual age in controlled breeding programmes is 8–12 weeks. If mice are kept in polygamous trios, that is, one male and two females per cage, litters may be expected every 3–4 weeks. Young mice should be weaned before the next litter appears. If mice from a single stock are inbred for 7 or 8 generations a fall in average litter size may occur.

Rats

Oestrous cycle	4–5 days
Gestation	21–23 days
Weaning age	28 days
Mating age	10–12 weeks
Litter frequency	7–9 yearly
Litter size (average)	7
Optimum temperature	21 °C
Humidity	45–55 per cent

Breeding

The animals should be 10–12 weeks old. Mating is usually carried out in polygamous trios. Females, when pregnant, should be removed to a separate cage with plenty of nesting material. The female and litter should be left undisturbed as far as possible until weaning. After weaning, the young should be kept in a high density population for at least a week.

Rabbits

Oestrous cycle	No definite cycle. Ovulation occurs ten hours after mating. There is evidence of rhythmic sexual activity
Gestation	28–31 days
Weaning age	6–8 weeks
Mating age	6–9 months
Litter frequency	4 yearly
Litter size (average)	4
Optimum temperature	18 °C
Humidity	40–45 per cent

Breeding

Nine-month-old rabbits should be mated when the female is on heat. It is advisable to introduce the doe to the buck's cage, not vice versa. The vulva of the doe becomes red, moist, and swollen when she is on heat. After 24 days the doe should be transferred to a large cage with a screened breeding compartment. Plenty of hay for nesting must be provided and the litter should not be disturbed for 10 days if cannibalism is to be avoided. The young should be weaned when they are 6–8 weeks old and the sexes separated into different runs.

Guinea-pigs

Oestrous cycle	16 days
Gestation	Usually around 63 days, but may vary between 59 and 72 days
Weaning age	14–21 days
Mating age	12–20 weeks
Litter frequency	3 yearly
Litter size (average)	3
Optimum temperature	21 °C
Humidity	45 per cent

Breeding

12 weeks is the usual minimum age. One male is placed in a pen with 5–10 females allowing at least 1 square foot (1000 cm^2) of floor space per animal. Pregnant females are placed in separate cages.

Food

In addition to pelleted food, guinea-pigs should be given two ounces (60 grams) of fresh food daily – carrots, cabbage, and swedes are suitable.

Hamsters

Oestrous cycle	4–5 days
Gestation	16–17 days
Weaning age	3–4 weeks
Mating age	7–9 weeks
Litter frequency	3 or 4 yearly
Litter size (average)	6
Optimum temperature	21 °C
Humidity	40–50 per cent

Hamsters must be kept in strong cages; they will very quickly gnaw their way out of wood, aluminium, or zinc cages. The polypropylene cages manufactured by North Kent Plastic Cages Ltd., Home Gardens, Dartford, Kent, are excellent. Hamsters must be provided with wood to gnaw, otherwise their incisor teeth will overgrow until feeding is not possible.

Mating should not be carried out until the female is 12–20 weeks old. Mating should be carried out on a table top under supervision as the female is liable to attack the male savagely after copulation, or if she is not receptive. The female comes on heat every fourth night, usually after sunset. If a female is on heat she will adopt a mating position when placed with the male.

Pelleted diets

For all school purposes the following are suitable:
Mice, Rats, and Mongolian gerbils: Diet FFG(M) or Diet 86
Rabbits and Guinea-pigs: Diet 18
Hamsters: Diet FFG(M), or 86, or 18, with some green supplement if possible

Diet RGP is suitable for rabbits, hamsters, and guinea-pigs without a green supplement. E. Dixon and Sons (Ware) Ltd., Crane Mead Mills, Ware, Herts. or Peter Fox (Scientific Animals) Ltd., Home Farm, Aldenham Park, Elstree, Herts., supply all the above diets in $\frac{1}{2}$ cwt (25 kg) sacks.

Xenopus

The African clawed toad has many advantages as a laboratory animal when compared with *Rana* and *Bufo*. Stocks are easily maintained in the laboratory and breeding can be carried out at any time of the year. *Xenopus* will not breed until they are two years old – it is therefore necessary to order a breeding pair when starting a colony.

Containers

Ordinary aquaria or plastic tanks are suitable. Use a large size, 20 in. × 16 in. × 6 in. (50 cm × 40 cm × 15 cm approx.) deep for stocks and a smaller size, 12 in. × 8 in. × 8 in. (30 cm × 20 cm × 20 cm approx.) for breeding. Plastic tanks of these dimensions are available from Messrs Thermo Plastics Ltd, Luton Road, Dunstable. The tanks should be firmly covered with transparent sheeting reinforced with cross wires, such as Claritex, which is available from ironmongers.

Temperature

Adult stocks can be maintained without heat as long as the ambient temperature does not fall below 10 °C. For breeding and rearing the larvae, the tank should be heated to 23 °C using an aquarium heater and thermostat.

Feeding

Earthworms are an ideal food for the adults as they are consumed without fouling the water. If earthworms are not available, finely chopped liver or heart may be given. The stock should be fed twice weekly, and the water changed immediately afterwards. Changing the water is facilitated if the plastic tanks are fitted with taps. Some workers prefer to feed the stock once a week until food is refused, and then change the water 24 hours later when the toads have defaecated.

Diseases

Two per cent mortality per year is to be expected in a large stock kept under clean conditions with no overcrowding. Adults should be kept at a population density of 10–15 *Xenopus* to 30 litres of water.

Marking

Xenopus may be marked by clipping one or more claws.

Breeding

Hormonal stimulation is necessary to ensure breeding in captivity. Two cm^3 disposable syringes fitted with No. 18 needles are used and both male and female toads are given a primer and a final injection.

	Primer	Final
Female	100 i.u. Pregnyl	300 i.u. Pregnyl
Male	50 i.u. Pregnyl	100 i.u. Pregnyl

A breeding pair are kept at 23 °C for 4 weeks before the primer injection is given. The final injection is given four days after the primer. Pregnyl is obtainable from Organon Ltd, Crown House, Morden, Surrey and is supplied in separate ampoules of Pregnyl and distilled water. The water is taken into the syringe and transferred to the ampoule of Pregnyl which dissolves instantly and can be drawn up into the syringe. **Air bubbles are removed by holding the syringe, needle uppermost, and pushing the plunger in until all air is expelled.** The injection is made by inserting the needle into the thigh, and passing it forward in the subcutaneous connective tissue to the dorsal lymph space which lies just beneath the skin, between the 'stitch marks' and the mid-line.

If, after injection, the toads pair without spawning the temperature should be raised to 28 °C.

After spawning, the adults should be removed and the eggs distributed between containers so that each egg is in direct contact with the surrounding water and no eggs are left enclosed in solid masses. Gentle aeration may be given by means of an aquarium pump fitted with a diffuser block.

The newly-hatched tadpoles may be fed on dried nettle powder which can be obtained from health food stores. Metamorphosis occurs approximately seven weeks after hatching. During the metamorphosis period the larvae may be fed on micro-worm, *Daphnia*, *Tubifex*, or white-worms. After metamorphosis, finely-shredded liver and heart may be given.

Xenopus in all stages are available from Phillip Harris Ltd.

Locusts

Locusts are particularly appropriate laboratory animals for schools as they can be used for a wide range of practical work. They are relatively easy animals to maintain as a permanent stock.

Cages

For stock colonies and breeding the most suitable cage is a glass fronted container approximately 15 in. × 15 in. × 20 in. high (38 cm × 38 cm × 50 cm), with a false floor of perforated zinc. The cage should be heated by means of two electric light bulbs, one below the false floor and one in the locust compartment. The false floor should be arranged so that containers of sand for egg laying may be placed underneath. The minimum size for the egg containers should be 4 in. (10 cm) deep by 1¼ in. (3 cm) diameter. An excellent cage of the pattern used by the Anti-Locust Research Centre, complete with egg tubes, is available from Phillip Harris Ltd. A cage of this type may be used to house up to 200 adults.

For raising locusts in isolation, an excellent container may be made by replacing the top of one of the new pattern Kilner jars with a disc of perforated zinc. The disc is best attached with Evo-Stik or Araldite. Several containers of this type may be heated by placing them around a 40 watt bulb at a distance of approximately eight inches.

Cylindrical perspex or cellulose acetate cages 15 in. × 6 in. (38 cm × 15 cm) are available from Griffin and George Biological Labs Ltd., and may house 20–30 adults.

Perches

It is essential to provide each cage with twiggy branches or a cylinder of large mesh wire netting to provide perching space. Unless the instars can hang freely when moulting a high proportion of deformed adults will occur.

Temperature

Locusts will tolerate a wide range of temperatures and can be kept in an unheated laboratory during the summer; development will be slow and mortality high under these conditions. Ideally locusts should be kept within the range 28–34 °C. Adjustments may be made by changing the bulbs in the cages. A suitable arrangement is a 25 watt bulb below the floor and a 40 watt bulb in the locust compartment. A temperature gradient will be set up in the cage and provided that plenty of perching space has been given, the locusts will bask in the optimum region.

Humidity

Excess humidity will favour disease. If condensation appears on the cages, or faeces are soft and wet instead of hard and dry, steps should be taken to lower the humidity. Assuming that the temperature of the cage is not too low, the most usual cause of excessive humidity is overfeeding with lush grass.

Feeding

Locusts may be fed entirely on grass. Sufficient fresh grass should be placed in the cage each day, the amount should be only very slightly more than can be consumed in the 24 hours before the next feed. Overfeeding makes cleaning unnecessarily difficult and raises the humidity. Grass need not be given daily if a plentiful supply of wheat bran, available from pet shops, is always present in the cage. Wheat bran is particularly useful in schools for overcoming the weekend feeding problems. When grass is not available, cabbage may be used as a substitute but has the disadvantage of resulting in very malodourous faeces.

Breeding

The female lays eggs in a frothy 'pod' in a hole which she excavates in moist sand. Each pod contains 30–100 eggs, each of which is the size of a rice grain. For laboratory use, sharp builders' sand should be sieved to remove particles over $\frac{1}{10}$ of an inch (3 mm), then washed with repeated changes of water

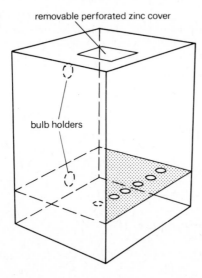

removable perforated zinc cover

bulb holders

Figure 4.1   A cage for locusts

until all dust-like particles have been removed. The sand should then be sterilised in an oven or autoclave. Once a stock of sand has been prepared it can be used repeatedly, provided that it is resterilised before each use. To obtain the correct moisture content, the dry, sterilised sand is mixed in the proportion of 100 parts of sand to 15 parts of water by volume. The sand is then placed in containers at least 4 inches (10 cm) deep, beneath holes in the floor of the cage. Porous containers should not be used as they dry out too rapidly. The containers should be replaced every one or two days, even if they have not been used. Gently tip off the top $\frac{1}{2}$ inch (12 mm) of dry sand to check for the presence of pods. Two or three pods in a standard tube are sufficient – if the numbers are greater than this there will be a sharp fall-off in the hatching percentage. After checking for the presence of pods, the containers should be loosely covered to prevent excessive evaporation, and set to incubate at 28–32 °C. At these temperatures *Locusta migratoria migratorioides* (African migratory locust) will hatch in 11–16 days, and *Schistocerca gregaria* (Desert locust) in 12–17 days.

Five hundred first stage hoppers may be kept in a standard cage and will produce 100–200 adults. Although the mortality rate is high, few corpses will be observed as cannibalism is prevalent.

Locust allergy

Some people, after prolonged exposure to high density populations of locusts, show allergy symptoms. The severity of the symptoms and the rapidity of onset depends on the degree of exposure to locusts and the individual's susceptibility to allergic conditions. Any person with a family history of allergies, hay-fever, or asthma should avoid regular handling of locusts or working for long periods in rooms where locusts are kept in large numbers. From the school point of view, the only people at risk are technicians responsible for feeding and cleaning, and staff who spend most of their time working in the same laboratory. At present, expert opinion suggests that there is no risk to children who only use the laboratory once or twice per week, but that it would be unwise to keep locusts in a classroom used by one class for long periods, unless the locusts were few in number. The author prefers to keep his locusts in a corridor where they are a source of considerable interest without exposing any individual to prolonged contact.

The symptoms of allergy start with an apparent head cold and sneezing, these symptoms often disappearing at night. Concurrently, or sometimes at a later date, a characteristic skin irritation of the hands and fore-arms appears. If these symptoms appear, the individual should stop working with locusts and the symptoms will disappear without trace. Once a person is sensitised to locusts, continued exposure may lead to respiratory conditions, such as bronchitis and asthma.

'Amoeba'

Shallow, glass containers capable of taking a two centimetre deep culture solution should be used. The tops of household pyrex dishes are ideal. Extreme cleanliness is essential to success; all glassware should be washed in Teepol and then rinsed with at least twenty changes of clear tap water, with a final rinse in glass distilled water (metal contaminants are lethal).

Culture medium

Chalkley's medium is excellent. It is prepared as a stock solution and diluted for use.

Chalkley's medium (Stock solution)

NaCl	16.0 g
NaHCO$_3$	0.8 g
KCl	0.4 g
NaHP$_4$	0.2 g

Make up to one litre with glass distilled water. De-ionised water contains phenols which are harmful to Protozoa.

For use, take 5 cm^3 of stock solution and dilute to 1 dm^3 (litre) with glass distilled water. Distribute the medium to the culture dishes, giving each a depth of 2 cm. Boil wheatgrains, place four in each dish to grow mould and bacteria, and finally *Colpidium* which forms the food for Amoebae.

Technique of culturing

Inoculate the culture with *Amoeba* using Pasteur pipettes for handling. Growth is erratic at low temperatures.

At 7 °C there is one division per week.

At 30 °C division is most rapid, but the culture produces small, underfed individuals.

17–23 °C is the most suitable range (many workers keep their cultures at 20 ± 1 °C).

At 17 °C the lifespan is 2–3 days and approximately two per cent of the culture are at the division stage at any one time.

If dividing *Amoebae* are transferred they may produce binucleate individuals. After inoculating the culture in a Pyrex dish with about 300 *Amoebae*, it should be covered and left for four weeks without disturbance. A harvest of 5000 cells may be expected. The culture should be kept in a fairly dark place to avoid the growth of algae.

Always transfer attached individuals; floating forms are not healthy.

Mass culture

If a large plastic sandwich box is charged with 2 cm depth of Chalkley's medium and inoculated with a four week culture (approximately 5000 cells), it is possible to harvest 5 cm^3 of packed cells (1 million) after centrifuging. For mass culture the *Colpidium* required for food should be cultured separately to ensure a sufficient supply (see below).

Notes

1 Wheat obtained from seed merchants may have been dressed with a mercuric fungicide and is unsuitable for culture media. (Undressed wheat is available from Carters Seeds Ltd and health food stores.) If undressed wheat cannot be obtained, polished rice grains may be substituted.

2 Aromatic compounds must be kept out of culture rooms. Xylene, methyl benzoate, and the solvents used in paints are particularly lethal to protozoan cultures.

3 It is almost certainly a waste of time to attempt to collect the proteus group of *Amoebae* from the wild.

Colpidium

Colpidia will usually appear in a suitable culture medium if it is left exposed to the air. An excellent medium is 1 per cent proteose peptone (available from Oxoid or Difco). If cultures are available, the medium should be sterilised and inoculated with *Aerobacter aerogenes*. Twenty-four hours later the culture should be cloudy with bacteria and ready for inoculation with *Colpidium*. Another excellent medium is 0.1 per cent milk in Chalkley's medium; when the culture clears, it is at its peak, and the *Colpidia* may be removed by means of a centrifuge. (2000 r.p.m. for 8 minutes will bring down the ciliates, leaving the bacteria in suspension.) If the *Colpidia* are required to feed *Amoebae*, decant off the supernatant liquid, rinse with Chalkley's medium, and re-centrifuge to remove the peptone or milk.

Stentor

Stentor grows well in Chalkley's medium. Feed with *Colpidium* every 2 or 3 days and keep the culture in good light. Beakers are suitable vessels.

Paramecium

Add 30 grains of boiled wheat to a litre of Chalkley's medium and inoculate with *Bacillus subtilis*. Twenty-four hours later, inoculate with *Paramecium*, 500 cm^3 flasks containing 300 cm^3 of medium, plugged with cotton wool, are ideal vessels. Subculture every three weeks.

Algae and green flagellates

Add 10 grains of wheat to 500 cm^3 Chalkley's medium and autoclave for 15 minutes at 15 lb in.$^{-2}$ (0.1 N m^{-2}) pressure. Inoculate from purchased culture, and place in a sunny window. The culture should be plugged with sterile cotton wool to keep out airborne spores.

Hydra

Hydra present two main difficulties when kept as cultures. Firstly, they require suitable pond water or a chemically defined medium. Tap water is usually lethal as *Hydra* are sensitive to the copper ions present. Tap water can be used if the copper ions are first chelated by the addition of 50 mg dm^{-3} of 'Versene' (di-sodium ethylenediamine tetra-acetate). The second difficulty is the provision of a suitable supply of small crustaceans for food. Unless young *Daphnia* are readily available, the most convenient food will be newly hatched brine-shrimps (*Artemia*).

Chemically defined medium

Stock solution A

Potassium chloride	3.75 g
Sodium hydrogencarbonate	42.00 g
2-amino-2-(hydroxy-methyl) propane-1 : 3-diol.	60.57 g

Make up to one dm^3 (one litre) with de-ionised or glass distilled water.

Stock solution B

Calcium chloride	41.6 g
Magnesium chloride (hydrous)	10.2 g

Make up to one dm^3 with de-ionised or glass distilled water.

The working solution is made up by adding 20 cm^3 of each stock solution to 10 dm^3 of de-ionised or glass distilled water, approximately 9 cm^3 N hydrochloric acid is then added to bring the pH within the range 7.5–7.8.

2-amino-2-(hydroxy-methyl) propane-1 : 3-diol is available from Harrington Bros Ltd, Weir Road, Balham, London S.W.12. It is also marketed under the name Tris Buffer.

Culture technique

Shallow culture dishes, such as domestic Pyrex lids, are filled with one centimetre depth of medium and kept at room temperature. *Hydra viridis* should be kept in good light, but for *Hydra fusca* a more shady position is to be preferred. *Hydra* should be fed twice per week with newly hatched *Daphnia* or *Artemia*. One hour after feeding, the culture dish should be gently tilted to pour off the medium and any surplus food, and refilled with fresh medium.

Sexual reproduction may be induced by placing the culture in a refrigerator for two to three weeks at 10–15 °C. The culture should be fed normally.

Artemia

Artemia (brine-shrimp) eggs are available from aquarium shops. They are hatched by placing them in a salt solution at room temperature. At 15–20 °C, hatching occurs in two to three days. If supplies are required at short notice, they will hatch in 24 hours at 30 °C. The culture solution is made by dissolving 360 g sodium chloride in 1 dm^3 of hot water and cooling. For use, the stock solution is diluted with 10 parts of water. *Artemia* must be rinsed in clean water to remove salt before feeding to *Hydra*.

Daphnia

Daphnia may be obtained from tropical aquarium shops who obtain supplies at intervals. In some localities *Daphnia* may be collected in ponds during the summer using fine nylon sweep nets.

Daphnia are difficult to maintain in continuous culture because of the difficulty of obtaining a satisfactory balance between food supply and oxygen supply. Place the *Daphnia* in an aquarium with a large surface to volume ratio and aerate if possible. Feed the *Daphnia* on infusoria, yeast, or a few drops of suspended Horlicks powder. Only give sufficient food to cloud the water for about an hour – if the *Daphnia* cannot clear the water, it will become foul, and the bacterial action which ensues deoxygenates the water and the *Daphnia* asphyxiate.

The temperature must be kept below 20 °C.

Woodlice

Collect woodlice from under stones, bark, and fallen logs. In towns, woodlice are usually common at the base of tombstones, and can be found by pulling back the grass.

Large plastic sandwich boxes make ideal containers. Place three centimetres of damp soil, or a mixture of damp sand and peat, in the bottom and at all times make sure that the soil is kept damp as the animals can only survive in high humidity. Keep below 20 °C in a very shady or dark place, and feed on small pieces of raw carrot or potato. If pieces of broken clay flower pot are placed with their convex side up on the surface of the soil, animals will tend to congregate underneath thus making collection easy. Ventilation may be provided by drilling a number of small holes in the box lid, or by a cover of muslin held in place by a large rubber band. Occasional feeding and damping will maintain the culture for long periods. Subculture every year to eighteen months.

'Microworm'

These are an easily bred species of freeliving nematode which can be obtained as a culture from aquarium shops. Place the culture on the surface of a plate of cold, not too thick, porridge. Within a few weeks the porridge will be a seething mass of worms which can be used to feed small carnivores in the aquarium, and fish fry. If flat ice-lollie sticks or unpainted plant labels are placed on the surface of the porridge, the worms will swarm on to them and can be rinsed off into the aquarium without introducing too much porridge.

Whiteworm

These are quick-breeding annelids which can be obtained as a culture from aquarium shops. The culture is placed on the surface of moist loam in a container and covered with slices of bread which have been soaked in water or milk. The slices of bread are lifted occasionally and the worms 'harvested' with coarse forceps.

Earthworms

Obtained by digging, or less laboriously by watering the playing field with a dilute solution (pale pink) of potassium permanganate. The solution is best applied by pouring a bucketful onto a small area and allowing it to spread naturally. In dry weather it may be necessary to add a second bucket of solution as the earthworms may burrow two feet or more. Rinse the worms which come to the surface before placing them in a wormery.

A large wormery can be made to act as a reconstruction of a typical soil profile. This will require two panes of glass 60 cm high and 45 cm wide, supported 10 cm apart by a wooden frame. The glass should be given a lightproof cover which can be removed for observation. Smaller wormeries for pupil experiments are fully described on page 25 of the *Nuffield Biology Teacher's Guide 1*.

Blowflies and blowfly larvae

The larvae are sold as 'gentles' by fishing shops during the coarse fishing season, and this is the most convenient source of supply. If they are placed in the refrigerator soon after purchase, pupation will be delayed for several weeks. Blowflies may also be cultured by exposing tinned dogmeat to adult flies and then keeping it in a light, airy place, covered by muslin. At 25 °C a continuous culture can be maintained. Tinned dogmeat is used in preference to fresh meat or liver as it is less smelly.

Caterpillars

A suitable cage for the rearing of caterpillars can be made by rolling a sheet of cellulose acetate into a cylinder to fit the base of a round biscuit or cake tin 12–20 cm in diameter. The edges of the cylinder are sealed with adhesive tape and the tin lid is used as a top. The caterpillars must be given a fresh spray of their food plant held in a small bottle of water. Some caterpillars pupate by spinning a coccoon on the food plant, others, such as the cabbage white butterfly pupate, in the soil; for such species a layer of moist, but not wet, peat should be provided in the cage base.

Mealworms

Mealworms are available from pet shops throughout the year.

Flour beetles (*Trilobium* species)

These are best obtained as genetic types from biological supply houses. Stock cultures are kept in jam jars containing five centimetres of medium and covered with muslin. Genetic crosses are kept in 10×2.5 cm tubes containing four centimetres of medium.

The medium used is a mixture composed of 1 part powdered yeast (not pelletted) with 12 parts wholemeal flour. Culture at 25–30 °C in a relative humidity

of 65 per cent. This humidity may be readily maintained by placing the cultures in a box which contains a beaker filled with a saturated solution of sodium nitrate.

Mites can cause problems, and it is best to heat-sterilise the medium before use.

To handle the flour beetles first chill the culture in a refrigerator to immobilise them, and then separate them from the flour with a kitchen flour sieve.

Drosophila

Preparation of the medium

Add 6 g agar, 35 g black treacle, and 72 g oatmeal to 560 cm^3 of water and bring to the boil, stirring continuously. Boil with continuous stirring for 15 minutes and add a pinch of Nipagin (methyl p-hydroxy-benzoate) which acts as a mould inhibitor. This quantity will be sufficient for sixty 3 in. × 1 in. (75 mm × 25 mm) specimen tubes or twelve $\frac{1}{3}$-pint milk bottles. The medium should be allowed to cool slightly before pouring it into the containers to a depth of 1 in. (2.5 cm); pouring is best done though a warmed filter funnel to help keep the sides of the containers clean. When the culture medium has set, add a few drops of bakers yeast suspension. After the steam has escaped, the container may be plugged with cotton wool or foam plastic (see Figure 4.2). A piece of folded filter paper, or non-medicated paper towel, should be placed in the container before the introduction of flies. It does not seem to be necessary to autoclave the containers each time they are used; a thorough wash seems to be all that is needed.

The optimum temperature for experimental matings is 25 °C, but room temperature is satisfactory if longer life-cycles are acceptable. Prolonged exposure at 10 °C or below will kill the flies, and a temperature of 28–30 °C will result in sterile progeny.

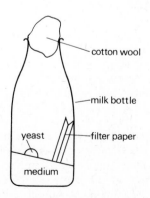

Figure 4.2   Preparation of a medium for Drosphila

Parasitic mites and moulds are the main problem, but they can be completely controlled by regular subculturing if the flies are kept at 25 °C because at this temperature the life cycle of the fly is shorter than that of the pest.

A full account of handling *Drosophila* for experimental work is given in *An Experimental Biology Manual* by G. D. Brown and J. Creedy (Heinemann Educational Books, 1968).

THE AQUARIUM

The keeping of aquaria often seems to have a 'mystique' for the uninitiated. It is however a very straightforward process provided that a few basic facts are understood.

Never use goldfish bowls; they contain a large volume of water for their very small surface area. A well designed aquarium is as wide as it is deep to allow adequate diffusion of oxygen from the air. Very small aquaria should be avoided as they are very difficult to manage; the most useful type for schools is probably the standard 60 cm × 30 cm × 30 cm tank. Tanks 37.5 cm deep are also available and may look more attractive but they will only house the same number of fish. The limiting factor is the surface area, **not** the volume of water.

Fishtanks should be sited near a power point so that electrical equipment can be used without trailing cables. The ideal site is well away from sunny windows which encourage the growth of green algae; lighting is best provided artificially so that the amount can be regulated by duration and bulb size.

Fish tanks with artificial light and heat are a potential source of danger and it is as well to consider safety from the outset. The aquarium frame and hood should be earthed, if made of metal, by brazing on an earth strap. Only a heater and thermostat with an integral earth wire should be fitted (the pattern developed in response to a C.L.E.A.P.S.E. Biology panel request is admirable, whereas the majority of heaters sold to the amateur market are very questionable from the electrical safety point of view). A heater will not be necessary when keeping British freshwater life or goldfish, unless the tank is situated in an unheated animal house when a heater and thermostat may be included on a low setting as a precaution against frost damage.

When the tank is in position, a layer of well-washed aquarium gravel is spread on the bottom; the depth of gravel should be about 8 cm at the back of the tank, sloping to 2 cm at the front. Any rocks are then placed in position (taking care not to use limestone or other soluble material) so that they will not form pockets to obstruct the gravitation of 'mulm' (which consists of fish droppings, unconsumed food, and bits of decaying plant), towards the front from where it can easily be removed by siphoning.

To fill the tank, place a few layers of paper over the gravel with a saucer on top. Water can now be run in gently with a length of tubing from the tap without disturbing the gravel. The paper is removed when the tank is full. Tap water is generally suitable for all but a few species, except in areas where the tap water is alkaline. If the local water is heavily chlorinated it is wise to allow the tank to stand for 24 hours before introducing plants and fish.

Plants are placed around the back and sides of the tank to leave a clear swimming area where the fish can be seen. Plants look best in groups of the same kind and should be planted with their crowns just above the gravel. Trailing plants, such as Myriophyllum, Elodes, and Fontinalis, should be broken into lengths of 10–15 cm and then inserted in the gravel until the first node is covered. Floating plants such as duckweed, Azolla, and Riccia are of botanical interest as well as providing a protective layer for young fry and small animals. The plant sold under the name of 'Indian fern' is of particular interest; it reproduces vegetatively by producing plantlets on its leaves, also producing totally different plant forms when grown as a rooted plant or as a floating plant.

If the tank is to hold tropical fish, it is advisable to allow the tank to settle for a week to ensure that the thermostat is functioning correctly. It is usual to set the thermostat to give an average temperature of 22 °C with a fluctuation of about 2 °C on the make and break.

When fish are introduced, the container in which they arrive should be floated in the aquarium until it has assumed the temperature of the surrounding water, and then gently sunk to allow the new fish to swim quietly into the tank.

Feeding with dried fishfood of a reliable make should occur once a day, but the fish can be safely left at week ends, and for periods of a week to ten days during the holiday, provided that the tank is well established and plenty of food, including live food, has been given just before the holiday. It is safer to leave fish for ten days than to allow inexperienced persons to feed them with the subsequent risks of overfeeding and foul water. If it *is* necessary for inexperienced persons to come in during the holiday, it is best to leave them a number of packets of food which are sufficient for just one feed.

In a well set up aquarium very little cleaning should be necessary, and it *should never be necessary to change the water*, although a routine should be established to draw off the mulm with a siphon once a fortnight. The mulm incidentally is often a superb culture of ciliates and frequently contains diatoms.

Aeration of the tank can be increased by the use of a pump and diffuser block. This will increase the fish holding capacity of the tank by 50 per cent, but it can be irritating in a teaching laboratory to have to listen to a pump running continuously. Some types of pumps are much quieter than others. Some aquarists use a pump to operate a continuous filtration system which keeps the water crystal clear. Algae on the front glass can be removed with a razor blade in a special holder sold by pet shops. For the tropical tank, catfish are available which will eat the algae.

Marine aquaria

Marine aquaria are very difficult to maintain. It is usually best to half fill the tank with sea water and mark the level carefully, so that the aquarium can be topped up to this mark with distilled water regularly to prevent a change in the concentration of salts present. Aeration is desirable, but the greatest problem of all is the maintenance of a low temperature. In technical colleges and university laboratories a continuous-flow heat exchanger may be available. At least one school known to the author has overcome the problem by using a discarded beer cooler from a public house! If possible the temperature should be kept below 18 °C.

Why keep aquaria?

In many schools a great deal of effort is involved in keeping aquaria with very little educational advantage other than the development of the interests of a few keen pupils. With a little forethought on the choice of material and some judicious display labelling, a great number of biological (and physical) teaching points can be made. A well balanced aquarium is a useful biological system from which to begin the discussion of food webs, nitrogen, oxygen, and carbon cycles. Courting behaviour, sexual dimorphism, and a wide range of breeding patterns may be observed. Most of the livebearers such as guppies, platies, and the Mexican swordtail have mouths placed in a position for surface feeding; the barbs, tetras, and zebra fish have central mouths for feeding at all levels; the catfish have ventral mouths for bottom feeding. Angel fish, besides being decorative, show pelvic fins modified as tactile organs – a vertically striped pattern which makes them conspicuous in the aquarium but is basically camouflage. Place them in a tank planted fully with Valisneria, a grasslike plant, and they will be hard to see. The chromatophores in the stripes are controlled by the nervous system and can be contracted suddenly, leaving the fish pale all over – they often respond this way when startled.

Fish diseases

Fungus is often seen on fish as white or opaque patches and may cause a condition known as fin-rot. Fungal spores are always present but do not attack unless the mucous membrane on the surface of the fish is damaged. For this reason fish should never be handled; when netting them take care not to abraid the skin. Fish affected by fungus should be transferred to a separate tank containing 40 cm^3 of 1 per cent phenoxetol per gallon of water. The fish should be left in this solution until cured.

Swim bladder disease results in fish having difficulty in maintaining their balance – some swim upside down, others vertically. There are probably a variety of causes and some fish recover naturally; the majority are incurable, although they may live for a considerable time in this condition. The disease is *not* infectious but in a school tank it is best to kill affected specimens. The most humane way to kill a fish is to throw it hard at the floor; needless to say, this should not be done in the presence of pupils.

Dropsy. Fluid accumulates in the tissues and the fish swells with the scales standing out from the body. The cause is believed to be a virus attack followed by bacterial infection. There is no cure at present, and the affected fish should be humanely killed.

White spot is a common disease of tropical fish which will quickly spread though a tank and kill all the fish. Fortunately the disease is curable. The first signs of attack are fish with white bubbles the size of a pinhead on their bodies. These bubbles are protozoan parasites within a cyst; shortly the cyst will burst releasing a large number of the parasites, *Ichthyophthiriasis multifilaris*, which then cause further infections. At the point when the parasites are released into the water they can be destroyed by drugs. Various proprietary drugs are available from aquarist shops; those based on mercury salts are best avoided as in the author's experience they invariably lead to sterility in livebearers. Some of the proprietary cures can be used in the tank without removing the plants. One of the best treatments is to transfer all the fish to another tank without plants and add 1 cm^3 of 1 per cent medical quality methylene blue per gallon of water. The fish should be cured within a week. The parasite cannot survive for long without fish and the original tank may have its fish returned in fourteen days. Another well tried method, but which some aquarists suspect of having undesirable long term affects, is to add half a grain of quinine hydrochloride per gallon of water on three successive days. This treatment is best carried out in a separate tank as the drug damages plants. Whatever treatment is employed, the temperature of the tank should be raised to 27 °C; this gives maximum development to the parasites and ensures their release into the treated water in the shortest possible time.

White spot usually breaks out after the introduction of new fish from pet shops. It is a wise precaution to float new fish in a jamjar in the aquarium so that they can be observed for a week before release. Outbreaks of the disease sometimes follow a lowering of the temperature after heater failures.

Netting fish

Never chase fish with the net. Place a square-framed net near the corner of the tank and drive the wanted fish into it with a stick.

AMPHIBIANS AND REPTILES

Vivaria

The common British amphibians and reptiles may be kept in classroom vivaria during the spring and summer months. A responsible attitude towards collecting with an eye on conservation problems is essential. All British amphibians and reptiles hibernate, and suitable hibernation conditions cannot be easy provided in the classroom. Amphibians and reptiles should only be kept during the summer term and should be very well fed to allow them to build up the necessary food reserves for hibernation. They should be released *in the area in which they were collected* at the end of the term. Imported species may be obtained from dealers, but they have to be maintained throughout the winter.

Commercially available vivaria usually have a sloping glass front but this is not essential. The size of the vivarium should be related to the habits of the species kept – climbing and jumping types need deep vivaria, whilst the slower crawling types need more floor area. Old aquaria which have developed leaks make very good vivaria if they are otherwise sound. Small reptiles are skilled escapists and it is essential to provide a well fitting lid, preferably made of perforated zinc. When setting up the vivarium, reproduce the basic essentials of the natural environment as far as the restrictions of space will allow. Amphibians

need a cool humid atmosphere as they depend upon a moist skin for the greater part of their respiration – this can be achieved by covering the floor of the aquarium with 10 cm of damp peat and embedding a dish of water at one end. Shady hiding places constructed from flat stones should be provided, and a length of stiff branch should be made available for climbing species. For amphibians, the temperature should not exceed 20 °C. If plants are required, ferns are suitable; the vivarium should be set in a shady part of the laboratory. If reptiles are to be kept, the floor of the vivarium should be covered with dry sand or gravel, with a dish of water embedded at one end, and suitable stones and branches provided to give both hiding places and basking spots; the vivarium should be set up where it receives sun, and succulent plants may be grown if wished. In addition to less humid conditions, reptiles also need slightly higher temperatures, 20–25 °C during the day and a night minimum of 13 °C. Drinking water must always be available.

Frogs

Spawn should be collected in early March if tadpoles are required – a maximum of 50 eggs is sufficient for a 24 in. × 12 in. × 12 in. (60 cm × 30 cm × 30 cm) tank which should be filled to a depth of about 15 cm. Rocks or floating wood should be provided to enable the young frogs to climb out of the water. A handful of filamentous algae from a pond will provide food during the early vegetarian stages, and algal growth on the glass of the aquarium will also provide grazing. Later, when the tadpoles become carnivorous they should be fed on small water fleas or, as a substitute, small pieces of lean meat or raw liver should be suspended in the water on cotton. Do not leave the meat in the water for more than an hour at a time or it will cause the water to become foul. The young frogs are best returned to their natural habitat but they may be fed for a week or so on aphids. Adult frogs should only be kept for short periods. They need 60 cm headroom for jumping and consume a large quantity of insects, white slugs, and spiders.

Common toad

Management considerations are as for frogs, but the eggs are to be found a little later, in late March and early April. The adults are much easier to keep in a vivarium and will become quite tame. They feed on a wide range of live food: slugs, spiders, earthworms, and a wide range of insects.

Tree frogs

They are actually toads, and can be kept successfully in the vivarium throughout the year provided that a suitable temperature and a continuous supply of flies for food can be maintained.

Newts

Newts emerge from hibernation at the end of March and enter the water for breeding. If brought into the aquarium at this time, courting and egg laying can be observed. After egg laying, the adults must be moved to another aquarium or they will eat the eggs. The adults remain aquatic until July when they crawl out onto rocks. During the aquatic stage the adults must be fed on waterfleas, whiteworm, chopped garden worm, or small pieces of fresh meat (take care not to foul the water by overfeeding). Newts are best returned to their natural environment during July as then they normally hide under stones all day and only emerge to feed at night. The tadpoles are carnivorous from hatching and may be fed on *Infusoria*, followed by *Daphnia*, as they grow. Tadpoles are ready to emerge as young adults in about 12 weeks, and rafts or rocks must be provided. The young newts should also be returned to the wild in July.

Common lizards and slow-worms

These need to be handled gently and should never be picked up by the tail as they shed this organ readily. Once shed, the tail will be regenerated but will remain stumpy. They may be successfully kept in the vivarium during the summer term and fed on small earthworms, slugs, and small insects – in particular mealworms and blowfly larvae.

Grass snakes

Grass snakes can be kept in the vivarium for short periods but are not easy to keep in captivity because they often refuse food. Their normal food in the wild consists of newts, lizards, small voles, mice, and young birds. When handled they may produce a malodorous fluid from between the scales as a defence. It is probably best to bring a grassnake into the laboratory for just one week to create interest and then to release it again at the place of capture.

Tortoises

Tortoises are best kept out of doors during the summer. They require a drinking container of adequate depth as they need to immerse their whole head when drinking. Tortoises are herbivores and need generous provision of cabbage and lettuce with the opportunity to graze on grass. If powdered cuttlefish bone is sprinkled on their lettuce, this will ensure an adequate supply of calcium for the shell. In the autumn tortoises hibernate; a tea-chest half filled with dried leaves or peat, and placed in a cool but frostproof outhouse will provide suitable conditions. The animals should not be disturbed from October until they are ready to emerge in the spring.

Terrapins

The small green baby terrapins which are often sold in pet shops come from North America, and it is nearly impossible to provide them with suitable conditions and food. Because of this they seldom survive and should never be kept in schools. The slightly larger and less colourful European terrapin is more suitable. The European terrapin is a good swimmer and usually feeds in the water, but it must be provided with a basking stone in its aquarium. Small terrapins can be fed on chopped earthworm or whiteworm; larger specimens should be fed three times a week on small earthworms. Terrapins can be hibernated in a tin of damp peat kept in a cool, frost-free place.

There has been some suggestion recently that imported terrapins may carry typhoid and salmonella food poisoning. Until the evidence is more certain, schools would be wise not to acquire new specimens from pet shops.

HATCHING THE EGGS OF THE DOMESTIC HEN

Fertile eggs should be stored for a maximum of seven days before incubation begins, the optimum temperature being 7–15°C. Storage for longer periods up to fifteen days is possible but viability falls off. Viability of the eggs is adversely affected by shaking prior to incubation.

Fertile eggs are incubated at an optimum temperature of 38 °C and an optimum humidity of 60 per cent. Uniform heating is just as successful as top heat. The incubator must be well-ventilated to allow oxygen to reach the egg, and carbon dioxide to leave. Eggs must be turned twice a day for the first 18 days of incubation to prevent adhesions forming in the embryonic tissues – a cross marked in pencil on one side of the egg will help in identifying which eggs need turning. From the 18th day until hatching on the 21st day (under optimum conditions), the eggs should remain unturned in the horizontal position.

HATCHING TROUT EGGS IN THE CLASSROOM

Trout eggs can be hatched with little difficulty or equipment in any classroom where running water is available.

The eggs are available from December until February or early March and may be obtained from any commercial hatchery. If a local fish hatchery can supply eggs this is the best arrangement. If no local hatchery exists, eggs may be obtained from a number of addresses which can be found in angling magazines. One firm, which has been providing eggs for school use for many years is:

The Surrey Trout Farm & United Fisheries Ltd.,
The Midland Fishery,
Nailsworth, Glos.
or
The Surrey Trout Farm,
Haslemere,
Surrey

The minimum order is 100 eggs, and they can only be supplied Cash with Order. As far as possible this firm will try to post for a specified date, but try to avoid despatch later than Wednesday in any week so that eggs do not spend the weekend in the post. The eggs are sent packed in a small tin and are supported on damp muslin or moss.

As soon as the eggs arrive they should be emptied gently into fresh water and any eggs showing signs of injury or discolouration should be removed. Healthy eggs are translucent and firm, with the dark eyes of the embryo fish showing through the egg membrane. Diseased or dead eggs usually look white, cloudy, and opaque (owing to attack by fungus and/or precipitation of proteins in a dead egg).

When the eggs have been picked over they should be placed in a glass tank with a constant circulation of water. This can be arranged by means of a slowly dripping tap which ensures that the water is changed steadily and that the supply will be well oxygenated (Figure 4.3). The tank should be placed in as cool a place as possible and shaded from direct sunlight. The rate of development of the eggs depends upon temperature, the best range of temperature being between 4 °C and 10 °C (40 °F and 50 °F).

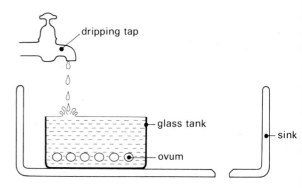

Figure 4.3 Tank arrangement for hatching trout eggs

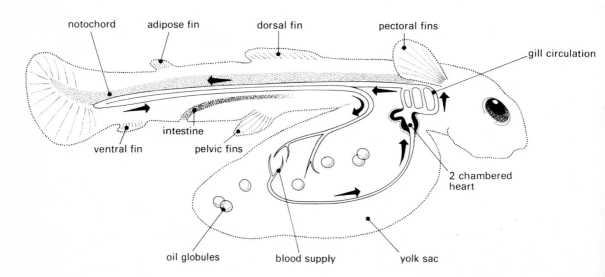

Figure 4.4 An alevin, 9 to 10 days after hatching

The eggs will begin to hatch a few days after arrival, the alevins wriggling free of the egg membrane to lie on the bottom of the tank, with occasional movements, whilst the still-large yolk sac is absorbed (Figure 4.4). After a few days the alevins become lively and are liable to swim over the edge of the tank. At this stage it is advisable to cover the tank with muslin or perforated zinc (copper gauze should not be used as young fish are highly susceptible to poisoning by copper ions).

Throughout the time that trout ova and young fish are being kept it is necessary to pick out any diseased or discoloured specimens at least once per day. After hatching there is usually quite a large mortality rate, but with care it is possible to raise some of the fry, which may be kept in an aquarium. Trout require a fairly high concentration of oxygen in the water (they naturally live in fast-flowing streams). The author has kept two trout about ten centimetres in length for over a year in a 2 ft × 1 ft × 1 ft (60 cm × 30 cm × 30 cm).

Figure 4.5 shows the stages in development which may be observed in school. The alevin soon after hatching is very transparent and the beating of the heart and gill circulation can be clearly seen. If a micro projector is available, the alevin may be placed in a 'pond life' cell to enable more detail to be seen by projections.

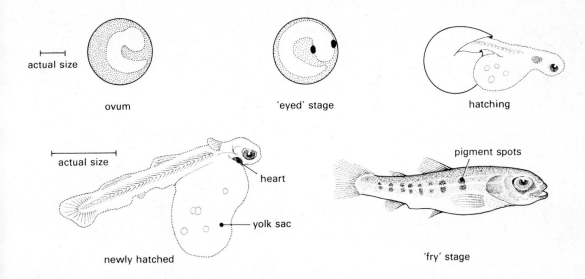

Figure 4.5 The stages of development of trout observable in school

5. Greenhouse Management

TYPE, DESIGN, AND PURPOSE

As greenhouses are often set up after a school itself has been built, teachers and technicians are sometimes consulted at the planning stage, so a few words on design will not be out of place here.

The design of greenhouses is a great deal more specialised and scientific than would appear at first sight. It is definitely *not* a job for the school's architect. The author has seen many architect-designed greenhouses, none of which are satisfactory from the horticultural and managerial point of view, and all of which cost far more than their commercial counterparts. In every case the extra cost incurred by the 'one-off' design would have paid for the installation of automatic heating, watering, and ventilation equipment.

Size is clearly related to function, but it should be born in mind that smaller glasshouses less than 15 ft × 10 ft (4.5 m × 3 m) are very difficult to control satisfactorily because of the rapid thermal changes caused by sunlight. The size chosen will reflect the degree to which the house is to provide plants for demonstration and experiment, and the amount of space required for pupil experiments at any given time. A 15 ft × 10 ft (4.5 m × 3 m) house, if well planned, may be large enough to house a demonstration collection and to raise some plants for demonstration experiments. If pupil experiments are contemplated, 30 ft × 15 ft (9 m × 4.5 m) should be regarded as the minimum. Wherever possible, 15 ft (4.5 m) should be regarded as a minimum width as this will provide sufficient space for side benches and/or beds as well as a centre bench. The centre-bench type of layout gives a more satisfactory space and better circulation for pupils than a side bench layout. At least one Education Authority has recognised the need for more space for pupil experiments and is planning 60 ft × 60 ft (18 m × 18 m) commercial span houses. If tropical plants are to be kept, consideration should be given to partitioning the house, so that a small section can be kept at a higher temperature. Always choose a commercially produced house with an aluminium frame as this will be free of maintenance costs, and perhaps more important, need never be taken out of commission for painting. Glass to ground houses are more versatile than the halfwalled plant houses. If money is particularly tight, dispense with benching rather than choose too small a house – many efficient nurseries grow 'on the ground'.

Siting, services, and layout should all be considered at an early stage. If a choice of site can be made, remember that the major problem in Britain is that of poor winter light, and for this reason a house sited east–west is to be preferred. Shelter from north and east winds will reduce heating costs, but shading from nearby trees and buildings must be avoided. In urban areas a rooftop greenhouse may be the only possibility, the only snag being that water pressure is sometimes too low to operate some automatic equipment.

Figure 5.1　A well-designed greenhouse frame, with guttering built into the top glazing bar

Figure 5.2　Painting wooden framed greenhouses is an expensive process

44

SERVICES

Adequate electrical supplies and water supplies are needed. Controls for both water and electricity should be placed near the door – it is desirable to be able to switch off overhead sprinklers before entering with a class! Waterproof power points, switch-boards, and light fittings *must* be used. The number of outlets will vary but it is better to plan generously as water and electricity are used in close proximity in a greenhouse and makeshift arrangements at a later date cannot be tolerated.

Figure 5.5 A well-designed electrical distribution box for a school greenhouse; note the low wattage heater, which is connected to a frost-stat to prevent condensation when the temperature falls

Figure 5.3 The 'Camplex' greenhouse control panel

Figure 5.4 Waterproof sockets are essential unless all equipment is permanently wired to the distribution box

Figure 5.6 Well-planned greenhouse services provide several points for water connections

Water

There should be a main stopcock followed by a short length of copper pipe fitted with stopcocks for each outlet required. If the outlet ends of these cocks are fitted with about ten centimetres of copper pipe, hoses can be connected with circlips. In a 30 ft × 15 ft (9 m × 4.5 m) greenhouse laid out as shown in Figure 5.7, six water points are desirable:

1 Tap (often forgotten in automatic houses but very necessary)
2 Overhead watering and damping line
3 Capillary bench
4 Mist propagation bench
5 Trickle line to bed
6 Spare point or trickle or capillary supply to the side bench if its use changes.

Electrical supply

In the same layout the following electrical services are included:

1 Tubular heaters
2 Extractor fan
3 Small air-circulation fan
4 Low voltage soil warming cable for propagation bench
5 Electrical controller for mist on propagation bench
6 Low voltage soil warming cable for bed
7 Waterproof lighting
8 Two waterproof power points, one at each end of the house, for use during experiments (such as supplementary lighting and temporary attachments such as soil sterilisers).

LAYOUT

The basic problem with the school greenhouse is that it is a multi-purpose unit, and any layout is of necessity a compromise. The example given in Figure 5.7 provides bench space for propagation, capillary benching for pupil experiments, a small potting bench area, a bed with soil warming for permanent subjects and for limited experiments with tomatoes, and 90 ft² (8.4 m²) slatted benching with overhead spray watering for experiments and for demonstration pot plants. Below the slatted side bench provides an ideal environment for ferns, mosses, selaginellas, and liverworts. It will be suitable also for some of the shade-loving flowering plants. The continuous path provides good pupil access.

Such a greenhouse equipped with the automatic heating, ventilating, and watering equipment outlined below can be left without attention throughout

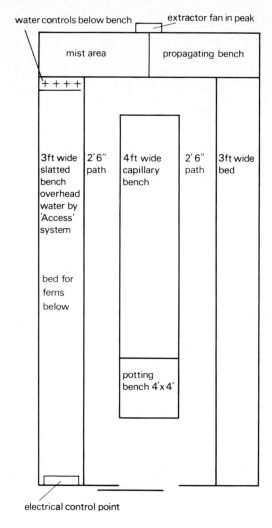

Figure 5.7 Possible layout for a 30 ft × 15 ft (9 m × 4.5 m) greenhouse for automatic watering and ventilation

school holidays if necessary. Ninety per cent of the work involved in running a school greenhouse is taken up by watering and adjusting ventilation – the expenditure of between £30 and £60 can completely automate this part of the work.

MANAGEMENT

Automation

Teachers' and technicians' time is far too valuable to spend on greenhouse management chores which can be carried out automatically. The provision of good growing conditions should be left to automatic equipment. Large plants, and material pupils will examine in the greenhouse, should be grown in permanent beds if possible.

Ventilation

Good management of ventilation is essential if good growing conditions are to be obtained. A well-designed glasshouse has peak ventilation equal to one sixth of the floor area of the house. Such a standard is rarely found in smaller houses, but reasonable control of ventilation may be achieved by fitting a device which varies the ventilator opening with temperature by mechanical means. Such devices depend upon the expansion and contraction of a material with a high coefficient of thermal expansion operating a lever mechanism – the 'Bayliss' and 'Thermofor' ventilator controls are of this type.

Figure 5.8 An automatic extractor fan, the 'Humex' Autofan

Figure 5.9 A rod-type thermostat, the 'Humex' C/5 soil/air model

The most appropriate control for school greenhouses is a 9 or 12 in. (23 cm or 30 cm) louvred extractor fan operated by a 24 in. (60 cm) rod-type thermostat. The thermostat should be placed so that it reacts to air temperature rather than solar radiation! A suitable fan and thermostat will cost around £60 for a house up to 30 ft × 15 ft (9 m × 4.5 m). Houses larger than this are usually controlled by geared motors which open the peak ventilators (Figure 5.10). Once chosen, an extraction fan should be placed near the peak in the end wall, opposite the door. Roof mounting types are available, but they produce unwelcome shading during the winter months. The ventilator thermostat should be set to operate at 5 °C above the operating temperature for the heating system during the spring and winter, and at 13 °C during the summer when the heating is not in use.

Air circulation

This is essential for healthy growth during the winter months; a stagnant atmosphere seems to encourage the growth of fungal diseases. A fan heater will produce the necessary air movement; with other types of heating, air movement may be achieved by placing small ventilators in the side walls below the heaters, or using a small fan such as that made by Simplex for the purpose (cost approximately £8).

Heating

With regard to efficiency, safety, and convenience, tubular electrical heaters of the type supplied by Simplex and Ecko are the most suitable form of heating for schools. It is essential, if the greenhouse is an integral part of the design of a new school, that its heating is independent of the school system which is often turned off at weekends and during holidays.

Modern rod-type thermostats give reasonable control at a cost of about £15, but much more accurate control can be achieved by the use of aspirated thermostats, which draw air over the thermostat by means of a fan. The cost, in the region of £25–35, can be made up by the saving in electrical costs in a couple of seasons. In a small house the aspirated thermostat fan will help to produce some air circulation. Aspirated thermostats are also manufactured which give differential day and night temperatures.

The loading of the heating system will depend on the type of house, the locality, and the degree of exposure. The Electricity Council publishes a free 88 page book called *Electrical Growing* which contains a detailed appendix on how to calculate the kilowatt

Figure 5.10 Geared motors used to give very precise
ventilation control on a large greenhouse

Figure 5.11 Electrical tubular heaters ('Humex')

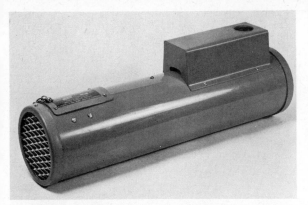

Figure 5.12 An aspirated thermostat gives the most accurate control of heating and ventilation equipment. This is the 'Camplex' HD 7074

heater loading needed. The system should be designed to maintain a temperature 12 °C above that outside.

Benching should always be fitted with small air space left at the back to allow warm air from the heaters to rise. If this is neglected, cold spots will occur on the benches and frosting of some plants near the glass may occur even in a heated house. Tubular heaters should be evenly spaced.

Soil-warming and air-warming cables

The permanent bed may be heated by a soil-warming cable placed at a depth of 40 centimetres. Soil-warming cables are also used in the propagating bench. The cables may be mains or low voltage types which run from a mains transformer, although the low voltage type is to be preferred for school use. In propagating beds and benches, the output of the cable is controlled through a rod-type thermostat which is usually set to give a temperature of 26.5 °C. In heated beds, control may be achieved by use of a thermostat, or by using a time switch to give an eight hour 'dose' each night. When setting up heated propagating benches, very even heating is obtained if the wire is covered with 3 cm of sand, a layer of chicken wire is placed on the sand to act as a heat distributor, and then a further 5 cm of sand are added.

Very few experiments in schools need the full height of the greenhouse, and it is possible to utilise frames and cloches in conjunction with soil and air warming cables to increase capacity at low capital cost.

Watering

Capillary benches are constructed as shown in Figure 5.13. Pots used on capillary benches should not be crocked as this prevents root penetration and capillary contact between the pot and the bench sand. Well designed plastic pots are the most suitable (Ward's plastic pots can be strongly recommended). Some coloured pots, particularly greens and yellows, seem to inhibit root growth; brown pots are the most satisfactory. For capillary bench growing it is essential to use an 'open' compost in the pot to allow air to permeate. Specially designed plastic seed trays are also made for capillary benches – they have long slots rather than holes in the base. When choosing seed boxes it is as well to pick a stackable pattern.

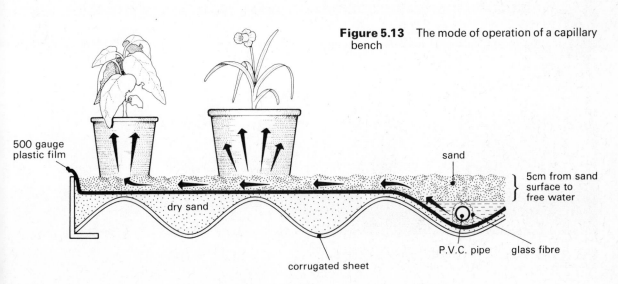

Figure 5.13 The mode of operation of a capillary bench

500 gauge plastic film

sand

5cm from sand surface to free water

dry sand

P.V.C. pipe   glass fibre

corrugated sheet

Figure 5.14 Commercial capillary trays for the small greenhouse

Figure 5.15 A plant lifted from a capillary bench to show rooting into sand. These roots can be removed without harm if the pot has to be brought in from the greenhouse

If mains water is available in the greenhouse, overhead mist watering is very easy to install. The 'Access' system (available from Access Ltd) is inexpensive; it depends for its operation on a simple electrical sensing device which operates a solenoid water valve. This controller will operate any number of mist nozzles, each covering 10 sq ft of bench.

Trickle or drip irrigation systems are inexpensive to install and some, such as the 'Access', do not require mains water.

Beds, and if desired capillary benches, may be supplied by a cheap perforated polythene tube, or by the trickle system supplied by Boil Home Garden Department. Such systems are also suitable for watering straw bales for growing tomatoes and cucumbers.

Figure 5.16 The 'Access' overhead mist watering system

Figure 5.17 The 'Access' drip irrigation system

Fertiliser distribution

If a Cameron dilutor filled with concentrated ferti-liser solution is placed in the water supply line of a drip feed (trickle) system, fertiliser can be applied continuously at a predetermined rate. A dilutor can be used with other systems but has the disadvantage of encouraging algal growth on the surface of capillary benches. The dilutor is most useful for applying fertiliser to beds and bales as well as for trickle irrigation of pots containing soil-less com-posts such as 'Levington'.

In commercial practice, control of water appli-cation rate and the amount of fertiliser in solution, is related to the amount of solar radiation on the previous day. Such control equipment is at the moment rather expensive but, as the electronics are not particularly complex, to build a system could well form a 'technology' project within the school.

Pest control

If the greenhouse is used purely to maintain a collection of teaching material which will be brought into the laboratory, continuous control of pests by the 'Aerovap' system is possible. The 'Aerovap' system depends on the continuous evaporation of pesticides by use of an electrical element; the vapour deposits itself on all plant surfaces in the house. If the house is to be used for pupil work, the author feels that it would be wiser to rely on other means of control such as the use of appropriate smoke cones which are available from horticultural sundriesmen. Smoke cones are best used at the end of the day in a closed greenhouse. When the cone has been ignited

the operator should leave immediately, *locking* the door behind him to make certain that no one enters. The following morning when the house is unlocked it should be thoroughly ventilated.

Shading

Shading is used to prevent scorching of delicate plants during the summer months. It must be elim-inated during winter as all possible light must be

Figure 5.18 An electrically operated pesticide evaporator

admitted. External wooden blinds on rollers are regarded as the best device, but they are both expensive and cumbersome. Green polythene blinds are now available for use inside the roof of the house, but they become brittle after a few seasons and are difficult to use if there are climbers in the house. The author prefers to use a wash of whiting or slaked lime on the sunny side of the roof and to leave the rest to efficient ventilation and damping down. In wet weather the lime becomes translucent and admits more light. Application is made with a soft household brush and only takes a few minutes; some seasons a second application is necessary. The same brush in conjunction with a hose will remove any remaining wash at the end of the season. Proprietary shading washes containing size are also available, but these are more difficult to remove at the end of the season.

Hygiene

The removal of dead leaves, flowers, and experiments which have come to an end will help to reduce the spread of fungal diseases and aphids. The greenhouse is an expensive item and consequently should be used for the purpose for which it was provided. It should not be used as a store for all the 'outdoor' paraphernalia which biologists accumulate.

Lighting

The greenhouse should be fitted with a waterproof light or a fluorescent tube to allow work to be carried out on winter afternoons.

Other lights may be required from time to time for horticultural or experimental purposes such as investigation of daylength effects and nightbreaks. Suitable lights are available from Simplex and other suppliers. In the school greenhouse, with a mixed collection, remember that increasing the daylength of one plant may also affect its neighbour which is being used in an allegedly controlled experiment. With a little ingenuity it is possible to screen plants with *heavy* black polythene (the lighter gauges of black polythene transmit some light).

6. Chromatographic Techniques

Table 6.1 Identification of spots on chromatograms

Substances	Solvents
Chlorophyll and carotenoids	100 parts petroleum ether (B.P. 100–120 °C), 12 parts propanone (acetone). For thin layer techniques, 100 parts of pet. ether (B.P. 60–80 °C) to 20 parts of propanone is better.
Plant acids	100 parts of butyl formate, 40 parts of 98 per cent methanoic (formic) acid, 10 parts distilled water. Take care with these substances. Add 0.5 g of sodium methanoate to each 100 cm³ of solvent and sufficient solid bromophenol blue to turn the solvent a pale orange colour.
Amino acids, sugars, and plant acids	28 g of pure crystalline phenol **(use rubber gloves for handling this substance)** should be placed in a separating funnel with 12 cm³ of distilled water. Add a little NaCl and shake thoroughly. Fill the funnel with coal gas to prevent oxidation and leave for the layers to separate out. This may taken an hour. The lower layer is the phenol saturated with water; reject the upper layer.
Amino acids, and sugars	120 parts butan-1-ol, 30 parts glacial ethanoic acid, 60 parts distilled water.
Amino acids (2-way)	For the second solvent use 180 parts ethanol, 10 parts ammonium hydroxide, 10 parts distilled water.
Anthocyanins	40 parts butan-1-ol, 10 parts glacial ethanoic acid, 50 parts distilled water. Make up the solvent in a separating funnel and reject the upper layer.

PHOTOGRAPHING CHROMATOGRAMS

Reflex contact document paper should be used; there are two grades, rapid and normal. This may be handled in a bright yellow safe light. Opening the door does not matter, unless light falls directly on to the paper.

Place the chromatogram in close contact with the photographic paper.

Expose the paper by holding it one metre from the lamp. On average exposure must be for 8 to 12 seconds for Whatman number 1 filter paper, and 3 minutes for Whatman number 3 filter paper. If the spots are very thick, it may be possible to make several exposures to resolve them. Exposure times are usually not critical. Very thick spots may be made less opaque by wiping with tetrachloromethane or ethyl ethanoate.

Develop the film in the following solution:

Metol	3 g
Na_2SO_4 (anhydrous)	50 g
Na_2CO_3 (anhydrous)	70 g
Hydroquinone	13 g
NaBr	1 g
Distilled water	1 litre

Thirty to sixty seconds is usually a suitable time for development.

Dip the print in water and then in an acid fixing bath made up as follows:

$Na_2S_2O_3$, $5H_2O$	250 g
$K_2S_2O_5$	25 g
Water	1 litre

The print should only be in this bath for a few seconds, then wash the print for a few minutes in running water, blot it dry, and hang it up to completely dry. A short wash in propanone (acetone) may be useful for the rapid drying of the print.

Table 6.2 **Extraction and development**

Substances	Methods of extraction from plant material	Development
Chlorophyll and carotenoids	Grind in acetone	No development necessary. Ultraviolet light is useful for viewing the spots. Store the chromatograms in the dark as these pigments are not stable in light.
Plant acids	Grind in 70% ethanol	Dry carefully in a fume cupboard. Sometimes further development is worth while by holding the chromatogram over a bottle of ammonium hydroxide. Do not let the paper get too blue. Sometimes the spots become more clear after a few days.
Amino acids	Grind in water or 80% ethanol	Dry in a fume cupboard. Spray with 2% ninhydrin in propanone (acetone). Heat the chromatogram strongly for the spots to appear.
Sugars	Grind in 50% ethanol	Dry the chromatogram in a fume cupboard. Spray with either a 3% solution of para-anisidine hydrochloride in butan-1-ol plus a few drops of hydrochloric acid or with a 10% solution of resorcinol in propanone (acetone) plus a few of hydrochloric acid.
Plant acids	Grind in water	Dry the chromatogram in a fume cupboard. Spray with bromothymol blue adjusted to pH 8.5 with sodium hydroxide
Amino acids	Grind in water or 80% ethanol	Dry carefully as butanol is highly inflammable. Spray with 2% ninhydrin in propanone (acetone). Heat strongly for the spots to appear.
Sugars	Grind in 0.1 M sodium acetate in water	Dry carefully as butanol is highly inflammable. Spray with 10% resorcinol in propanone (acetone) plus a few drops of hydrochloric acid. Heat strongly for the spots to appear.
Amino acids	(as above)	Dry carefully as ethanol is highly inflammable. Spray with 2% ninhydrin in propanone (acetone). Heat strongly for the spots to appear.
Anthocyanins	Grind in 80% ethanol	No development is necessary. Treatment with ammonia vapour may intensify the spots.

Table 6.3 Table of R_f values [after W. M. M. Baron, *Organization in Plants* (Edward Arnold, 1967)]

Identification of spots

Group of substances	Thin layer chromatography			Paper chromatography		
	Substance	R_f value	Colour	Substance	R_f value	Colour
Chlorophyll and carotenoids	Chlorophyll b	0.10	Yellow-green	Chlorophyll b	0.45	Green
	Chlorophyll a	0.11	Blue-green	Chlorophyll a	0.65	Blue-green
	Xanthophyll	0.22	Yellow	Xanthophyll	0.71	Yellow-brown
	Phaeophytin	0.28	Grey	Phaeophytin	0.83	Grey
	Carotene	0.90	Yellow	Carotene	0.95	Yellow
	When using petroleum ether (60–80 °C):					
	Chlorophyll b	0.17	Yellow-green			
	Xanthophyll	0.19	Yellow			
	Chlorophyll a	0.23	Blue-green			
	Xanthophyll	0.35	Yellow			
	Phaeophytin	0.44	Grey			
	Carotene	0.96	Yellow			
Plant acids	Tartaric acid	0.32 ⎫		Tartaric acid	0.20 ⎫	
	Citric acid	0.50 ⎪		Citric acid	0.25 ⎪	
	Malic acid	0.61 ⎬ Yellow against		Oxalic acid	0.32 ⎬ Yellow against	
	Pyruvic acid	0.85 ⎪ purple background		Malic acid	0.37 ⎪ purple background	
	Butanedioic (succinic) acid	0.92 ⎭		Butanedioic (succinic) acid	0.57 ⎭	
Amino acids	Glycine	0.29	Orange-red	Glutamic acid	0.38	Orange-red
	Arginine	0.32	Deep red-purple	Glycine	0.50	Brown-purple
	Cystine	0.32	Pink	Tyrosine	0.66	Deep purple
	Valine	0.41	Red-purple	Arginine	0.70	Red-purple
	Phenyl-alanine	0.52	Brown	Alanine	0.72	Blue-purple
				Leucine	0.91	Deep purple
				Proline	0.95	Yellow
Sugars				Glucose	0.31	Red with resorcinol, yellow-brown with anisidine
				Sucrose	0.35	
				Fructose	0.46	
Plant acids				Tartaric acid	0.23	Yellow against a blue background
				Citric acid	0.32	
				Ethanedioic (oxalic) acid	0.35	
				Malic acid	0.43	
				Pyruvic acid	0.59	
				Butanedioic (succinic) acid	0.60	
Amino acids	Glycine	0.20	Orange-red			
	Arginine	0.23	Deep red-purple			
	Tyrosine	0.25	Purple			
	Cystine	0.26	Pink			
	Valine	0.39	Red-purple			
Sugars	Sucrose	0.24	Yellow-brown			
	Fructose	0.31	Yellow-brown			
	Glucose	0.43	Yellow-brown			
Anthocyanins				Delphinidin	0.59	Blue-purple
				Pelargonidin	0.73	Bright red
				Peonidin	0.74	Magenta
				Cyanidin	0.79	Mauve-purple

7. Laboratory Solutions

PURITY OF REAGENTS AND ACCURACY OF SOLUTION

Many of the reagents in science teaching are used dissolved in water or in some other solvent or mixture of solvents. In the case of most bench reagents, a high degree of accuracy is not essential when the solutions are made up, but in other cases, for example the preparation of standard solutions, extreme accuracy is required; both teacher and technician should be fully aware of the standards of accuracy which are sufficient for a given purpose. At the present time there is much confusion in educational establishments about the standards of purity of reagents used, and as a result enormous sums of money are wasted each year by the use of *Analytical grade* reagents for purposes which demand no greater purity than is found in the less expensive *Technical grade* reagents. *General Purpose (G.P.R., G.C.R., or G.L.A.) grades* may be used if higher purity than *Technical grades* is required, but without incurring the expense of using *Analytical grades.*

Most reagents are available in the highly-purified form intended for analytical work, indicated by trade names such as *Pronalys* or *AnalaR*. The label on the bottle of *AnalaR* grade reagents gives the percentage of the reagent present and lists the maximum limits of any impurities which might be present. In the case of a few reagents *special grades* are available for specific reactions which require very low concentrations of a particular impurity which would otherwise interfere with the reaction; needless to say, such chemicals are expensive and should only be purchased for the specific reaction for which they are intended.

General Purpose Reagents (G.P.R.s) with a lower standard of purity are available with a similar type of specification on their label; these are sufficiently pure for most school purposes.

Many reagents are used in pharmaceutical work and dispensing and the letters *B.P.* following the name of the reagent indicate that the substance complies with the standard of purity laid down in the *British Pharmacopoeia.*

Technical grades of many reagents are available which have much lower standards of purity, the bottle label usually bearing no specification.

The efficient teacher and technician will always select the cheapest grade of reagent suitable for the reaction to be carried out. **The only exceptions to this rule are certain organic reagents such as diphenylamine which may carry carcinogens as impurities** – where such impurities exist it is advisable to purchase only AnalaR grades.

Volumetric reagents

For titrimetric work there is a growing tendency for manufacturers to produce *certified standard solutions* ready for use, or ampoules of accurately prepared concentrated solutions which only need to be diluted according to the instructions on the label. The use of such prepared chemicals may represent an overall saving if technicians' time is taken into account, as well as rendering less expensive chemical balances acceptable for school purposes.

DANGEROUS REAGENTS

These will be found listed on page 10. Some dangerous materials such as sodium, potassium, and phosphorus are now available in small packs for educational use. There is no sense in, or excuse for, storing 500 g and 1000 g quantities. Bromine is now available in 1 cm^3, 2 cm^3, and 2.5 cm^3 phials packed in dozens; purchased this way the risk of a serious accident is considerably reduced.

DEFINITION OF BASIC TERMS USED IN RELATION TO SOLUTIONS

Solutions of which the precise concentrations are known are referred to as **standard solutions** and are usually used in quantitative work or very precise experimental work. Substances which are sufficiently pure to be weighed out accurately and can be dissolved in distilled or de-ionised water to give solutions of accurately known concentration are called *primary standards.*

All reagents which are used as standard solutions, but which do not meet the criteria necessary for listing as primary standards, are called *secondary standards*. They are made up as accurately as possible by weighing and then *standardized* by titration against a solution of known concentration. (Very accurate secondary standard solutions are available as volumetric solutions from chemical suppliers – these are often standardized with great precision by electrochemical methods.)

The mole

By international agreement, the *mole* is now the chemists' unit for *amount of substance*. The mole is a *physical quantity*, not a number, and is one of the base units of the SI system of units (see page ix). The definition of this quantity is:

> The amount of substance of a system which contains as many elementary entities as there are atoms in 0.012 kilogram of carbon-12 ($^{12}_{6}C$). When the mole is used, the elementary entities must be specified, e.g. atoms, molecules, ions, electrons.

The actual number of atoms in 0.012 kilogram of $^{12}_{6}C$ has been determined and the accepted value is 6.02×10^{23}. This is known as the *Avogadro constant, L*.

Thus the *amount of substance* is *proportional to* the number of (specified) elementary particles of that substance, and the proportionality factor is $1/L$. Alternatively, one mole of any substance is that amount of the substance which contains 6.02×10^{23} particles of that substance.

Now *relative atomic (or molecular) mass* is defined as:

> The mass of one atom (or molecule) of the element (or compound) compared with the mass of an atom of $^{12}_{6}C$ which is arbitrarily assigned as 12.000.

The relative atomic (or molecular) mass has *no units*, it is simply a number. Because of this definition however, the relative atomic (or molecular) mass of a substance expressed in *grams* always contains the Avogadro number of particles of that substance, and thus the amount of substance present is *one mole*.

It should be borne in mind that the elementary entity may be any *specified* group of particles, whether or not that specified group has any real separate existence (e.g. a mole of NaCl(s), or a mole of C—C bonds).

Thus the mass of *one mole* of a substance can always be determined by expressing the *formula mass in grams* (see table of relative atomic masses, page 225). The formula mass of a substance is the sum of the relative atomic masses of all the atoms present in its conventional formula. For example,

Substance	Conventional formula	Formula mass	One mole weighs
Sodium metal	Na	22.99	22.99 g
Nitrogen gas	N_2	28.02	28.02 g
Nitric acid molecule	HNO_3	63.02	63.02 g
Phenylamine (aniline)	$C_6H_5NH_2$	93.14	93.14 g

Concentration of solutions

The term *concentration* is now understood by I.U.P.A.C. to mean *amount of substance per unit volume*, and in this sense concentration will be measured in units such as $mol\,dm^{-3}$. Thus the word *molarity* is, strictly speaking, merely a synonym for concentration, and I.U.P.A.C. recommends that it should not be used.

The word *molar* is now restricted to the meaning 'divided by the amount of substance' (e.g. molar volume) and it is inappropriate to talk of a 'molar solution' when 'a solution having a concentration of one $mol\,dm^{-3}$' is meant. However, the abbreviation 0.1 M for 'having a concentration of 0.1 $mol\,dm^{-3}$' will undoubtedly continue to be used, and for reasons of brevity will occur frequently in this manual.

Some other related quantities which are often applied to solutes and solutions are shown below, together with their recommended usage:

Physical quantity	Usage	Coherent SI units	Units for general use
Molar mass	$\dfrac{mass}{amount\ of\ substance}$	$kg\,mol^{-1}$	$g\,mol^{-1}$
Molality (of solute X)	$\dfrac{amount\ of\ X}{mass\ of\ solvent}$	$mol\,kg^{-1}$	
Concentration (of solute X)	$\dfrac{amount\ of\ X}{volume\ of\ solution}$	$mol\,m^{-3}$	$mol\,dm^{-3}$ $(mol\,l^{-1})$
Mass concentration (of solute X)	$\dfrac{mass\ of\ X}{volume\ of\ solution}$	$kg\,m^{-3}$	$g\,dm^{-3}$ $(g\,l^{-1})$

Solution	Concentration	Formula mass of solute	Mass concentration
Dilute hydrochloric acid (HCl) (bench)	$2 \, mol \, l^{-1}$ (2M)	36.46	$72.9 \, g \, l^{-1}$
Sodium hydroxide (NaOH) (bench)	$2 \, mol \, l^{-1}$ (2 M)	40.00	$80.0 \, g \, l^{-1}$
Calcium chloride ($CaCl_2$) (bench)	$0.25 \, mol \, l^{-1}$ (0.25 M)	110.98 (anhydrous)	$27.8 \, g \, l^{-1}$
Calcium ion solution	$0.1 \, mol \, l^{-1}$ (M/10)	40.08	$4.01 \, g \, l^{-1}$
Sodium thiosulphate solution ($Na_2S_2O_3.5H_2O$)	$0.1 \, mol \, l^{-1}$ (M/10)	$158.10 + 5 \times 18.02$ $= 248.20$	$24.82 \, g \, l^{-1}$

Note particularly that the term mass concentration must be used for the quantity *mass of solute per unit volume* to avoid confusion with *amount of substance per unit volume*. Some examples are shown above.

Other expressions of strength of solution

The I.U.P.A.C. recommendations are probably not yet fully implemented in schools, and there may well be instances where established practice is superceded only with reluctance; for these reasons the technician ought to be aware of some other quantities occasionally used to express the strength of a solution.

Biochemists, who often work with substances with very high formula masses and require their solutions in a very dilute state (comparable with those found in living tissue), often prefer to express the concentration of their solutions in millimoles per litre (one thousandth of a mole per litre).

In physiological work very dilute solutions of such substances as hormones are used, often in serial dilutions, and these may be expressed by means of indices; for example,

1 g of acetylcholine per litre is expressed as acetylcholine solution 1 in 10^3,
0.1 g per litre as acetylcholine solution 1 in 10^4,
0.01 g per litre as acetylcholine solution 1 in 10^5,
0.001 g per litre as acetylcholine solution 1 in 10^6, etc.

In *percentage solutions* the number refers to the number of grams of the substance dissolved in 100 g of solvent. For example 10 per cent ethanol means 10 g of ethanol dissolved in 100 g of water, a 0.25 per cent Neutral Red in ethanol solution contains 0.25 g Neutral Red dissolved in 100 g ethanol.

Occasionally the strength of a solution is expressed in terms of its *specific gravity* (or density). The most common example of this is '880 ammonia' which refers to the specific gravity of concentrated aqueous ammonia solution, 0.880.

Normalities

The expression of concentration in moles per litre has superceded the older system of expressing concentrations of solutions used for volumetric analysis in *normalities*. A normal solution contains the *gram equivalent mass* of the solute in one litre of solution. The gram equivalent mass of a substance is the mass in grams which combines with or displaces 8 g (strictly 7.9997 g) of oxygen, 1.008 g of hydrogen, or the gram equivalent mass of any other substance.

The expression of concentrations in normalities has been extremely widespread for some time and is unlikely to disappear totally from use in the laboratory during the lifespan of this manual; indeed some suppliers of laboratory chemicals and solutions still express the strength of solutions in normalities rather than moles per litre. For these reasons, a number of the tables in this chapter give the concentration of solutions in terms of normality as well as in moles per litre.

In the tables of solutions set out below the relationships between concentration in moles per litre (molarity) and normality will be shown as M = N or M = 2N etc. Although the solutions are often still labelled as multiples or fractions of normality (e.g. N, 2N, 0.1N or N/10), it is desirable to use only the molar system to avoid confusion.

Units of volume

Throughout this book the cubic centimetre has been used in preference to the millilitre, in line with the SI system of preferred units, although in many instances the litre has been retained in preference to the cubic decimetre as the litre is probably the more commonly used unit. Synonymous units are:
cm^3; ml; c.c.
dm^3; l.

CLEANLINESS

The accurate preparation of solutions depends upon the use of accurately graduated glassware which is free from grease and otherwise chemically clean. If the glassware is not clean the purity of the final solution will be affected and any grease on the glass will make accurate reading of the meniscus impossible because the curve of the meniscus will be distorted.

If glassware is in regular use, washing with Teepol or one of the special laboratory glassware detergents should be sufficient to keep it clean. If detergent alone does not succeed, and grease or other deposits remain, the following solutions should be tried in the order given, washing thoroughly with water and distilled water between each.

1 To 10 g sodium dichromate in 15 cm³ of water, add concentrated sulphuric acid slowly (with cooling) until the solution is made up to 100 cm³. Treat the solution with *the same care* as concentrated sulphuric acid; if necessary leave the glassware immersed overnight.
2 Sodium hydroxide; 41 g of pellets in 1 litre of water.
3 Potassium hydroxide in ethanol (56 g in 1 litre of industrial spirit).
4 Potassium permanganate; 10 g of solid in 1 litre of water. After immersion in this solution the glassware should be rinsed in tapwater and any brown stains removed with concentrated hydrochloric acid.
5 Any remaining stains may be treated as follows:
   White deposit – soak in sodium metasilicate (5 per cent aqueous).
   Carbon deposits – soak in a solution of 6 g trisodium phosphate and 3 g sodium oleate in 100 cm³ water.

Indelible pencil – wash with acetone.
Iodine – soak in sodium thiosulphate solution (25 g litre⁻¹).
Iron stains – soak in hydrochloric acid.
Sulphur deposits – soak in a bench solution of ammonium sulphide.

Once the glassware is clean it can be rinsed with a little propanone (acetone) to remove water and then dried in a cabinet **(narrow necked vessels rinsed in acetone must not be brought near a naked flame as they may explode).**

VOLUMETRIC GLASSWARE

Graduated flasks, *pipettes*, and *burettes* are normally used for making up accurate solutions, and *measuring cylinders* for making up bench solutions. Volumetric glassware is manufactured in three standards; A and B, for which there are British Standards for accuracy, and ungraded. The British Standard tolerances for volumetric glassware are given in Table 7.1. Class B glassware is approximately one third of the price of class A, and is suitable for all educational use.

It will be seen that the accuracy of class B glassware is better than 0.2 per cent and is more than adequate for the limits of experimental accuracy found in schools.

Markings on volumetric glassware

C or IN means 'contains'
D or EX means 'delivers'
A means A standard of accuracy
B means B standard of accuracy

Thus a flask marked '250 cm³ C 20 °C B' means 'contains 250 cm³ at 20 °C plus or minus 0.30 cm³'. A pipette marked '50 cm³ D 20 °C A' means 'will deliver 50 cm³ plus or minus 0.04 cm³ at 20 °C'.

Table 7.1 British Standard maximum permitted errors (tolerances) for volumetric glassware

Size	5 cm³	10 cm³	20 cm³	25 cm³	50 cm³	100 cm³	250 cm³	500 cm³	1000 cm³
Pipette A	0.015	0.02	0.03	0.03	0.04	–	–	–	–
Pipette B	0.03	0.04	0.06	0.06	0.08	–	–	–	–
Burette A	0.02	0.01	–	0.03	0.05	–	–	–	–
Burette B	0.04	0.02	–	0.06	0.10	–	–	–	–
Flask A	–	–	–	0.03	0.05	0.08	0.15	0.25	0.40
Flask B	–	–	–	0.06	0.10	0.15	0.30	0.50	0.80

The letters D and C are found on older laboratory glassware and the letters EX and IN on glass of more recent manufacture.

The standards of accuracy given above apply only if the apparatus is used correctly.

On all volumetric glassware the reading is taken at the base of the meniscus.

The pipette

The pipette is a glass tube with a bulb partway along its length (except graduated pipettes which are tubes of even bore used for measuring very small quantities – the divisions may be as small as 0.01 cm^3). One end of the tube is drawn out to a point, the other left wide for sucking by mouth, or by means of a safety bulb. Between the mouth end and the bulb there is a gradation mark. **Safety bulbs must always be used when pipetting poisons, concentrated acids, corrosive liquids, strong alkalis, radioactive solutions, bacterial suspensions, and anaesthetics!**

The routine for using a pipette is:

1   Fill with distilled or deionised water, allow to drain, blow out the last drop of water and wipe the outside dry.
2   Draw about 3 cm^3 of solution into the pipette, place the forefinger over the mouth end, and hold the pipette horizontally to run the solution over the entire inside surface to a point just above the gradation mark; drain and repeat.
3   Fill the pipette to a point just above the gradation line and place the forefinger over the mouth end with a firm pressure. Keep the top of the pipette dry or accurate adjustment of the meniscus will be impossible.
4   With the eye on an exact level with the gradation mark, release the pressure with the forefinger without removing it from the end until the meniscus just touches the gradation mark.
5   Touch the tip of the pipette against the beaker to remove any hanging drip of solution.
6   Allow the solution to drain into the required vessel. **Do not blow.**
7   If the pipette is marked IN or C, hold its tip against the vessel for three seconds after the last drop has run out. If the pipette is marked EX or D it should be held against the vessel for fifteen seconds after the last drop has run out freely (these were the times used when the pipette was graduated).

The burette

The burette is a graduated glass tube with a tap to deliver any quantity of liquid accurately. The gradations marked are in 1 cm^3 and 0.1 cm^3 steps, with the zero at the top. The tap may be of glass, of plastic, or may be replaced by a spring clip or a pinch valve. Ground glass taps are given the faintest smear of petroleum jelly or a silicone burette grease after cleaning and held in place with a rubber band looped across the barrel of the tap. (Some makes have a stainless steel retaining clip or spring.) Plastic taps do not require vaseline or burette grease. If a ground glass tap becomes blocked with petroleum jelly it should be placed in hot water to melt the obstruction, which should then be flushed out with the solution. Should this be unsuccessful, a suction pump may be used. Glass taps sometimes become stuck, particularly if not thoroughly cleaned after use with alkalis. Pouring hot water over the tap causes the barrel to expand before the stopcock and may free the tap. If this fails, use penetrating oil, or soak in hot water. Clean thoroughly once the tap is free.

The routine for use is:

1   Place about 5 cm^3 distilled water in the burette and rinse by inverting several times with the thumb held over the end; drain through the tap. Repeat.
2   Repeat **1** using solution.
3   Fill to a point just above the zero mark (1–2 cm above).
4   Open the tap to allow the tip to be filled with liquid and the meniscus to touch the zero mark.
5   Check that the burette is vertical in its stand by sighting it with two vertical corners of the room.
6   The best technique for using the tap is to curl the left hand around the stem of the burette, turning the tap with the thumb and index finger. This method automatically presses the stopcock into its barrel to prevent leaks and also leaves the right hand free for mixing the solution in the receiving vessel.
7   The meniscus should be read with the eye on an exact level to avoid parallax errors (see Figure 7.1). A second decimal place can be estimated and recorded in brackets.

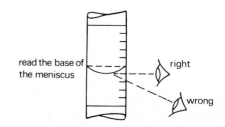

read the base of the meniscus — right / wrong

Figure 7.1   Reading a meniscus

Measuring cylinders

These are less accurate than pipettes and burettes and their use is usually limited to the making up of bench solutions or measuring poisonous or corrosive substances if other means are not available. Stoppered cylinders are very useful when preparing mixtures of solvents for chromatography.

Tall glass cylinders of the traditional design are very easily knocked over and broken. Glass cylinders with a plug-in plastic base cost only half as much as the traditional pattern and replacement bodies are available. Square section sponge rubber rings called cylinder protectors or lifeboys are available, and will prevent breakage in most cases when a cylinder is knocked over on the bench. They should be fitted about 3 cm below the lip. Polythene cylinders are cheap and unbreakable but they are less accurate than glass and more difficult to clean after use with some reagents.

The routine for use is:

1 Rinse with distilled water.
2 Rinse with the liquid to be measured.
3 Place on a level bench and fill up to the mark, keeping the eye level with the meniscus.

Volumetric flasks

These are available with groundglass stoppers or unstoppered. The stoppered type are preferred for making up standard solutions but the unstoppered type, if provided with a rubber bung, are suitable for student use and are less prone to damage as well as being cheaper.

Balances

The ideal balance for school use is a direct readout single pan or top pan balance. If the third decimal place is considered necessary, the balance will cost in the region of £200, but only half this sum need be spent if two decimal places are considered satisfactory. One direct readout balance of good design will service a class of thirty. It is very rare in new schools to find solid balance benches because of the expense. In the rare cases where vibrations from traffic, trains, or vibrations inherent in the design of the building are a serious problem, the purchase of an antivibration table will solve the problem. These tables are used in industrial laboratories, which are often situated near heavy machines.

PREPARATION TECHNIQUES

Making bench and other approximate solutions

1 Weigh out the formula mass in grams of the substance if solid; use a measuring cylinder if liquid (assuming that the density is known).
2 Dissolve in distilled water (or deionised water) in a beaker, heating and stirring as necessary.
3 Make up to the volume required.
4 Shake thoroughly to ensure that the concentration is the same throughout the solution.

Concentrated acids should be added to water in the beaker, never the reverse. Substances such as sulphuric acid and sodium hydroxide which produce heat on solution should be added to water in small quantities with frequent stirring and cooling. Thick-walled vessels can be cracked by the heat of solution if this is produced rapidly and locally. Solutions which have been made by heating, or which produce heat on solution, should always be cooled before dilution to final volume.

Making Primary Standard Solutions

1 Carefully weigh out the required amount of solid on a watchglass, scoop, or weighing bottle.
2 Small amounts of substance are washed into a graduated flask with distilled or deionised water from a wash bottle, using a funnel. Larger amounts of substance, and substances which need heating, are first washed into a beaker and made into a solution.
3 If prepared in a beaker, the solution should be poured into the graduated flask using a glass rod to prevent spillage. Use distilled water to wash the remnants of solution from the beaker to the flask.
4 Water is now added to bring the meniscus up to the gradation mark so that the meniscus just sits on or touches the mark. (The final adjustment is made at 20 °C and the last few drops should be added with a teat pipette to prevent overshooting the gradation mark.)
5 Label the flask with the name of the dissolved substance, the solvent unless this is water (aqueous solutions are only labelled as such if the solute is commonly dissolved in other solvents), and the date of preparation.

Making Secondary Standard Solutions

Proceed as described above, and then standardize by titration. If the secondary standard substance is a liquid, use a burette to deliver the required volume.

Table 7.2 Primary standards
Always use AnalaR or Pronalys grades.

Substance	Formula mass	M :N	Use
Potassium dichromate, $K_2Cr_2O_7$	294.19	M = 6N	Direct titration and standardization of sodium thiosulphate
Potassium iodate, KIO_3	214.01	M = 6N	As above
Sodium carbonate (anhydrous), Na_2CO_3	105.99	M = 2N	Standardization of solutions of strong acids
Sodium chloride, NaCl	58.44	M = N	Standardization of silver nitrate solution
Sodium oxalate, $(COONa)_2$	134.00	M = 2N	Standardization of potassium permanganate and ceric sulphate solutions

All the above are anhydrous salts which can be dried between 120 and 180 °C. To make a solution of the required concentration the correct mass of the substance is calculated from the formula mass. This mass of dry substance is then carefully weighed out and dissolved in the correct volume of distilled or deionised water. Always use AnalaR or Pronalys grades.

The substances listed below may also be used as primary standards if they are stored carefully. All may be weighed directly but unless a solution of exact concentration is required, its easier to weigh the approximate amount of the salt accurately, make it up to the required volume and calculate the precise concentration of the solution.

Substance	Formula mass	M :N	Use
Aminosulphonic acid (sulphamic acid), NH_2SO_3H	97.09	M = N	Standardization of alkaline solutions
Ammonium iron(II) sulphate, $(NH_4)_2SO_4.FeSO_4.6H_2O$ (make up in dilute sulphuric acid)	392.14	M = N	Standardization of oxidizing agents
Antimony potassium tartrate, $KSbO.C_4H_4O_6$	324.92	M = 2N	Iodine titrations
Arsenic(III) oxide, As_2O_3	197.84	M = 4N	Standardization of oxidizing agents
Benzoic acid, C_6H_5COOH	122.13	M = N	Standardization of alkaline solution
Guanidine carbonate, $[(NH_2)_2.C.NH]_2.H_2CO_3$	180.17	M = 2N	Standardization of solutions of strong acids
Hydrazine sulphate, $NH_2.NH_2.H_2SO_4$	130.12	M = 4N	Standardization of oxidizing agents
Oxalic acid, $(COOH)_2.2H_2O$	126.07	M = 2N	Standardization of oxidizing agents. Standardization of alkaline solutions
Potassium hydrogen phthalate, $COOH.C_6H_4.COOK$	204.23	M = N	Standardization of alkaline solutions
Potassium hydrogen tartrate, $KHC_4H_4O_6$	158.16	M = N	Standardization of alkaline solutions
Silver nitrate, $AgNO_3$	169.88	M = N	Precipitation reactions
Sodium tetraborate (borax), $Na_2B_4O_7.10H_2O$	381.37	M = N	Standardization of solutions of strong acids

Table 7.3 Secondary standards
Use laboratory grade chemicals unless AnalaR grade is indicated

Substance	Formula mass	M:N	Special notes and standardization
Boric acid, H_3BO_3	61.83	M = N	Dissolve 61.83 g in water and make up to one litre. Standardize against standard sodium hydroxide with phenolphthalein as indicator
Citric acid, $C_3H_5O(COOH)_3.H_2O$	210.00	M = 3N	Dissolve 70.00 g in water and make up to 1 litre. Standardize with standard sodium hydroxide with phenolphthalein as indicator
Ethanoic (acetic) acid, CH_3COOH	60.05	M = N	Dissolve 58.0 cm³ of AnalaR grade glacial ethanoic acid and make up to one litre. Standardize against standard sodium hydroxide with phenolphthalein as indicator
Hydrochloric acid, HCl	36.46	M = N	AnalaR concentrated acid has a density of 1.18, and contains 36% acid. Concentration = 11.7 moles per litre. Add 86 cm³ of concentrated acid to 500 cm³ of distilled water and make up to 1 litre. Standardize with standard sodium carbonate and methyl orange indicator or standard sodium hydroxide with methyl red
Nitric acid, HNO_3	63.01	M = N	AnalaR concentrated acid has a density of 1.41 and contains 70% acid. Concentration = 15.8 moles per litre. Pour 64 cm³ into 500 cm³ of water and make up to 1 litre. Standardize as for HCl
Phosphoric(V) acid, H_3PO_4	98.00	M = 3N	AnalaR concentrated acid has density of 1.75 and contains at least 88% acid. Concentration = 15.7 moles per litre. Pour 66 cm³ of the syrupy acid into 500 cm³ of distilled water and shake the flask vigorously. Make up to 1 litre. Standardize with standard sodium hydroxide using methyl orange as indicator
Succinic acid (butanedioic acid), $(CH_2)_2(COOH)_2$	118.00	M = 2N	Dissolve 118 g in water and make up to 1 litre
Sulphuric acid, H_2SO_4	98.08	M = 2N	AnalaR concentrated acid has density of 1.84 and contains at least 97% acid. Concentration = 18.5 moles per litre.

			Pour 55 cm³ of concentrated acid very slowly into 500 cm³ of distilled water. Make up to 1 litre. **Caution.** Standardize as for HCl
Tartaric acid, $CH(OH)_2(COOH)_2$	150.00	M = 2N	Dissolve 75 g in water and make up to 1 litre
Ammonium hydroxide, NH_4OH	17.03	M = N	AnalaR grade contains at least 30–35% ammonia. Specific gravity = 0.88, concentration = 18 moles per litre. Standardize with standard HCl with methyl red as indicator
Ammonium chloride, NH_4Cl	53.50	M = N	53.5 g per litre
Barium hydroxide (baryta), $Ba(OH)_2.8H_2O$	315.47	M = 2N	Use water which has been boiled to remove dissolved CO_2. Molar or Normal solutions are not possible, as solubility is only 38.9 g per litre at 20 °C. Absorbs carbon dioxide readily from the air. Used in quantitative respiration experiments
Potassium hydroxide, KOH	56.11	M = N	Weigh about 56.5 g in a closed vessel (to minimize CO_2 absorption). Dissolve in freshly boiled and cooled distilled water. Store in bottle with rubber or polythene stopper or a cork. Standardize with standard hydrochloric or sulphuric acid, with methyl red as indicator
Sodium hydroxide, NaOH	40.00	M = N	As for potassium hydroxide, but using 41 g of solid
Sodium hydroxide solution (carbonate free)			(i) Dissolve 41 g of solid in water. Precipitate carbonate by adding barium chloride solution drop by drop. Filter and make up to 1 litre (ii) Dissolve 41 g of solid in a little ethanol (carbonate is insoluble in ethanol). Filter and dilute with boiled distilled water

N.B. If burettes are used for sodium or potassium hydroxide, the taps must be cleaned immediately after use.

Potassium carbonate, K_2CO_3	138.21	M = 2N	A solution can be made up with sufficient accuracy by weighing. No standardization
Potassium hydrogencarbonate, $KHCO_3$	100.12	M = N	As above
Sodium hydrogencarbonate, $NaHCO_3$	84.01	M = N	As above

Table 7.4 Solutions used in precipitation reactions

Substance	Formula mass	M :N	Notes and standardization
Ammonium thiocyanate, NH_4CNS	76.12	M = N	Dissolve 8 g of crystals in water and make up to 1 litre. Standardize with silver nitrate solution, acidified with 20 cm³ of 2 M nitric acid for each 25 cm³. Use ferric alum as indicator
Potassium thiocyanate, KCNS	97.18	M = N	As for ammonium thiocyanate but using 10 g of crystals made up to one litre
Silver nitrate	169.88	M = N	(i) Dry AnalaR crystals at 110 °C for a couple of hours. Use 16.988 g made up to one litre for M/10 solution (ii) It is considerably cheaper to dissolve 17 g of laboratory grade crystals, and make up to 1 litre. Standardize with sodium chloride solution with potassium chromate indicator (or screened indicator)
Sodium chloride	58.44	M = N	Use 5.845 g per litre for M/10 solution

Table 7.5 Miscellaneous titration solutions

Ammonium chloride, NH_4Cl	Dissolve 53.5 g in 1 litre of water for M or N solution
Barium chloride, $BaCl_2.2H_2O$	Dissolve 122.0 g in 1 litre of water for M/2 or N solution
Calcium ion solution	Add a little distilled water to 1 g of calcium carbonate (AnalaR grade) followed by sufficient 2 M hydrochloric acid, added one drop at a time, to dissolve all the carbonate. Boil, and when cool make up to 1 litre. 1 cm³ = 0.004 g of calcium ion
Copper(II) sulphate, $CuSO_4.5H_2O$	Dissolve 249.7 g in 1 litre of water for M or N solution.
Iodine, I_2	Dissolve 126.9 g in 1 litre of water for M/2 or N solution. Must be dissolved in KI solution
Lead(II) ethanoate, $Pb(OCO.CH_3)_2.2H_2O$	Dissolve 189.6 g in one litre of water for M/2 or N solution
Magnesium sulphate, $MgSO_4.7H_2O$	Dissolve 123.0 g in 1 litre of water for M/2 or N solution
Potassium bromate, $KBrO_3$	Dissolve 27.83 g in 1 litre of water for M/6 or N solution
Potassium bromide, KBr	Dissolve 119.0 g in 1 litre of water for M or N solution
Potassium chlorate, $KClO_3$	Dissolve 20.44 g in 1 litre of water for M/6 or N solution
Potassium iodide, KI	Dissolve 166.0 g in 1 litre of water for M or N solution
Soap solution	Dissolve 10 g of Castile soap in 100 cm³ of ethanol by warming in a beaker of hot water. Add a further 500 cm³ of ethanol and then distilled water to make 1 litre.
Soap solution (Boutron-Boudet)	Dissolve 100 g of Castile soap in a solvent made by adding 1400 cm³ of ethanol to 1100 cm³ of distilled water. Titrate against 40 cm³ of barium nitrate (0.59 g dm⁻³).

Dilute so that after dilution 2.4 cm³ of the soap solution gives a permanent lather with 40 cm³ of the barium nitrate.

2.4 cm³ = 22 French degrees of hardness, or 220 parts per million of calcium carbonate.

Soap solution (Clark's)	Dissolve 100 g of Castile soap in 800 cm³ of ethanol and 200 cm³ of distilled water. Stand for 24 hours, decant the liquid, and titrate against standard calcium solution. Dilute with 80% ethanol until 1 cm³ = 1 mg calcium carbonate, i.e. when a permanent lather is just obtained
Zimmermann-Reinhardt solution for the titration of iron(II) ions in the presence of HCl, using permanganate solution	Add 70 g of hydrated manganese(II) sulphate crystals to 500 cm³ of water, add with stirring 125 cm³ of concentrated sulphuric acid followed by 125 cm³ of syrupy phosphoric(V) acid (85%) and dilute to 1 litre

SOLUTIONS USED IN REDOX REACTIONS

Table 7.6 Oxidizing agents or oxidants

Ceric sulphate hydrated, $Ce(SO_4)_2.4H_2O$ anhydrous, $Ce(SO_4)_2$	Formula mass = 404.3 $$M = N$$ Formula mass = 332.24 Ceric sulphate is very expensive in AnalaR grade but technical grades are readily available and much less expensive. Below are two methods of preparing solutions from the technical grade. *Method (i)* Dissolve 56 g of ceric sulphate (technical grade) in 800 cm³ of M(2N) sulphuric acid, boil, cool, and filter to give a clear solution which is diluted to 1 litre. *Method (ii)* Add 50 g of ceric oxide in small quantities to 100 cm³ of sulphuric acid, which has been added to 100 cm³ of distilled water. Do not let the temperature exceed 100 °C and after ten minutes stirring at 100 °C, cool the solution and dilute to 800 cm³. Filter and make the solution up to one litre with M(2N) sulphuric acid. Standardize solutions made by either method with arsenic(III) oxide solution using ferroin as an indicator.
Iodine, I_2	Formula mass = 253.81 M = 2N Dissolve 1.27 g of iodine crystals in a solution of 40 g of potassium iodide in 300 cm³ distilled water. When all the iodine has dissolved, make up to one litre. Standardize with arsenic(III) oxide solution diluted with an equal volume of water using starch as an indicator.
Potassium bromate Potassium dichromate Potassium iodate	See *primary standards*
Potassium permanganate, $KMnO_4$	Formula mass = 158.04 M = 5N Add 3.2 g of AnalaR grade crystals to 300 cm³ of water. Heat and stir until the crystals have dissolved. Cool and make up to 1 litre. Filter through mineral wool after standing for 24 hours. Standardize with sodium oxalate solution above 70 °C.
Sodium hypochlorite, NaClO	Formula mass = 52.47 M = 2N Commercial solutions contain 10–14% sodium hypochlorite. Dilute 220 cm³ of 10–14% solution to 1 litre. Standardize with arsenic(III) oxide solution with Bordeaux indicator. **Unstable** – store in the dark for a maximum of one week, and standardize before use.

Table 7.7 Reducing agents, or reductants

Ammonium iron(II) sulphate, $(NH_4)_2SO_4 \cdot FeSO_4 \cdot 6H_2O$	Formula mass = 392.14	M = N

Dissolve 39.214 g of crystals in 200 cm³ of M sulphuric acid and dilute to 1 litre. H_2SO_4 reduces atmospheric oxidation.

Arsenic(III) oxide — Formula mass = 197.84 — M = 4N

Dissolve 4.945 g in 20 cm³ of 2 M sodium hydroxide solution. Add M or 2 M sulphuric acid until the solution is just acid. Dilute to 1 litre.

Mercury(I) nitrate, $Hg_2(NO_3)_2 \cdot 2H_2O$ or $HgNO_3 \cdot H_2O$ — Formula mass (second formula) = 280.61 — M = N

Dissolve 28.061 g in 1 litre of 0.5 M nitric acid.
Alternatively, use slightly more mercury(I) nitrate and standardize with ferric alum using potassium thiocyanate as an indicator.

Sodium thiosulphate, $Na_2S_2O_3 \cdot 5H_2O$ — Formula mass = 248.20 — M = N

Dissolve 25 g of crystals in freshly boiled and cooled distilled water and make up to one litre. Some users advocate the addition of 0.1 g of sodium hydrogencarbonate.
Standardize with 25 cm³ of standard potassium dichromate solution to which 10 cm³ of 10% potassium iodide and 10 cm³ of M sulphuric acid have been added. Use starch as the indicator.

Titanium chloride, $TiCl_3$ — Formula mass = 154.26 — M = N

Add 100 cm³ of the commercial solution to 100 cm³ of concentrated hydrochloric acid and boil. Cool and make up to one litre

SOLUTIONS FOR TITRATIONS

Certain solutions are used regularly in most laboratories and are usually kept in the laboratory or its immediate storage area. These 'bench solutions' require regular servicing by the technician to ensure that they are always available in good condition. It is a growing (and desirable) practice in school laboratories to regard the laboratory as a general purpose one, wheeling in on a trolley the apparatus and reagents needed for a particular class. A well-organized technician will welcome such a method of working as it means that every reagent bottle passes through his or her hands after every session of use. If stock bottles of the bench reagents used in the laboratory are always kept full, it is a simple matter to recharge the working bottle as it is returned to the store. If this method of working is adopted rather than waiting until bottles run out, the operation of the laboratory becomes much smoother and a teaching session is never disrupted by the technician being caught 'on the hop' for reagents.

The reagents listed below should include all those in common use, and the concentrations given are those most generally accepted. Many reagents are prepared in batches of one litre, but some which are used frequently, may be prepared in greater amounts – usually $2\frac{1}{2}$ litres, which is the volume held by a winchester bottle. If a permanent mark, such as a ring of paint, is put on the neck of a winchester at the exact $2\frac{1}{2}$ litre point much time will be saved in making up bench solutions. For many bench solutions accuracy matters less than clean working, but this should never be the excuse for lack of care or for guess-work.

Tables 7.8, 7.9, and 7.10 give recipes for 1 litre of reagent in most cases, but for the more commonly used materials, appropriate figures are given in the $2\frac{1}{2}$ litre column. Technicians who have occasion to make up winchesters of the reagents listed only in the 1 litre column have only to perform a simple calculation and record the quantities in the appropriate column.

BENCH SOLUTIONS
Table 7.8 Bench acids

Acid	To make 1 litre	To make 2½ litres	Comments
Aqua Regia	Add one part of concentrated nitric acid to three parts of concentrated hydrochloric acid		Make as required. **Do not store**
Ethanoic acid, glacial			As purchased
Ethanoic acid, concentrated	290 cm³ of glacial ethanoic acid made up to 1 litre		5 M solution. Occasionally used
Ethanoic acid, dilute	58 cm³ of glacial ethanoic acid made up to 1 litre	145 cm³	M solution
Hydrochloric acid, concentrated			As purchased. Always stand on porcelain or a plastic tray
Hydrochloric acid, dilute (2 M)	Add 172 cm³ of concentrated acid to water and make up to 1 litre	430 cm³	
dilute (4 M)	344 cm³ of concentrated acid made up to 1 litre	860 cm³	
Nitric acid, concentrated			As purchased. Always stand on porcelain or a plastic tray and store in brown bottles
Nitric acid, dilute, 2 M (2 N)	Add 128 cm³ of concentrated acid to water and make up to 1 litre	320 cm³	
dilute, 4 M (4 N)	256 cm³ of concentrated acid made up to 1 litre	620 cm³	
Phosphoric(V) acid, M (3 N)	Add 66 cm³ of syrupy acid to water, mix well, and make up to 1 litre	165 cm³	
Sulphuric acid, concentrated			As purchased. Always stand on porcelain or a plastic tray. **Never add water to concentrated sulphuric acid**
Sulphuric acid, dilute, M (2 N)	Add 55 cm³ of concentrated acid to water very slowly in a large beaker. Make up to 1 litre	137.5 cm³	**Always add acid to water – never the reverse.** The dilution is best carried out in a large beaker as the heat of dilution has been known to crack thick glass such as winchesters
dilute, 2 M (4 N)	110 cm³ of concentrated acid per litre of solution	275 cm³	**Caution**

Concentrated acids should always stand in saucers.

Concentrated acids should be brought into laboratories in schools only when needed for a particular experiment. Plastic bottles should not be used as they tend to 'concertina' if dropped, squirting the contents towards the face.

Table 7.9 Bench alkalis

Alkali	To make 1 litre	To make 2½ litres	Comments
Ammonium hydroxide, 0.880 (aqueous ammonia)			As purchased. 0.880 is the specific gravity of the concentrated solution. Only used in this strength for a few tests
Ammonium hydroxide, 2 M (2 N)	Add 112 cm³ of 0.880 ammonia to water, and make up to 1 litre	270 cm³	Avoid breathing the fumes, and any risk of ammonia getting into the eyes
Ammonium hydroxide, 4 M (4 N)	224 cm³ in 1 litre	540 cm³	As above
Barium hydroxide (baryta water), 0.2 M (0.4 N)	Add 70 g of the hydrated hydroxide to 1 litre of boiled water. Shake well to form a saturated solution, and leave to settle	175 g	Absorbs atmospheric carbon dioxide very rapidly. Best stored in an aspirator with a soda-lime tube at the inlet
Calcium hydroxide (lime water), 0.02 M (0.04 N)	Solubility is approximately 1.5 g per litre. The usual method is to place 75 g in a winchester, shake with 2 litres of water, allow to settle, and decant the clear liquid for use, refilling the bottle. Protect from atmospheric carbon dioxide – best stored in an aspirator with the air inlet protected with a soda-lime tube (the self-indicating type is excellent as it changes colour when exhausted).		
Lithium hydroxide, M (approx.)	20 g of lithium hydroxide and 50 g of potassium nitrate dissolved in water and made up to 1 litre		
Potassium hydroxide, 2 M (2 N)	Dissolve 58 g of pellets in 800 cm³ of water and make up to 1 litre	145 g	**Stir constantly** as the heat of solution can crack glass vessels if care is not taken
Potassium hydroxide, 4 M (4 N)	Dissolve 116 g	290 g	Use rubber-stoppered or polythene-stoppered bottles
Potassium hydroxide, alcoholic (in ethanol)	Dissolve 56 g of pellets in ethanol and make up to 1 litre with ethanol	140 g	
(in methanol)	Dissolve 56 g in methanol and make up to 1 litre with methanol	140 g	
Sodium hydroxide, 2 M (2 N)	Dissolve 41 g of pellets in 800 cm³ of water and make up to 1 litre	100.5 g	Precautions and stoppers as for KOH

Strong alkalis should be kept on tiles or saucers, as they will dissolve wax and varnish on benchtops.

Bench indicator solutions

The following are often kept readily available in the laboratory:

Litmus solution   Used as purchased.

Litmus paper   Red and blue litmus paper is often put out in widenecked jars with screw caps or ground glass stoppers.

Methyl orange   2 g dissolved in 1 litre of ethanol. Use small dropper bottles.

Phenolphthalein   1.75 g dissolved in 1 litre of ethanol. Use small dropper bottles.

Table 7.10 Inorganic bench reagents

Reagent	To make 1 litre	To make 2½ litres
Ammonium carbonate, M (2 N)	Dissolve 160 g of solid in 500 cm³ water, add 120 cm³ of 0.880 ammonia, and make up to 1 litre	
Ammonium chloride, 3 M (3 N)	Dissolve 160 g in water and make up to 1 litre	400 g
Ammonium chloride, 5 M (5 N)	Dissolve 231 g in water and make up to 1 litre	567 g
Ammonium ethanoate, 3 M (3 N)	Dissolve 231 g in water and make up to 1 litre	
Ammonium molybdate, 0.5 M	Dissolve 75 g in 50 cm³ of 0.880 ammonia and 25 cm³ of water. Dilute to 500 cm³ and then add slowly, with continuous stirring, to 500 cm³ of 5 M nitric acid	Unstable solution best made frequently
Ammonium nitrate, M (N)	Dissolve 80 g of crystals in water and make up to 1 litre	200 g
Ammonium mercuri-thiocyanate, 0.3 M	9 g of ammonium thiocyanate are dissolved in 50 cm³ of water. Add 8 g of mercury(II) chloride and make up to 1 litre	
Ammonium oxalate, 0.25 M (0.5 N)	Dissolve 34 g of crystals in water and make up to 1 litre	95 g
Ammonium sulphide, a 10% solution is 3 M (6 N)	Sold as a colourless solution, or as a yellow solution which contains sulphur. Somewhat tedious to prepare and is best purchased. Commercial solutions contain 7–10% by mass of H_2S. To prepare the colourless solution, cool 200 cm³ of 0.880 ammonia on ice and saturate it with H_2S. Add a further 200 cm³ of 0.880 ammonia and make up to 1 litre. The yellow solution is prepared by saturating 200 cm³ of ammonia with H_2S. 10 g of flowers of sulphur are then dissolved in the solution. Add another 200 cm³ of 0.880 ammonia and make up to 1 litre	
Ammonium thiocyanate, 0.5 M (0.5 N)	Dissolve 38 g of solid in water and made up to 1 litre	95 g
Barium chloride, 0.25 M (0.5 N)	61 g made up to 1 litre in water	152.5 g
Barium nitrate, 0.25 M (0.5 N)	65 g made up to 1 litre in water	162.5 g
Bromine water	Make up in small quantities and store in a dark bottle. Use 5 cm³ of bromine in 100 cm³ of water. **Use the fume cupboard and extreme caution when handling bromine**	
Calcium chloride, 0.25 M (0.5 N)	Dissolve 55 g of crystals, or 27 g of dry anhydrous salt, in water and make up to 1 litre	127.5 g of crystals or 67.5 g of the anhydrous salt
Calcium hypochlorite (bleaching powder)	Shake 125 g of powder vigorously with 1 litre of water. Allow to stand for several hours, then filter	312 g
Calcium nitrate, 2 M (4 N)	Dissolve 470 g of dried crystals in water and made up to 1 litre	1175 g
Calcium sulphate; saturated solution is approx. 0.03 M (0.06 N)	Grind 3 g of the salt to a fine powder and add to 1 litre water. Allow to stand for 24 hours and then filter	Grind 8 g
Chlorine water; a saturated solution is approximately 0.09 M	Prepare chlorine by allowing concentrated hydrochloric acid to drop on crystals of potassium permanganate.	

	Bubble the resultant chlorine gas through water until saturated. **Use a fume cupboard.** Store in a dark bottle	
Chlorine in carbon tetrachloride	Prepare chlorine as above, bubble through water to wash, and then through carbon tetrachloride until saturated	
Cobalt nitrate, 0.01 M (0.02 N)	2.9 g dissolved in water and made up to 1 litre	7.25 g
Cobalt nitrate (5 %)	Dissolve 50 g in water, add 200 cm³ of glacial ethanoic acid, and make up to 1 litre	
Cobalt nitrate (10 %)	Dissolve 100 g in water, add 200 cm³ of glacial ethanoic acid, and make up to 1 litre	
Copper(II) sulphate, 0.5 M (N)	125 g of hydrated crystals made up to 1 litre in water	312 g
Copper(II) sulphate (0.005 M, used in mercuri-thiocyanate test for zinc)	1.2 g made up to 1 litre in water	3 g
Copper(I) chloride (acid)	Shake 100 g of copper(II) chloride and 50 g of copper turnings with 200 cm³ of concentrated hydrochloric acid	
Copper(I) chloride (alkaline)	Neutralise the above solution with ammonium hydroxide until it smells of ammonia. Leave the metallic copper in the solution	
Iron(III) chloride, 0.05 M	Dissolve 135 g of solid in water, acidify with 12 cm³ of concentrated hydrochloric acid, and make up to 1 litre	332.5 g 30 cm³ of HCl
Iron(II) sulphate, 0.05 M	Dissolve 140 g of solid in water containing 10 cm³ of concentrated sulphuric acid, and dilute to 1 litre	350 g 25 cm³ of H_2SO_4
Hydrogen peroxide, 10 volume solution	Available in 20 and 100 volume solutions commercially. **Refer to Table 1.2 before ordering** Bench solution is 10 volumes and is prepared by diluting 100 cm³ of 100 volume with 900 cm³ of water	250 cm³ of 100 volume solution 2250 cm³ of water
Hydrogen sulphide solution; a saturated solution is approximately 0.1 M	Bubble washed hydrogen sulphide through water to obtain a saturated solution. **Use a fume cupboard**	
Iodine solution, 0.05 M (0.1 N)	Dissolve 20 g of potassium iodide crystals in 600 cm³ of water, and add 12.7 g of iodine crystals. Make up to 1 litre	50 g of KI 31.75 g of I_2
Iodine, tincture of	Dissolve 50 g of potassium iodide in water. Add 70 g of iodine crystals and dilute to 1 litre with ethanol	
Lead ethanoate, 0.1 M (0.2 N)	Dissolve 38 g of crystals in water and make up to 1 litre	95 g
Lead nitrate, 0.1 M (0.2 N)	33 g in water made up to 1 litre	82.5 g
Lime water – see calcium hydroxide (Table 7.9)		
Magnesia mixture (phosphate and arsenate test)	Dissolve 55 g of magnesium chloride crystals and 100 g of ammonium chloride in water. Add 50 cm³ of 0.880 ammonium hydroxide and make up to 1 litre	Prepare just before use, or add ammonia just before use
Magnesium nitrate reagent (phosphate and arsenate test)	Dissolve 128 g of magnesium nitrate crystals and 240 g of ammonium nitrate in water and add 20 cm³ of 0.880 ammonium hydroxide. Make up to 1 litre	
Magnesium sulphate, 0.25 M	62 g crystals in water made up to 1 litre	155 g

Magnesium uranyl ethanoate	Dissolve 12 g of magnesium uranyl ethanoate in 200 cm³ of M ethanoic acid. *or* Boil 6 g of magnesium ethanoate in 100 cm³ of M ethanoic acid. Boil 6 g of uranyl ethanoate in another 100 cm³ of M ethanoic acid. Mix the hot solutions and stand for 24 hours before filtering	
Manganous chloride, 0.5 M	99 g in water made up to 1 litre	247.5 g
Mercury(II) chloride (0.1 M) Mercury(II) chloride (saturated)	Dissolve 27 g in water and make up to 1 litre 40 g per litre	67.5 g 100 g
Mercury(I) nitrate, 0.2 M $HgNO_3.H_2O$	50 g of the salt dissolved in 40 cm³ of concentrated acid and made up to 1 litre with distilled water	125 g of salt 100 cm³ of acid
Nessler's reagent (ammonia test)	Dissolve 35 g of potassium iodide in water and add this solution to 12.5 g of finely powdered mercury(II) chloride in a large mortar. Grind until dissolved. Pour into a beaker and add saturated mercury(II) chloride solution until a faint precipitate forms. Add 120 g of sodium hydroxide pellets and make up to 1 litre	**Store in a dark bottle**
Potassium antimonate	Boil 22 g of the solid in 1 litre of water, cool quickly, and add 30 cm³ of 2 M potassium hydroxide	
Potassium chromate, 0.1 M (0.2 N)	Dissolve 20 g in water and make up to 1 litre	50 g
Potassium cyanide, 0.5 M (0.5 N)	Dissolve 3.3 g in 100 cm³ of water. The solution deteriorates on storage. **Extremely poisonous – make up only small quantities as required.** This reagent must **not be mouth pipetted.**	
Potassium ferricyanide, 0.133 M (0.5 N)	55 g made up in 1 litre of water. The solution deteriorates, and some workers prefer to make fresh solutions by washing a small crystal in water and then dissolving it for use	
Potassium ferrocyanide, 0.1 M	Dissolve 40 g in water and make up to 1 litre	100 g
Potassium iodide, 0.2 M (0.2 N)	Dissolve 33 g in water and make up to 1 litre	79.5 g
Potassium permanganate, 0.02 M (0.1 N)	Heat 3.2 g in water and make up to 1 litre. Filter with glass wool	8 g
Potassium oxalate, 0.5 M (N)	Dissolve 138 g in water and make up to 1 litre	335 g
Potassium thiocyanate, 0.1 M (0.1 N)	Dissolve 10 g in water and make up to one litre	25 g
Silver nitrate, 0.1 M (0.1 N)	Dissolve 17 g in water and make up to 1 litre. Store in a brown bottle	42.5 g
Silver nitrate (alcoholic)	1.7 g in 100 cm³ of methanol. Store in a brown bottle	
Silver sulphate, approximately 0.025 M (0.05 N)	Dissolve 8 g in 1 litre of water to form a near saturated solution	
Sodium ethanoate, M (N)	Dissolve 136 g in water and make up to one litre	340 g
Sodium ethanoate, 3 M (3 N)	408 g made up to 1 litre	1020 g
Sodium carbonate, M (2 N)	Dissolve 286 g of the hydrated salt and make up to 1 litre in water *or* Dissolve 106 g of the anhydrous salt in hot water and make up to 1 litre	715 g 265 g
Sodium chloride, M (N)	Dissolve 58.5 g in water and make up to 1 litre	146.25 g

Sodium chloride (normal). *This is a physiological solution and the name does not imply normal in the chemical sense.*	Dissolve 9 g of sodium chloride in freshly boiled distilled water	22.5 g
Sodium cobaltinitrate, 0.16 M (0.32 N)	Dissolve 30 g of cobalt nitrate in 200 cm³ of water, and 240 g of sodium nitrite in 200 cm³ of water. Mix the two solutions and add 60 cm³ of glacial ethanoic acid. Make up to 1 litre and filter *or* Dissolve 68 g of the solid in water and make up to 1 litre. **An unstable solution – maximum shelf life 3 weeks**	
Sodium hypochlorite, M (2 N)	Available commercially as a solution containing 10–14% hypochlorite. A bench solution is prepared by diluting the commercial solution with an equal volume of water. A weaker solution is sometimes used in which 100 cm³ is diluted to 1 litre *or* A solution may be prepared in the laboratory by bubbling chlorine (conc. HCl on potassium permanganate) through 200 cm³ of 12.5 M sodium hydroxide solution to which 500 g of crushed ice have been added. When the solution has increased in weight by 71 g it is diluted to 1 litre	
Sodium hypobromite, 2 M (4 N)	Place a beaker containing 90 cm³ of 5 M sodium hydroxide solution on ice in a trough in a fume cupboard. Add 5 cm³ of bromine slowly with stirring. The solution is unstable and should be freshly prepared	
Sodium nitroprusside, 0.03 M	1 g is dissolved in 100 cm³ of water. Prepare freshly when required	
Sodium perchlorate, 0.2 M	Dissolve 100 g of solid in 250 cm³ of water, then add 250 cm³ ethanol	
Sodium phosphate (disodium hydrogen phosphate), 0.33 M (N)	Dissolve 120 g in water and make up to 1 litre	300 g
Sodium sulphide, 0.5 M (N)	Dissolve 120 g of the solid in water and make up to 1 litre *or* Pass hydrogen sulphide through 500 cm³ of M sodium hydroxide solution until it is saturated. Add 500 cm³ of M sodium hydroxide	300 g
Tin(II) chloride, 0.5 M (N)	Heat 113 g of the solid in 200 cm³ of concentrated hydrochloric acid, add some pieces of metallic tin, and make up to 1 litre with water	282.5 g 500 cm³ of acid
Uranyl zinc(II) ethanoate	Dissolve 10 g of uranyl ethanoate and 30 g of zinc(II) ethanoate in 12 cm³ of 5 M ethanoic acid. Dilute to 100 cm³, add 0.1 g of sodium chloride, stand for 24 hours, then filter	
Zinc nitrate, 0.25 M (0.5 N)	Dissolve 74 g of crystals in water and make up to 1 litre	183 g
Zirconium nitrate (for phosphate separation)	Boil 10 g of solid in 100 cm³ of M nitric acid, stand for 24 hours, and filter through mineral wool	
Zirconium nitrate (for fluoride test)	Boil 1 g in 200 cm³ of concentrated hydrochloric acid and make up to 1 litre with water	

INDICATOR SOLUTIONS

Indicator solutions may be purchased ready-made or may be prepared in the laboratory from dyes. If an indicator is used in large quantities, there is a considerable saving if the solution is made up on the premises.

pH indicators

These are used for the determination of the pH of a solution or to detect changes in pH during acid-alkali titrations. The colour of the indicator changes with the pH of the solution; the pH change which gives the full colour range of the indicator is called its *pH interval* or *pH range*.

Some indicators may have their colour change altered by the addition of a second dye and are known as *screened indicators*. This second dye does not alter in colour with pH, but absorbs some of the light which would otherwise be transmitted through the solution. For example, methyl red changes from red to yellow over the range pH 4.2–6.3; if methyl red is *screened* by the addition of methylene blue, the colour change over the same pH range is from mauve to green, which is easier to detect.

For some purposes a mixture of two indicators is used which alters the colour change of each and results in a single solution with a wider pH range than that of a single indicator. If several indicators are mixed, the range of pH which can be detected with a single solution is greatly increased and the mixture is called a *universal indicator*, or *wide-range indicator*.

If pH papers are required, they can be made by drying sheets of filter or chromatography paper which have been dipped in the appropriate indicator, or universal indicator, solution. (This method is only justified for occasional use; it is more economical to purchase rolls of commercially prepared papers for regular use.)

Where the colour, or turbidity, of titration solutions prevents the use of the usual indicators, *luminescent indicators* may be used. The colour change may be observed in ultra violet light in a dark room.

When making solutions of pH indicators, the correct amount of dye is weighed and dissolved in ethanol, the solution then being diluted to 1 litre with de-ionised water. When solutions must not contain ethanol, they can be made up from sodium salts of the indicator which may be purchased in phials ready to dilute to 1 litre, or they may be prepared by the Clark and Lubs method which depends on dissolving the dye in a precise ammount of 0.01 M sodium hydroxide solution, and then diluting to 1 litre with deionised water.

The number of pH-sensitive dyes is legion, but the selection displayed in the following tables will be more than adequate for most educational laboratories.

Table 7.11 Water soluble indicators (Clark & Lubs)

When the dye has dissolved in the sodium hydroxide solution, make up to 1 litre with deionised water.

Indicator	cm³ of 0.01 M NaOH	pH range
Cresol red	131.0	0.2–1.8
m-cresol purple	104.8	0.6–2.4
Bromophenol blue	59.9	2.8–4.6
Bromocresol green	57.5	4.0–5.4
Chlorophenol red	94.4	5.2–6.6
Bromocresol purple	74.0	5.2–6.8
Bromophenol red	78.0	5.2–7.0
Bromothymol blue	64.0	6.0–7.6
Thymol blue	86.0	8.0–9.6

Table 7.12 General pH indicators

Dissolve the dye in the volume of ethanol indicated and make up to 1 litre with deionized water.

Indicator	Mass of dye in g	Ethanol in cm³	Colour change	pH range
Methyl violet	0.5	0	yellow-violet	0.1–3.2
Cresol red	0.5	200	red-yellow	0.2–1.8
m-cresol purple	0.4	200	red-yellow	0.6–2.4
Thymol blue	0.4	200	red-yellow	1.2–2.8
Quinaldine red	0.4	1000	colourless-pink	1.4–4.2
Bromophenol blue	0.4	200	yellow-blue violet	2.8–4.6
Methyl orange	0.4	200	red-yellow	2.8–4.6
Congo red	0.2	200	blue-red	3.0–5.0
Bromocresol green	0.4	200	yellow-blue	4.0–5.4
Methyl red	0.2	200	red-yellow	4.4–6.0
Chlorophenol red	0.4	200	yellow-violet red	5.2–6.6
Bromocresol purple	0.4	200	yellow-purple	5.2–6.8
Bromothymol blue	0.4	200	yellow-blue	6.0–7.6
Phenol red	0.2	200	yellow-red	6.8–8.4
Cresol red	0.2	200	yellow-purple	7.2–8.8
m-cresol purple	0.4	200	yellow-violet blue	7.6–9.2
Thymol blue	0.4	200	yellow-violet	8.0–9.6
Phenolphthalein	1.0	600	colourless-violet red	8.4–10.0
Phenolthymophthalein	1.0	600	colourless-violet	8.3–11.0
Thymolphthalein	2.0	1000	colourless-blue	9.4–10.4
Brilliant orange	0.4	200	yellow-red	10.5–12.0
Alizarin	1.0	0	pink-violet	11.0–13.0
Titan yellow	1.0	500	yellow-red	12.0–13.0

Table 7.13 Fluorescent (luminescent) indicators for acid-alkali titrations in ultraviolet light

Indicator	Mass in g	cm³ of ethanol	cm³ of water	Colour in ultraviolet light	pH range
Eosin	1.0	700	300	red-yellow	2.0–3.5
(or the sodium salt in water)	1.0	—	1000		
Erythrosine B	1.0	700	300	colourless-light green	3.0–5.0
Fluorecein	1.0	1000	—	colourless-green	0.5–1.0
(or the sodium salt in water)	1.0	—	1000		

Table 7.14 Screened indicators

Indicators	Mass in g	cm³ of ethanol	cm³ of water	Colour change	pH range
Dimethyl yellow Methylene blue	1.0 0.5	1000	—	pink-yellow-green	2.9—4.0
Methyl orange Xylene cyanol	1.0 2.6	—	1000	mauve-green	2.9—4.6
Methyl orange Analine green	0.5 0.5	—	1000	violet-green	2.9—4.6
Methyl red Methylene blue	1.0 0.5	1000	—	mauve-green	4.2—6.3
Chlorophenol red (sodium salt) Aniline blue (methylene blue)	0.5 0.5	—	1000	green-violet	4.8—6.4
Neutral red Methylene blue	0.5 0.5	1000	—	blue-violet green	6.8—8.0
Phenolphthalein Methyl green	0.33 0.66	1000	—	green-violet	8.0—9.8

Table 7.15 Universal indicators

Indicator	Composition		Colour	pH
Yamada's universal indicator	Bromothymol blue	0.25 g	red	4
	Thymol blue	0.025 g	orange	5
	Methyl red	0.0625 g	yellow	6
	Phenolphthalein	0.5 g	green	7
	Dissolve in 500 cm³ of ethanol and		blue	8
	dilute to 1 litre with water		dark blue	9
			violet	10
van Urk's universal indicator	Tropaeolin 00	0.35 g	orange red	2
	Methyl orange	0.5 g	red orange	3
	Methyl red	0.4 g	orange	4
	Bromothymol blue	2.0 g	yellow orange	5
	α-naphthol phthalein	2.5 g	orange yellow	6
	o-cresol phthalein	2.0 g	yellow	6.5
	Phenolphthalein	2.5 g	green yellow	7
	Alizarin yellow R	0.76 g	green	8
	Dissolve in 700 cm³ of ethanol and		blue green	8.5
	make up to 1 litre with water		green blue	9
			violet blue	9.5
			violet	10
			violet-violet red	11
			violet red	12

Table 7.16 Mixed indicators

Indicators	Mass in g	cm³ of solvent	Colour change	Median pH
Methyl orange	0.66	1000 water	orange-blue-green	4.3
Bromocresol green (Na salt)	0.33			
Methyl red	0.4	1000 ethanol	wine-red-green	5.1
Bromocresol green (Na salt)	0.6			
Chlorophenol red (Na salt)	0.5	1000 water	yellow-green-violet	6.1
Bromocresol green (Na salt)	0.5			
Bromothymol blue (Na salt)	0.5	1000 water	yellow-blue-violet	6.7
Bromocresol purple (Na salt)	0.5			
Bromothymol blue (Na salt)	0.66	1000 water	violet-blue	6.9
Azolitmin	0.33			
Neutral red	0.5	1000 ethanol	rose-green	7.2
Bromothymol blue	0.5			
Cyanine	0.66	500 ethanol	yellow-violet	7.3
Phenol red	0.33	500 water		
Phenol red	0.5	1000 ethanol (or Na salts in	yellow-violet	7.5
Bromothymol blue	0.5	1000 water)		
Cresol red	0.33	1000 ethanol	pale pink-violet	8.3
Naphtholphthalein	0.66			
Cresol red (Na salt)	0.25	1000 water	yellow-violet	8.3
Thymol blue (Na salt)	0.75			
Thymol blue	0.25	Dissolve in 500 ethanol and	yellow-violet	9.0
Phenolphthalein	0.75	add 500 water		
Thymolphthalein	0.5	1000 ethanol	colourless-violet	9.9
Phenolphthalein	0.5			
Thymolphthalein	0.66	1000 ethanol	yellow-violet	10.2
Alizarin yellow	0.33			

Bench indicators

See end of section on *Bench Solutions*

Table 7.17 Miscellaneous titration indicators

Indicator	Preparation
Starch	Mix 10 g of AnalaR soluble starch into a paste with cold water and pour, with constant stirring, into 1 litre of boiling water. Make freshly as needed. Used for iodine titrations
Sodium starch glycollate	Dissolve 10 g in 1 litre of hot water. Cool
Iron(III) alum (for Volhard's method)	Dissolve 100 g of iron(III) alum crystals in 1 litre of water and add sufficient 2 M nitric acid to clear the solution. Use 1 cm³ per titration – colour change from white precipitate to orange red
Potassium chromate	Dissolve 50 g of potassium chromate in 1 litre of water *or* dissolve 42 g of potassium chromate and 7 g potassium dichromate in 1 litre of water for buffer action in slightly acid or alkaline titrations. Use 1 cm³ of indicator per 50 cm³ of final solution

Table 7.18 Absorption indicators

Indicator	g l⁻¹	Solvent	Colour change	Determination of
Alizarin	1.0	water		nitrate
Alizarin red S	1.0	water	yellow-pink	molybdate and ferrocyanide (hexacyanoferrate(II)) with lead(II) nitrate fluoride
Bromophenol blue	0.4	200 cm³ ethanol 800 cm³ water	colourless-blue	bromide chloride thiocyanate
Dibromo(R)fluorescein	1.0	ethanol	yellow green-pink	chlorides bromides borates phosphates
Dichloro(R)fluorescein	1.0	ethanol	yellow green-pink	borate bromide chloride iodide
Di-iodo(R)dimethyl (R) fluorescein	1.0	ethanol	orange pink-red	iodide in the presence of chloride
Di-iodo(R)fluorescein	1.0	ethanol	yellow-pink	chloride iodide
Diphenyl carbazone	1.0	ethanol	orange pink-violet	cyanide
Eosin	1.0	700 cm³ ethanol 300 cm³ water	pink-red purple	bromide iodide thiocyanate
Erythrosin B	1.0	700 cm³ ethanol 300 cm³ water	pink-blue pink	iodides
Fluorescein (Na salt)	1.0	water	green-pink	sulphate oxalate cyanate chloride bromide iodide
Mercurochrome	1.0	water	pale orange-pink	chloride thiocyanate
Phenosafranine	1.0	water	red-blue	bromide chloride
Rhodamine 6G	1.0	water	orange pink-violet	bromide silver
Solochrome red B	2.0	water	colourless-orange red precipitate blue pink-white precipitate	molybdate ferrocyanide (hexacyano-ferrate(II))
Tartrazine	1.0	water	colourless-yellow	thiocyanate bromide iodide silver
Titan yellow	1.0	500 cm³ ethanol 500 cm³ water	yellow-blue	chloride & bromide with mercury(I) nitrate

Indicators for redox reactions

Redox indicators change colour on reduction and the change may be reversed by oxidation. A small and specific change in the redox potential of the solution gives a complete colour change. The point at which the oxidized and reduced states of the indicator are in equilibrium is called the *normal* or *transition potential* of the indicator and is represented by the symbol E_0.

When used for a redox titration, the E_0 of the indicator must match the potential of the solution when equivalent quantities of reducing and oxidizing agents have been added.

The E_0 is affected to some extent by the pH of the solution.

The choice of indicator depends upon the reaction concerned. The redox potential of a chemical or biological system can be measured by the use of a series of redox indicators.

Table 7.19 Redox indicators

Indicator	$E^{\ominus}$ or E_0	Mass in g	Solvent	Colour		Determination of
				Oxidized	Reduced	
Barium diphenylamine-4-sulphonate	0.84	0.2	100 cm³ water	red violet	colourless	iron(II) ion
3,3-dimethyl naphthidine	0.80	0.1	100 cm³ glacial ethanoic acid	purple red	grey green	zinc and other metals by ferrocyanide (hexacyanoferrate (II))
3,3-dimethylnaphthidene-4-sulphonic acid	0.84	1.0	100 cm³ water	red violet	colourless	zinc and other metals by ferrocyanide (hexacyano-ferrate(II))
Diphenylamine	0.76	1.0	100 cm³ conc. sulphuric acid	violet	colourless	dichromate titrations
Diphenylbenzidene	0.76	1.0	100 cm³ conc. sulphuric acid	violet	colourless	zinc by ferrocyanide (hexacyano-ferrate(II))
2,2-dipyridil (α-α-dipyridil)	0.97	1.172	100 cm³ water. Add 0.695 g of FeSO₄ crystals	pale blue	red	ceric sulphate titrations
Erioglaucine A	1.0	0.1	100 cm³ water	orange	green yellow	ceric sulphate titrations
Erio-green B	1.01	0.1	100 cm³ water	orange	green	ceric sulphate titrations
p-ethoxychrysoidine	0.76	0.02	100 cm³ conc. sulphuric acid	salmon pink	grey green	zinc by ferricyanide (hexacyano-ferrate(III))
		0.1	100 cm³ ethanol	orange yellow	red	arsenic by bromate or iodate
α-naphth-flavone (7,8-benzo-flavone)	0.8	0.5	100 cm³ ethanol	brown orange	yellow	arsenic by bromate
1,10-phenanthroline hydrate	1.06	1.485	100 cm³ water	blue	red	iron(II) ions by ceric sulphate
1,10-phenanthroline iron(II) sulphate complex	1.06	1.485	0.695 g of FeSO₄ crystals in 100 cm³ water	blue	red	iron(II) ions by ceric sulphate

N-phenylanthanilic acid	1.08	0.1	Dissolve in 5 cm³ 0.1 M NaOH and dilute to 100 cm³	purple red	colourless	iron(II) ions by ceric sulphate or dichromate
Setopaline	1.07	0.1	100 cm³ water	orange	yellow	ferrocyanide (hexacyano-ferrate(II)) and ceric sulphate titrations
Setoglaucine	1.0	0.1	100 cm³ water	yellow red	yellow green	reduced iron(III) ion by permanganate
Sodium-diphenylamine-4-sulphonate	0.85	0.2	100 cm³ water	red violet	colourless	iron(II) ion

Table 7.20 Irreversible indicators

The following react irreversibly with local concentrations of oxidizing agents and fresh indicator may have to be added before the end-point is reached. All are made up by dissolving 0.2 g of powder in 100 cm³ of water.

Indicator	Colour change	Use
Amaranth	red-colourless	titrations with hypochlorite (chlorate(I)) and iodide
Brilliant ponceaux 5R	orange-colourless	titrations with iodate
Bordeaux	pink-pale yellow green	titrations with hypochlorite (chlorate(I))
Naphthol blue black	green-very pale pink	titrations with iodate and bromate

Table 7.21 Indicators for EDTA (diamino-ethane-tetra-ethanoic acid disodium salt) titrations

Indicator	Preparation	Colour change	Determination of
Murexide (ammonium pupurate)	Grind 1.0 g of murexide with 100 g of sodium chloride; use 0.2 g of the powder for each titration	salmon pink-blue	calcium
Murexide with naphthol green	Grind together 0.2 g murexide, 0.5 g naphthol green B, and 100 g sodium chloride; use 0.2 g of powder	green-red-blue	calcium
Bromopyrogallol red	Dissolve 0.05 g in 50 cm³ of ethanol and add 50 cm³ of water	blue-red	bismuth, cobalt, nickel
Pyrogallol red	Dissolve 0.05 g in 50 cm³ of ethanol and add 50 cm³ of water	red-yellow	bismuth, cobalt, nickel
Pyrocatechol violet (catechol violet)	0.1 g in 100 cm³ of water	blue-red	bismuth nickel
		red-yellow	gallium thorium

Calcon	1.0 g in 100 cm³ of ethanol with 0.8 g Na_2CO_3 added		calcium
Gallocyanine	1.0 g in 100 cm³ of glacial ethanoic acid	blue-red	gallium
Phthalein purple	0.18 g phthalein + 0.02 g naphthol green B in 100 cm³ of water rendered slightly alkaline with ammonia	red-grey	calcium
		red-green	barium strontium sulphate
Eriochrome black T (solochrome)	1.0 g in 100 cm³ of ethanol *or* as for murexide	red-blue	magnesium calcium lead manganese zinc
Solochrome black 6B	0.5 g in 100 cm³ of ethanol *Screening:* to each titration add 6 drops followed by 2 drops of dimethyl yellow solution (0.25 g in 100 cm³ of ethanol)	red-blue red-green	magnesium calcium lead manganese lime
Sulphosalicyclic acid	2.0 g in 100 cm³ of water	green blue-pale yellow	iron
Tiron	2.0 g in 100 cm³ of water	green blue-pale straw or colourless	iron
Variamin blue B	1.0 g in 100 cm³ of water	blue-pale yellow	iron(III)
Pan 1-(2-pyridolazo)-2-naphthol	0.5 g in 100 cm³ of ethanol	pink-yellow	cadmium copper
Xylenel orange	0.1 g in 100 cm³ of water	red purple-lemon yellow	cadmium bismuth lead mercury zinc

BUFFER SOLUTIONS

A buffer system is a solution to which it is possible to add either hydrogen ions or hydroxide ions without greatly affecting the pH of the solution. Buffer systems usually consist of a weak base and its salt with a strong acid, or a weak acid and its salt with a strong base. Let us consider a buffer system containing ethanoic acid and sodium ethanoate. In solution the components of this system will behave thus:

$$NaEth \leftrightharpoons Na^+ + Eth^-$$
$$HEth \leftrightharpoons H^+ + Eth^-$$

the acid being partially dissociated and the salt completely dissociated.

If additional H^+ ions are added to such a system, they will combine with the negatively charged anions to form undissociated acetic acid. If additional hydroxyl ions are added to the system, they will combine with H^+ ions and be removed as water. The hydrogen ions removed from the system are then replaced by further dissociation of the acid.

Such a system therefore has the ability to 'mop-up' both hydrogen ions and hydroxide ions, thus resisting any change in pH.

Buffer solutions may be obtained commercially either as a concentrated solution or as tablets. These preparations are standardized electrometrically and adjusted within ± 0.02 pH unit of the nominal value, at 20 °C.

Commercial buffer solutions of definite pH are available from Hopkins and Williams in the following range:

pH 3.0; 3.2; 3.4; 3.6; 3.8 – containing potassium, chloride, and phthalate ions.

pH 4.0; 4.2; 4.4; 4.6; 4.8; 5.0; 5.2; 5.4; 5.6; 5.8; 6.0; 6.2 – containing potassium, sodium, and phthalate ions.

pH 6.4; 6.6; 6.8; 7.0; 7.2; 7.4; 7.6; 7.8; 8.0 – containing potassium, sodium, and phosphate ions.

pH 8.2; 8.4; 8.6; 8.8; 9.0; 9.2; 9.4; 9.6; 9.8; 10.0 – containing potassium, sodium, borate, and chloride ions.

Commercial buffer tablets are available from most chemical wholesalers. British Drug Houses and Hopkins and Williams sell boxes of 50 tablets, each tablet making 100 cm^3 of solution. The tablets are available for pH 4.0, pH 7.0, and pH 9.2. When the buffer solution is to be used for standardizing a pH meter the most accurate results are obtained if a buffer is used from the approximate range expected in the experiment, i.e. if working in acid conditions use the pH 4 buffer.

G. T. Gurr, and Hopkins and Williams, also sell a range of tablets packed in boxes of 72 to make 100 cm^3 of buffer solution in the following values: pH 4.2, 4.8, 6.4, 6.8, 7.0, 7.2, 7.4, 7.6, 8.0, 8.4, 9.0. The same suppliers also stock boxes of 72 tablets, each of which will make one litre of buffer solution in the following range: 4.8, 6.4, 6.8, 7.0, 7.2, 7.4, 9.0.

Commercial Universal buffer mixture is available from B.D.H. and Hopkins and Williams. It consists of a mixed salt. If the contents of the purchased tube are dissolved to make 1 litre of solution, the pH value of this solution will be 3.1. Buffer solutions with pH values ranging from 2.7 to 11.4 can be made from this by the simple addition of 0.2 M HCl or NaOH. When these additions have been made, the pH value will be $3.1 \pm 0.1185V$ where $V =$ the number of cm^3 of 0.2 M HCl or NaOH added to 100 cm^3 of solution with the subsequent dilution to 200 cm^3 with water. The NaOH being (+) and the HCl being (−). De-ionised or freshly distilled water must be used, i.e. the addition of each cm^3 of HCl or NaOH will change the pH by ± 1.185 respectively.

pH	Na_2HPO_4 (cm^3)	Citric acid (cm^3)	pH	Na_2HPO_4 (cm^3)	Citric acid (cm^3)
2.2	2.00	98.00	5.2	53.60	46.40
2.4	6.20	93.80	5.4	55.75	44.25
2.6	10.90	89.10	5.6	58.00	42.00
2.8	15.85	84.15	5.8	60.45	39.55
3.0	20.55	79.45	6.0	63.15	36.85
3.2	24.70	75.30	6.2	66.10	33.90
3.4	28.50	71.50	6.4	69.25	30.75
3.6	32.20	67.80	6.6	72.75	27.25
3.8	35.50	64.50	6.8	77.25	22.75
4.0	38.55	61.45	7.0	82.35	17.65
4.2	41.40	58.60	7.2	86.95	13.05
4.4	45.10	55.90	7.4	91.85	9.15
4.6	46.75	53.25	7.6	93.65	6.35
4.8	49.30	50.70	7.8	95.75	4.25
5.0	51.50	48.50	8.0	97.25	2.75

pH 3.0–6.2 Citrate buffer

Make up a 0.1 M solution of citric acid (19.21 g per 1000 cm^3) and a 0.1 M solution of sodium citrate (29.41 g per 1000 cm^3).

To make the appropriate buffers add the quantities of each of these solutions indicated below and make the solution up to 100 cm^3 with de-ionised water.

Citric acid (cm^3)	Sodium citrate (cm^3)	pH	Citric acid (cm^3)	Sodium citrate (cm^3)	pH
46.5	3.5	3.0	23.0	27.0	4.8
43.7	6.3	3.2	20.5	29.5	5.0
40.0	10.0	3.4	18.0	32.0	5.2
37.0	13.0	3.6	16.0	34.0	5.4
35.0	15.0	3.8	13.7	36.3	5.6
33.0	17.0	4.0	11.8	38.2	5.8
31.5	18.5	4.2	9.5	40.5	6.0
28.0	22.0	4.4	7.2	42.8	6.2
25.5	24.5	4.6			

BUFFER SOLUTIONS WHICH CAN BE PREPARED IN THE LABORATORY

pH range 2.2–8.0

To make up 100 cm^3 of buffer add 0.2 M Na_2HPO_4 to 0.1 M citric acid in the proportions shown.

pH 3.6–5.6 Ethanoate (acetate) buffer

Make up a solution of ethanoic acid (0.2 M, 11.55 cm^3 in 1000 cm^3) and a solution of sodium ethanoate (0.2 M, 16.4 g of CH_3COONa per 1000 cm^3 or 27.2 g of $CH_3COONa, 3H_2O$ per 1000 cm^3).

Add the amounts of these indicated below and make the solution up to 100 cm^3.

Ethanoic acid (cm³)	Sodium ethanoate (cm³)	pH	Ethanoic acid (cm³)	Sodium ethanoate (cm³)	pH
46.3	3.7	3.6	20.0	30.0	4.8
44.0	6.0	3.8	14.8	35.2	5.0
41.0	9.0	4.0	10.5	39.5	5.2
36.8	13.2	4.2	8.8	41.2	5.4
30.5	19.5	4.4	4.8	45.2	5.6
25.5	24.5	4.6			

pH 5.8–8.0 Phosphate buffer

Make a solution of sodium dihydrogen phosphate (31.2 g of $NaH_2PO_4, 2H_2O$ per litre) and a solution of disodium hydrogen phosphate (28.39 g of Na_2HPO_4 per litre or 71.7 g of $Na_2HPO_4, 12H_2O$ per litre). Each of these solutions will be 0.2 M solutions. To make up the required buffers, add the indicated amounts of the solutions together and make up to 100 cm³ with de-ionised water.

NaH_2PO_4 (cm³)	Na_2HPO_4 (cm³)	pH	NaH_2PO_4 (cm³)	Na_2HPO_4 (cm³)	pH
46.0	4.0	5.8	19.5	30.5	7.0
44.0	6.2	6.0	14.0	36.0	7.2
40.7	9.2	6.2	9.5	40.5	7.4
36.7	13.2	6.4	6.5	43.5	7.6
31.2	18.7	6.6	4.2	46.7	7.8
25.5	24.5	6.8	2.6	47.3	8.0

pH 7.6–9.2 Boric acid-Borax buffer

Make up a solution of boric acid (12.4 g per litre) and one of borax (19.05 g per litre). The acid solution is 0.2 M and the borax 0.5 M (0.2 M in terms of sodium borate).

To 25 cm³ of the acid add the indicated amount of borax solution and make the solution up to 100 cm³ with de-ionised water.

Borax solution (cm³)	pH	Borax solution (cm³)	pH
1	7.6	8.7	8.6
1.55	7.8	15.0	8.8
2.45	8.0	29.5	9.0
3.6	8.2	57.5	9.2
5.7	8.4		

pH 9.2–10.6 Carbonate-Hydrogencarbonate buffer

Make up a solution of anhydrous sodium carbonate (21.2 g per litre), and a solution of sodium hydrogencarbonate (16.8 g per litre).

Both these solutions will be 0.2 M solutions.

To make up the required buffers, add the indicated amounts of the solutions together and make the solution up to 100 cm³ with de-ionised water.

Na_2CO_3 (cm³)	$NaHCO_3$ (cm³)	pH	Na_2CO_3 (cm³)	$NaHCO_3$ (cm³)	pH
2.0	23.0	9.2	13.7	11.2	10.0
4.7	20.2	9.4	16.5	8.5	10.2
8.0	17.0	9.6	14.2	5.7	10.4
11.0	14.0	9.8	21.2	3.7	10.6

A simply prepared Universal buffer

Prepare accurately 0.1 M solutions of sodium dihydrogenphosphate and disodium hydrogenphosphate. The following table indicates the proportions necessary for the preparation of buffer solutions between pH 4 and pH 9.

pH	NaH_2PO_4 (cm³)	Na_2HPO_4 (cm³)	pH	NaH_2PO_4 (cm³)	Na_2HPO_4 (cm³)
4.0	10	0	6.75	4	6
6.0	8	2	7.2	2	8
6.5	6	4	9.0	0	10

pH 0.65–5.20 Ethanoate – HCl buffer

Make up a M solution of sodium ethanoate (136.08 g in one litre) and a M solution of hydrochloric acid (86 cm³ concentrated acid in one litre).

To use, add 50 cm³ of the M sodium ethanoate solution to the volume of hydrochloric acid indicated and dilute to 250 cm³ with deionised water.

pH	HCl (cm³)	pH	HCl (cm³)	pH	HCl (cm³)
0.65	100	1.99	52.5	3.75	44.0
0.75	90	2.32	51.0	3.95	41.5
0.91	80	2.64	50.0	4.19	37.5
1.09	70	2.72	49.75	4.39	32.0
1.24	65	3.09	48.5	4.58	26.5
1.42	60	3.29	47.5	4.76	21.0
1.71	55	3.49	46.25	4.92	15.0
1.85	53.5	3.61	45.0	5.2	9.0

pH 2.2–3.8 Phthalate – Hydrochloric acid buffer

Make up a 0.2 M solution of potassium acid phthalate (40.24 g in 1 litre) and a 0.2 M solution of hydrochloric acid (1.72 cm^3 of concentrated acid in 1 litre). To use, add 50 cm^3 of the 0.2 M phthalate to the quantity of HCl indicated and dilute with de-ionised water to 250 cm^3.

pH	HCl (cm^3)	pH	HCl (cm^3)	pH	HCl (cm^3)
2.2	46.6	2.8	26.5	3.4	9.95
2.4	39.6	3.0	20.4	3.6	6.00
2.6	33.0	3.2	14.8	3.8	2.65

pH 4.0–6.2 Phthalate – Sodium hydroxide buffer

Make up a 0.2 M solution of potassium acid phthalate (40.24 g in 1 litre) and a 0.2 M solution of NaOH.

To use, add 50 cm^3 of the phthalate to the quantity of NaOH indicated and dilute to 250 cm^3 with deionised water.

pH	NaOH (cm^3)	pH	NaOH (cm^3)	pH	NaOH (cm^3)
4.0	0.4	4.8	17.5	5.6	39.7
4.2	3.65	5.0	23.65	5.8	43.1
4.4	7.35	5.2	29.75	6.0	45.4
4.6	12.0	5.4	35.25	6.2	47.0

pH 1.2–2.2 Potassium chloride – Hydrochloric acid buffer

Make up a 0.2 M solution of KCl (14.91 g per litre) and a 0.2 M HCl solution (1.72 cm^3 of concentrated acid in 1 litre).

To use, add 50 cm^3 of the KCl solution to the quantity of HCl solution indicated and dilute to 250 cm^3.

pH	HCl (cm^3)	pH	HCl (cm^3)
1.2	64.5	1.8	16.6
1.4	41.5	2.0	10.6
1.6	26.3	2.2	6.7

pH 1.04–4.96 Citrate – Hydrochloric acid buffer

Make up 0.1 M solution of citric acid (21.008 g of citric acid crystals per litre) by dissolving the crystals in approximately 500 cm^3 of water and then adding 200 cm^3 of M NaOH and making up to 1 litre. Prepare a 0.1 M solution of HCl.

To use, mix the solutions in the proportions indicated to obtain 100 cm^3 of buffer.

pH	Citrate (cm^3)	HCl (cm^3)	pH	Citrate (cm^3)	HCl (cm^3)
1.04	0.0	100.0	3.69	50.0	50.0
1.17	10.0	90.0	3.95	55.0	45.0
1.42	20.0	50.0	4.16	60.0	40.0
1.93	30.0	70.0	4.45	70.0	30.0
2.27	33.3	66.7	4.65	80.0	20.0
2.97	40.0	60.0	4.83	90.0	10.0
3.36	45.0	55.0	4.89	95.0	5.0
3.53	47.5	52.5	4.96	100.0	0.0

pH 4.96–6.69 Citrate – Sodium hydroxide buffer

Make 0.1 M citric acid solution as for the previous buffer. Prepare a 0.1 M NaOH solution (4.1 g per litre). Make up 100 cm^3 of working solution as indicated.

pH	Citrate (cm^3)	NaOH (cm^3)	pH	Citrate (cm^3)	NaOH (cm^3)
4.96	100.0	0.0	5.57	70.0	30.0
5.02	95.0	5.0	5.98	60.0	40.0
5.11	90.0	10.0	6.34	55.0	45.0
5.31	80.0	20.0	6.69	52.5	47.5

pH 9.2–12.38 Borate – Sodium hydroxide buffer

Dissolve 12.404 g of boric acid in 500 cm^3 of water, add 100 cm^3 of M NaOH, and make up to 1 litre. Prepare a 0.1 M NaOH solution (4.1 g per litre). Make up 100 cm^3 of working solution as indicated.

pH	Borate (cm^3)	NaOH (cm^3)	pH	Borate (cm^3)	NaOH (cm^3)
9.24	100.0	0.0	9.97	60.0	40.0
9.36	90.0	10.0	11.08	50.0	50.0
9.50	80.0	20.0	12.38	40.0	60.0
9.68	770.0	30.0			

pH 7.61–9.23 Borate – Hydrochloric acid buffer

Make up a borate solution as described for the previous buffer. Prepare a 0.1 M hydrochloric acid solution. Use by making up 100 cm³ of working solution as indicated.

pH	Borate (cm³)	HCl (cm³)	pH	Borate (cm³)	HCl (cm³)
7.61	52.5	47.5	8.79	75.0	25.0
7.93	55.0	45.0	8.89	80.0	20.0
8.13	57.5	42.5	8.99	85.0	15.0
8.27	60.0	40.0	9.07	90.0	10.0
8.49	65.0	35.0	9.15	95.0	5.0
8.67	70.0	30.0	9.23	100.0	0.0

pH 1.04–3.68 Glycine – Hydrochloric acid buffer

Dissolve 7.507 g of glycine (aminoethanoic acid) and 5.844 g of sodium chloride in water and make up to 1 litre. Prepare a 0.1 M solution of hydrochloric acid.

Make working solutions by mixing in the proportions indicated.

pH	Glycine (cm³)	HCl (cm³)	pH	Glycine (cm³)	HCl (cm³)
1.04	10	100	2.28	60	40
1.15	20	90	2.61	70	30
1.25	30	80	2.92	80	20
1.42	40	70	3.34	90	10
1.65	50	60	3.68	95	5
1.93	50				

pH 8.53–12.9 Glycine (aminoethanoic acid) – Sodium hydroxide buffer

Make up glycine (aminoethanoic acid) and sodium chloride solution as described for the previous buffer. Prepare a 0.1 M NaOH solution (4.1 g per litre). Make working solutions by mixing in the proportions indicated.

pH	Glycine (cm³)	NaOH (cm³)	pH	Glycine (cm³)	NaOH (cm³)
8.53	95	5	11.25	50	50
8.88	90	10	11.51	49	51
9.31	80	20	12.04	45	55
9.66	70	30	12.33	40	60
10.09	60	40	12.60	30	70
10.42	55	45	12.79	20	80
11.01	51	49	12.90	10	90

pH 7.8–10.0 Boric acid – Potassium chloride – Sodium hydroxide buffer

Make a solution of 12.366 g of boric acid and 7.456 g of potassium chloride in 1 litre of deionised water. Prepare a 0.2 M solution of NaOH. (8.2 g per litre.)

To use, add the volume of NaOH solution indicated to 50 cm³ of the boric acid/potassium chloride solution, and dilute to 250 cm³ with water.

pH	NaOH (cm³)	pH	NaOH (cm³)	pH	NaOH (cm³)
7.8	2.65	8.6	12.00	9.4	32.00
8.0	4.00	8.8	16.40	9.6	36.85
8.2	5.90	9.0	21.40	9.8	40.80
8.4	8.55	9.2	26.70	10.0	43.90

pH 7.19–9.1 Tris – Hydrochloric acid buffer

Make a 0.2 M solution of Tris(hydroxymethyl)aminomethane (24.228 g per litre) and prepare a 0.1 M solution of hydrochloric acid (0.86 cm³ of concentrated acid per litre). To use, add the amount of HCl indicated to 25 cm³ of Tris and dilute to 100 cm³.

pH	HCl (cm³)	pH	HCl (cm³)	pH	HCl (cm³)
7.19	45.0	7.96	30.0	8.51	15.0
7.36	42.5	8.05	27.5	8.62	12.5
7.54	40.0	8.14	25.0	8.74	10.0
7.66	37.5	8.23	22.5	8.92	7.5
7.77	35.0	8.32	20.0	9.10	5.0
7.87	32.5	8.51	17.5		

pH 6.8–9.6 Veronal (barbitone-sodium) – Hydrochloric acid

Dissolve 20.60 g of barbitone-sodium (sodium 5,5-diethylbarbiturate) in 1 litre of water. Prepare a 0.1 M solution of HCl (0.86 cm³ of concentrated acid per litre). Make working solutions as indicated.

pH	Barbitone (cm³)	HCl (cm³)	pH	Barbitone (cm³)	HCl (cm³)
6.8	5.22	4.78	8.4	8.23	1.77
7.0	5.36	4.64	8.6	8.71	1.79
7.2	5.54	4.46	8.8	9.08	0.92
7.4	5.81	4.19	9.0	9.36	0.64
7.6	6.15	3.85	9.2	9.52	0.48
7.8	6.62	3.38	9.4	9.74	0.26
8.0	7.16	2.84	9.6	9.85	0.15
8.2	7.69	2.31			

In a well-equipped school or college laboratory, a pH meter should be available. If so, all the above buffer solutions need be made up with only modest accuracy when preparing stock solutions; the working solutions can be adjusted *to any required pH* using the meter.

The meter must be calibrated on a suitable commercial buffer of known pH (see page 82).

REAGENTS USED IN QUALITATIVE ANALYSIS

Some of the reagents listed can also be used for quantitive tests, but since their main use is in spot tests or semi-micro qualitative analysis, the quantities given below are for the preparation of 100 cm^3 of test solution. If larger quantities are required it is only necessary to multiply by ten and make up the final solution to 1 litre.

Table 7.22 Reagents used in qualitative analysis

Reagent	Preparation of working solution	Test, or reagent, for
Alizarin	Saturated solution in ethanol	Aluminium
Alizarin S	0.1 g in 100 cm^3 of water	Aluminium
Aluminon (Ammonium aurine tricarboxilate)	0.1 g in 100 cm^3 of water	Aluminium
1-amino-4-hydroxy-anthroquinone	0.1 g in 100 cm^3 of ethanol	Lithium
Aniline sulphate (phenylammonium sulphate)	0.1 g in 100 cm^3 of water	Chlorate
Benzidene (4-4'-biphenyldiamine sulphate) (**Carcinogenic** – use is illegal in English schools)	Saturated solution in cold 2 M ethanoic acid	Ferricyanide
	0.05 g in 100 cm^3 of 1.5 M ethanoic acid	Lead
	2 g in 100 cm^3 of 2 M ethanoic acid	Persulphate
Benzidene-copper(I) ethanoate (**carcinogenic** – use is illegal in English schools)	(a) 4.75 cm^3 of saturated benzidene in glacial ethanoic acid, diluted to 100 cm^3 (b) 0.29 g copper(I) ethanoate in 100 cm^3 of water. For use, mix 5 parts (a) with 1 part (b).	Cyanide
Benzidene hydrochloride (**carcinogenic** – use is illegal in English schools)	0.8 g in 12 cm^3 of M hydrochloric acid and dilute to 100 cm^3.	Sulphate
S-benzyl-thiuronium chloride	10 g in 100 cm^3 of water	Nitrogen Sulphur Chlorine
Brucine	0.1 g in 100 cm^3 of ethanol	Bismuth Nitrate
Cacotheline (nitro derivative of brucine)	0.25 g in 100 cm^3 of water. Poor shelf life – prepare as needed	Tin(II) ion
Cadion 2B (4-nitro-naphthalene-diazoamino-azobenzine)	0.02 g plus cm^3 of 2 M KOH in 100 cm^3 of ethanol	Cadmium
Carmine	0.5 g in 100 cm^3 of ethanol	Boron
Catechol violet	0.1 g in 100 cm^3 of water	Aluminium Bismuth Tin Titanium Vanadium
Chloride test	Dissolve 1.7 g of silver nitrate in water, add 25 g of potassium nitrate followed by 17 cm^3 of 0.88 ammonia, and make up to 1 litre	Chloride

Chrome azurol S	0.1 g in 100 cm³ of water	Beryllium
Chromotrope 2B	0.005 g in 100 cm³ of concentrated sulphuric acid	Boron Chromium Titanium Nitrate Formaldehyde
Cinchonine	2.5 g in 100 cm³ of 0.25 M nitric acid	Bismuth
Cupferron	0.1 g of ammonium carbonate and 2 g of cupferron in 100 cm³ of water. Shelf life about 3–4 days	Aluminium Bismuth Iron Molybdenum Tin Tungsten Uranium Vanadium Zirconium
Cupron (α-benzoin-oxime)	5 g in 100 cm³ of ethanol	Copper Molybdenum Tungsten
Curcumin	0.1 g in 100 cm³ of ethanol	Boron
o-dianisidene (**carcinogenic** – illegal to use in English schools)	2 g in 100 cm³ of 6 M HCl	Chromium Gold Phosphate
4-dimethylamineazobenzine-4′-arsonic acid	1 g dissolved in 5 cm³ of concentrated HCl and diluted to 100 cm³ with ethanol	Zirconium
4-dimethylaminobenzylidenerhodanine	0.03 g in 100 cm³ of ethanol	Copper Gold Mercury Silver
Dimethylglyoxime	1 g in 100 cm³ of ethanol	Bismuth Cobalt Iron Nickel Palladium Silver
Di-β-naphthol	0.05 g in 100 cm³ of concentrated sulphuric acid	Tartrates
4,4′-dinitrodiphenylcarbazide	0.1 g in 100 cm³ of ethanol	Cadmium
2,4-dinitro-1-naphthol-7-sulphonic acid	0.5 g in 100 cm³ of formic (methanoic) acid	Caesium
Diphenylamine	0.5 g in 100 cm³ of concentrated H_2SO_4	Nitrate
Diphenylbenzidene	0.02 g in 100 cm³ of special grade, nitrogen free, concentrated sulphuric acid	Nitrate Nitrite
Diphenylcarbazide	1 g in 100 cm³ of ethanol	Cadmium Chromium Mercury Arsenate
Diphenylcarbazone	1 g in 100 cm³ of ethanol	Mercury
2,2′-dipyridol	Dissolve 0.146 g of iron(II) sulphate crystals and 0.2 g of dipyridol in 50 cm³ of water, then add 10 g of potassium iodide and dilute to 100 cm³	Cadmium

Dithiol(toluene-3,4-dithiol)	0.5 g in 100 cm³ of 0.25 M NaOH	Tin
Dithio-oxamide	0.5 g in 100 cm³ of ethanol	Bismuth Cobalt Copper Iron Nickel
Dithizone	0.005 g in 100 cm³ of tetrachloromethane	Cadmium Gold Indium Mercury Silver Zinc
Ethylene-diamine	Add a saturated solution in water to 0.1 M copper(II) sulphate until the colour no longer intensifies	Mercury
Fluorescein	Mix 50 cm³ each of water and ethanol and add fluorescein to saturate	Bromide
Formaldoxime (formaldehyde oxime hydrochloride)	2.5 g in 100 cm³ of water	Manganese
α-Furil-dioxime	10 g in 100 cm³ of ethanol	Nickel Palladium Rhenium
Gallic acid (3,4,5-trihydroxybenzoic acid)	0.02 g in 100 cm³ of water	Cerium
Gallocyanine	1 g in 100 cm³ of water	Lead
Hexanitro-diphenylamine	Heat 1 g in 100 cm³ of 0.05 M sodium carbonate and filter	Potassium
8-hydroxy-7-quinoline-5-sulphonic acid (ferron)	0.2 g in 100 cm³ of water	Aluminium Iron
4-hydroxy-3-nitro-phenyl-arsenic acid	Saturated aqueous solution	Cadmium Calcium Tin
8-hydroxyquinaldine	0.2 g in 100 cm³ of water	Lead Magnesium Titanium Zinc
8-hydroxyquinoline (oxine)	0.2 g in 100 cm³ of water	Aluminium Cadmium Cobalt Indium Molybdenum Tin Titanium Zinc
Indigo	Dissolve 0.4 g in 5 cm³ of hot, concentrated, sulphuric acid. Stand for 5 hours and *pour into* 95 cm³ of water	Hydrosulphites
Indole	0.02 g in 100 cm³ of water	Nitrite
Iron(III) periodate	Dissolve 2 g in 10 cm³ of 2 M KOH, dilute to 50 cm³ and add 3 cm³ of 10% iron(III) chloride. Dilute to 100 cm³ with 2 M KOH	Lithium
Magenta (basic fuschin)	0.02 g in 100 cm³ of water	Sulphites

2-mercapto-benzothiazol	2 g in 100 cm³ of 0.1 M KOH	Bismuth Cadmium Lead
Methyl violet	0.1 g in 100 cm³ of water	Antimony
α-Naphthylamine (**carcenogenic** – illegal in English schools)	0.5 g in 100 cm³ of ethanol	Gold
Nickel ethylene-diamine	Make a 0.5 M solution of nickel nitrate and add ethylene-diamine to give a violet-blue colouration	Thiosulphate
Nioxime	0.8 g in 100 cm³ of water	Nickel Palladium
Nitron	10 g in 100 cm³ of 0.1 M ethanoic acid	Chlorate Nitrate Perchlorate Rhenate Tetra-fluorborate Tungsten
Magneson 11	0.001 g in 100 cm³ of M NaOH	Magnesium
4-(p-nitrophenylazo)-orcinol	0.025 g in 100 cm³ of M NaOH	Beryllium
1-nitroso-2-naphthol	1 g in 100 cm³ of propanone (acetone)	Cobalt Copper Iron Lead Nickel Palladium Uranium Zinc
Nitroso-R-salt	1 g in 100 cm³ of water	Cobalt Iron
Morin (pentahydroxyflavonol)	5 g in 100 cm³ of ethanol	Aluminium Beryllium Boron Uranium Zirconium
Phenazone	1 g in 100 cm³ of water	Antimony Cobalt
Phenylarsonic acid	10 g in 100 cm³ of water	Bismuth Tin Zirconium
Phenylfluorone	0.1 g in 100 cm³ of water	Germanium Molybdenum
Phenylhydrazine	1.5 g in 50 cm³ of water and 50 cm³ of glacial ethanoic acid	Molybdate
Phosphomolybdic acid	5 g in 100 cm³ of water	Antimony
Phthalein purple (metalphthalein)	0.18 g, with 0.01 g of naphthol green (acts as screen), in 100 cm³ of water with 2 drops of ammonia	Barium Calcium Magnesium Strontium
Picrolinic acid	Saturated aqueous solution	Calcium Lead Thorium

Potassium periodate	Saturated aqueous solution	Manganese
PAN (1-(2-pyridylazo)-2-naphthol)	0.5 g in 100 cm³ of ethanol	Cadmium Cobalt Indium Nickel Vanadium Zinc
PAR (4-(2-Pyridylazo)-resorcinol)	0.5 g in 100 cm³ of ethanol	Bismuth Calcium Cobalt Cadmium Mercury Nickel Vanadium Zinc
Pyrocatechol	10 g in 100 cm³ of water. Must be prepared just before use	Titanium
Pyrogallol (benzene-1,2,3-triol)	As for Pyrocatechol	Bismuth
Pyrrole	1 g in 100 cm³ of aldehyde-free ethanol	Selenite
Quinaldic acid	Neutralise 1 g with M NaOH and dilute to 100 cm³	Cadmium Copper Palladium Zinc
Quinoline	Dissolve 2 g in 80 cm³ of hot water and add 2.5 cm³ of concentrated HCl. Filter and make up to 100 cm³ with water	Silicone
Resorcinol (benzene-1,3-diol)	1 g in 100 cm³ of water	Platinum
Rhodamine B	0.01 g in 100 cm³ of water	Antimony Gallium Gold Tungsten
Saltzmann's reagent	Dissolve 0.02 g of N-(1-naphthyl)-ethylene-diamine hydrochloride and 5 g of sulphanilic acid (4-aminobenzenesulphonic acid) in water containing 140 cm³ of glacial ethanoic acid. Make up to 1 litre	Nitrite
Silver periodate	Dissolve 2 g of potassium periodate and 10 g of silver nitrate in 80 cm³ of water. Add 2 cm³ of concentrated nitric acid and dilute to 100 cm³	Ethanoate
Sodium azide – iodine reagent	3 g of sodium azide in 100 cm³ of 0.05 M iodine solution	Sulphide Thiocyanate Thiosulphate
Sodium carbonate – phenolphthalein	Add 10 cm³ of 0.05 M sodium carbonate solution to 20 cm³ of phenolphthalein indicator solution, and dilute to 100 cm³	Bicarbonate
Sodium diethyldithiocarbamate	0.1 g in 100 cm³ of water	Cadmium Chromium Copper Cobalt Manganese Nickel Uranium Zinc

Sodium rhodizonate	0.5 g in 100 cm³ of water	Barium Lead Strontium
Sodium tetraphenylborate	0.5 g in 100 cm³ of water	Ammonium Caesium Potassium Rubidium
Sodium tungstate	10 g in 100 cm³ of water	Vanadate
Sulphanilic acid-1-naphthylamine reagent (**carcinogenic** – illegal in English schools)	(a) 1 g of sulphanilic acid (4-aminobenzenesulphonic acid) in 100 cm³ of 5 M ethanoic acid (b) 0.3 g of 1-naphthylamine boiled in 100 cm³ of 5 M ethanoic acid	Nitrites
Sulphosalicylic acid	25 g in 100 cm³ of water	Bismuth Iron
Tannic acid	1 g in 100 cm³ of water	Aluminium Beryllium Germanium Tin
4,4'-tetramethyl-diamino-diphenyl-methane (tetra-base)	Dissolve 0.5 g in 20 cm³ of glacial ethanoic acid and dilute to 100 cm³ with ethanol	Iodine Lead Manganese
Thioacetamide	7.5 g in 100 cm³ of water	Antimony Arsenic Bismuth Cadmium Cobalt Lead Nickel Zinc
Thiourea	10 g in 100 cm³ of water	Antimony Bismuth Cadmium Osmium Tin
Tiron (catechol-3,5-disulphonic acid, disodium salt)	0.5 g in 100 cm³ of water	Cerium Iron Molybdenum Titanium
Titan yellow	0.1 g in 100 cm of water	Magnesium
o-toluidene (**carcinogenic** – illegal in English schools)	0.1 g in 100 cm³ of 3 M HCl	Chlorine Gold Manganese

REAGENTS USED IN BIOCHEMISTRY AND ORGANIC CHEMISTRY

Biologists and biochemists use many solutions which are identical to those used in chemistry laboratories. Where a reagent is not listed as a biochemical reagent, it will probably appear in one of the previous sections on titrimetric or bench solutions. Where the biological use of a common reagent requires a solution of different strength from that used in chemical laboratories, the dilution and use are given below.

Table 7.23 Reagents used in biochemistry and organic chemistry

Reagent	Preparation of working solution	Use
Acetic acid	See *ethanoic acid*	
Acetone (propanone)	—	Fat solvent Plant pigment solvent Chromatography
Alkaline iodide (Winkler)	500 g of sodium hydroxide and 135 g of sodium iodide made up to 1 litre in water *or* 700 g of potassium hydroxide and 150 g of potassium iodide in 1 litre of water	Oxygen determination
Ammoniacal silver nitrate (Tollen's reagent)	Add 3 drops of 2 M NaOH to 5 cm³ of 0.2 M silver nitrate and then add 2 M ammonium hydroxide until the precipitate just disappears	Test for reducing sugars and aldehydes
Alcoholic silver nitrate	Dissolve 4 g in 10 cm³ of water and add 90 cm³ of ethanol	
Ammonium carbonate	Saturated aqueous	
Ammonium chloride	Saturated aqueous	
Ammonium molybdate	As bench solution	Phosphate determination and detection
Ammonium oxalate	Saturated aqueous	Calcium determination Precipitation of calcium in blood
Ammonium sulphate	56 g made up to 100 cm³ in water (saturated)	Colorimetric method for determination of protein in plasma Precipitation of casein in milk Coagulation of proteins and proteoses
Ammonium thiocyanate	13 g in 1 litre of water	Chloride determination in plasma and urine
Arsenomolybdate solution (Nelson's reagent)	25 g of ammonium molybdate in 450 cm³ of water. Add 21 cm³ of concentrated H_2SO_4 followed by 3 g of sodium arsenate dissolved in 25 cm³ of water. Stand at 37 °C for 2 days and store in a dark, glass-stoppered bottle	Colorimetric determination of blood glucose
Bang's reagent	Dissolve successively in 800 cm³ of water, 100 g of potassium carbonate, 66 g of potassium chloride, and 160 g of potassium hydrogencarbonate. Add 4.4 g of copper(II) sulphate crystals and make up to 1 litre. Stand for 24 hrs, and then dilute 300 cm³ of the solution to 1 litre with a saturated solution of potassium chloride. Stopper closely to prevent entry of air and stand for a further 24 hours before use	Estimation of glucose; 10 mg of glucose = 50 cm³ of Bang's reagent

Laboratory Solutions

Barfoed's reagent	13.3 g of copper(II) ethanoate and 2 cm³ of glacial ethanoic acid made up to 200 cm³ in water	Test for glucose
Baryta water	38.9 g per litre for saturated solution	Respiration experiments
Benedict's solution	*(a) for qualitative work;* 17.3 g of copper(II) sulphate crystals, 173 g of sodium citrate, and 100 g of anhydrous sodium carbonate made up to 1 litre in water	Detection of reducing sugars
	(b) for quantitative work; 100 g of anhydrous or 200 g of crystaline sodium carbonate, 200 g of sodium or potassium citrate, and 125 g of potassium thiocyanate are dissolved in water. To this solution, add 18 g of copper(II) sulphate which have been dissolved in 100 cm³ water, rinsing the flask twice to ensure that all the copper(II) sulphate goes into the final solution. Finally, add 6 cm³ of 0.1 M potassium ferrocyanide solution and make up to 1 litre at 20 °C	50 cm³ of this solution is reduced by 100 mg of glucose
Bial's reagent	Dissolve 6 g of resorcinol (benzene-1,3-diol) in 200 cm³ of 95% ethanol and add 40 drops of 10% iron(III) chloride	Test for pentoses and glycuronic acid
Biuret reagent	0.75 g of copper(II) sulphate crystals in 1 litre of 2 M potassium hydroxide	Test for urea and proteins
Bromine water	Saturated solution	Knoop's test for histidine
Brucke's reagent	Dissolve 50 g of potassium iodide in 200 cm³ of water and then saturate with 120 g of mercury(II) iodide. Filter and dilute with water to 1 litre	Protein test
Copper(I) chloride, ammoniacal	Dissolve 1 g in 10 cm³ of 2 M ammonium hydroxide solution. Add hydroxylamine-hydrochloride, drop by drop until the solution is colourless	Ethyne test
Calcium hypochlorite	Saturated aqueous	
Chloral hydrate iodine (2,2,2-trichloroethanediol iodine)	0.4 g of chloral hydrate in 100 cm³ of water and add a few iodine crystals	Starch test
Chlorine water	Saturated aqueous	
Cobalt(II) chloride	5% solution used to soak filter paper, which is then oven dried	Moisture detection (e.g. transpiration experiments)
Collodion	Dissolve in 50% mixture of ethanol and ether	Osmotic experiments and preparation of dialysis membranes
Copper(II) sulphate (Harding's)	70 g in 1 litre of water	Colorimetric estimation of blood glucose
	1% bench solution	
Copper(II) tungstate	Saturated aqueous	Precipitation of proteins for protein-free blood filtrate
Cross & Bevan's reagent	25 g of zinc(II) chloride crystals dissolved in 45 cm³ of concentrated HCl	Cellulose test
Denigè's reagent	Dissolve 50 g of mercury(II) oxide by heating in 200 cm³ of concentrated sulphuric acid. Dilute to 1 litre *by adding slowly* to 500 cm³ of water and then making up	Test for citrates

Dichromate solution	Add 250 cm³ of concentrated sulphuric acid *slowly to 750 cm³ of water* and then dissolve 100 g of sodium dichromate	Oxidation
Dimedone	0.3 g of 5,5-dimethylcyclohexane (dimedone) in 50 cm³ of ethanol; make up to 100 cm³ with water	Test for aldehydes
2,4-dinitrochlorobenzene	20 g in 100 cm³ of hot ethanol	Test for mercaptans
Diphenylamine	20 g in 95 cm³ of ethanol and 5 cm³ of water	Gives intense blue colouration with fructose
	0.1 g in 10 cm³ of nitrogen-free sulphuric acid	Molisch's test for nitrates
Disodium phenyl phosphate	1.09 g in 1 litre of water	Determination of alkaline phosphates in blood
Ethanoic acid (3%)	27 cm³ of glacial ethanoic acid made up to 1 litre with water	
Ethanoic acid/sodium ethanoate reagent	Mix equal volumes of saturated aqueous sodium ethanoate and glacial ethanoic acid	
Ether	—	Fat solvent and narcotising agent
Esbach's reagent	2 g of citric acid and 1 g picric acid (2,4,6-trinitrophenol) in 100 cm³ of water	Albumin test
Fehling's solution	*Fehling's A.* 69.28 g of copper(II) sulphate crystals in 1 litre of water. *Fehling's B.* 352 g of sodium potassium tartrate and 154 g of sodium hydroxide in 1 litre of water. To use: mix 5 cm³ of each solution	Test for reducing sugars. Can be used quantitatively — 10 cm³ of mixed solution = 0.05 mg glucose
Fenton's reagent	Saturated iron(II) sulphate solution. Add 10 volume hydrogen peroxide, followed by excess bench NaOH, when carrying out the test	Tartrates
Fischer's reagent	40 g of iodine (dried over concentrated sulphuric acid) in 315 cm³ of anhydrous methanol. Add 126 g of dry pyridine and pass through dry SO_2 until weight has increased by 32 g	Test for water in organic solvents
Folin's solution	Dissolve 500 g of ammonium sulphate, 5 g of uranyl ethanoate, and 6 cm³ of glacial ethanoic acid in water and make up to 1 litre	Uric acid test
Folin-Wu alkaline copper solution	Dissolve 40 g of anhydrous sodium carbonate (AnalaR) in 400 cm³ of water and add 7.5 g of tartaric acid. Add 4.5 g of copper(II) sulphate crystals (AnalaR) and make up to 1 litre	Folin-Wu colorimetric estimation of blood glucose
Fouchet's reagent	Dissolve 25 g of trichloroethanoic acid in 100 cm³ of water and add 10 cm³ of 10% iron(III) chloride	Fouchet's test for bile in urine
Glyoxalic acid (oxoethanoic acid)	Cover 10 g of magnesium powder with water and add 250 cm³ of saturated oxalic acid (ethanedioic acid) solution. Make the addition slowly, and cool as necessary. Filter and add ethanoic acid until just acid. Make up to 1 litre	Protein test. Specific for tryptophan and tryptophan radicals
Guaiac reagent	0.5 g of Guaiac resin in 30 cm³ of 95% ethanol	Test for blood in extreme dilution

Gunzberg's reagent	2 g of phloroglucinol and 1 g of vanillin in 100 cm³ of 95% ethanol	Test for free hydrochloric acid in gastric juices
Hager's reagent	Saturated aqueous solution of picric acid (2,4,6-trinitrophenol)	Test for alkaloids
Hanus's solution	Dissolve 13.2 g of iodine in 1 litre of glacial ethanoic acid using some heat. Cool, and add an equivalent quantity of bromine	Iodine number determination
Haupt solution	Boil 150 g of sodium carbonate in 500 cm³ of water and add 70 g of calcium hydrate. Continue boiling for 15 minutes, cool, and filter	Leaves are boiled in the solution for 1 hour to produce leaf skeletons
Hessler's solution	Dissolve 200 g of zinc(II) chloride in 4 litres of water. Add 100 cm³ of glycerine and 100 cm³ of methanal (formaldehyde)	Preservation of fruit to retain colour
Hubl's iodine reagent	3 g of mercury(II) chloride and 2.5 g of iodine in 100 cm³ of 95% ethanol	Test for unsaturated fats
Hydroxylamine hydrochloride	Saturated solution in methanol	Test for esters
Hydroxylamine reagent (0.05 M)	Dissolve 35 g of hydroxylamine hydrochloride in 160 cm³ of water. Make up to 1 litre with ethanol	
Iodine in potassium iodide	Bench solution	Starch test
Iodine in phosphoric(V) acid	Dissolve 0.5 g of potassium iodide in 25 cm³ of concentrated phosphoric(V) acid. Add a few crystals of iodine	Cellulose test – stains violet
Iodo-potassium iodide	6 g of potassium iodide and 2 g of iodine in 100 cm³ of water	Tests for aldehydes and alcohols
Iron(III) alum	Saturated aqueous	Test for chloride in urine
Iron(III) chloride	Bench solution	Test for lactic acid in muscle
Iron(III) chloride, neutral	Add drops of NaOH solution to 0.5 M iron(III) chloride solution until a faint precipitate forms. Filter	Test for salts of organic acids
Lactic acid (0.1 M, 0.9%)	10.5 cm³ of syrupy (85%) acid made up to 1 litre with water	
Lactic acid (0.5%)	6 cm³ of syrupy acid made up to 1 litre with water	
Lactic acid (0.1%)	1.2 cm³ of syrupy acid made up to 1 litre with water	
Lindt reagent	5 mg of vanillin, 3 cm³ of concentrated HCl, 0.5 cm³ of ethanol, and 0.5 cm³ of water	Test for glucosidic tannins and phloroglucinol
Locke's solution	Dissolve 0.42 g of potassium chloride, 9 g of sodium chloride, 0.48 g of calcium chloride, 0.2 g of glucose and 0.1 g of methylene blue in water and make up to 1 litre	Demonstration of tissue respiration. Change of colour indicates oxygen consumption. Approximates to the inorganic ionic composition and osmotic pressure of plasma
MacLean's reagent	Dissolve 5 g of iron(III) chloride in 100 cm³ of 0.2 M mercury(II) chloride, and add 1.5 cm³ of concentrated hydrochloric acid	Test for lactic acid

Reagent	Preparation	Use
Magnesium sulphate	Saturated aqueous solution	Fat test – forms an insoluble soap
Manganese(II) sulphate (Winkler)	480 g of $MnSO_4 \cdot 4H_2O$ dissolved and made up to 1 litre	Winkler titration for dissolved oxygen
Marme's reagent	33 g potassium iodide and 16 g cadmium iodide in 50 cm³ water. Make up to 100 cm³ with saturated solution of potassium iodide	Test for alkaloids
Marquis's reagent	10 cm³ methanal (formaldehyde) added to 50 cm³ concentrated sulphuric acid	Test for alkaloids
Mayer's reagent	1.36 g mercury(II) chloride in 50 cm³ water. Add 5 g potassium iodide and dilute to 100 cm³	Test for alkaloids
Mercury(II) chloride	70 g made up in water to 1 litre	Protein precipitation
Mercury(II) sulphate	10 g in 100 cm³ sulphuric acid (5%).	
Metaphosphoric acid	60 g in 1 litre water	Protein precipitant
Methylamine hydrochloride	5 g made up to 1 litre in water	Maltose test. Add 5 drops to 5 cm³ of solution under test and boil for 30 seconds. Add 5 drops 20% NaOH solution. Yellow colour which turns carmine on cooling indicates maltose
Millon's reagent	Dissolve 100 g mercury in 200 cm³ concentrated nitric acid by heating in a fume cupboard. When cool add 400 cm³ water	Test for proteins and tryptophan
Molisch's reagent	5 g α-naphthol in 100 cm³ ethanol	Test for carbohydrates. By adding a few drops of sodium hypochlorite and sodium hydroxide it can also be used as a test for arginine and arginine radicals in proteins
Nessler's reagent	Bench solution	Ammonia test
Nickel oxide (ammoniacal)	Dissolve 5 g of nickel sulphate in 100 cm³ of water and add excess 2 M NaOH. Filter and dissolve the precipitate in 25 cm³ of 0.88 ammonia and dilute to 50 cm³ in water	Silk test
Ninhydrin	0.1 g of indane-trione hydrate (ninhydrin) dissolved in 100 cm³ of water	Test for amino acids
	0.2 g of indane-trione hydrate dissolved in 100 cm³ of propanone (acetone) *Keep solutions in a refrigerator*	Location of amino acids on chromatograms
Nylander's reagent	4 g of sodium potassium tartrate and 2 g of bismuth subnitrate (bismuth nitrate, basic) in 100 cm³ of 2 M sodium hydroxide	Glucose estimation and tests for carbohydrates
Obermeyer's reagent	0.4 g of iron(III) chloride in 100 cm³ of concentrated HCl	Test for indoxyl
Orcin	0.5 g in 100 cm³ of 90% ethanol	Test for inulin
Osmic acid	Purchase as a 1% solution. **Do not inhale or make skin contact**	Test for fats
Pavey's solution	Mix 60 cm³ of Fehling's A with 60 cm³ of Fehling's B and add 300 cm³ of 0.88 ammonia. Add water to make 1 litre	Glucose reagent

Laboratory Solutions

Phenol solution	1 g in 1 litre of 0.1 M HCl	Estimation of alkaline phosphatase in blood
o-phenyldiamine	5 g in 100 cm³ of hot ethanol	Test for diketones and quinones
Phloroglucinol	3 g in 100 cm³ of ethanol	Pentosan test
Phosphomolybdic acid (Folin-Wu)	Boil 35 g of phosphomolybdic acid with 5 g of sodium tungstate, 200 cm³ of 10% NaOH, and 200 cm³ of water for 30 minutes. Cool and add 125 cm³ of syrupy (85%) phosphoric(V) acid and dilute with water to 500 cm³.	Folin-Wu determination of glucose in blood
Picric acid (2,4,6-trinitrophenol)	Saturated aqueous solution Saturated aqueous solution (AnalaR grade)	Precipitation of protein Determination of creatine and creatinine in urine
Potassium carbonate	Saturated aqueous solution	
Potassium dichromate	2.5% aqueous. Used with dilute H_2SO_4	Test for acetaldehyde
Potassium oxalate	18 g in 100 cm³ of water	Anti-coagulant for blood
Propanone (acetone)		Fat solvent plant pigment Chromatography solvent
Pyrogallol (alkaline) (benzene-1,2,3-triol)	(a) 28 g pyrogallol in 100 cm³ of water (b) 50 g potassium hydroxide in 100 cm³ of water Mix equal volumes of the two solutions just before use	Absorption of oxygen
Robert's reagent	Add 80 cm³ of concentrated nitric acid to 400 cm³ of saturated magnesium sulphate solution	Protein test
Salicylanilide ('Shirlan')	0.4 g in 1 litre of water	Antiseptic. Prevents growth of moulds in solutions
Schiff's reagent	Dissolve 0.5 g of fuschin in 500 cm³ of water and decolourise by bubbling through sulphur dioxide or Dissolve 0.5 g of fuschin in 500 cm³ of water and add 9 g of sodium hydrogensulphite followed by 20 cm³ of 2 M nitric acid	Test for aldehydes, lignin, and cuticle
Schimper's solution	Dissolve 160 g of chloral hydrate (2,2,2-trichloroethanediol) in 100 cm³ of water. Add sufficient iodine solution to just colour	
Schultze's reagent	Saturated solution of zinc chloride, potassium iodide, and iodine in water	Cellulose test
Schweitzer's reagent	Boil 5 g of copper(II) sulphate in 100 cm³ of water and add NaOH solution until no further precipitation occurs. Filter, and dissolve the washed precipitate in the smallest possible amount of 4 M ammonium hydroxide	Solution to dissolve cellulose
Seliwanoff's reagent	Add 0.5 g of resorcinol (benzene-1,3-diol) to 1 litre of 3 M hydrochloric acid	Test for fructose (glucose gives a false positive after prolonged boiling)
Silver nitrate	Bench solution	Chloride test
Silver nitrate	2.905 g (AnalaR) made up to 1 litre	Whitehorn procedure – 1 cm³ is equivalent to 0.001 g of NaCl
Sodium hydrogensulphite	380 g made up to 1 litre in water	Test for ketones and aldehydes
Sodium carbonate (1 M)	84 g in 1 litre of water	Pauli's test for histidine

Sodium carbonate (0.5 M)	42 g in 1 litre of water	Kept in laboratory as an antidote for mouth pipetted acid
Sodium carbonate (0.1–0.5 M)	Dilute range from molar solution containing 84 g per litre	Investigations into enzyme activity
Sodium carbonate (saturated)	Saturated aqueous solution	Blood alkaline phosphatase estimation
Saline solutions	See page 99	
Sodium citrate	Saturated aqueous solution	Anticoagulant for blood
Sodium hydroxide (10 M)	410 g in 1 litre of water	Saponification of fats Biuret test
Sodium hydroxide (0.5 M)	20.5 g per litre of water	Blood creatine estimations
Sodium nitroprusside (sodium pentacyanonitrosylferrate(II))	1 g in 100 cm³ of water. Solution does not keep; make as required	Test for aldehydes and ketones. Test for creatine and acetone in urine. Test for – SH group in muscle – use with drop of ammonium hydroxide and ethanoic acid; blue colouration with pyruvic acid
Sodium picrate (1%)	Dissolve 10 g (AnalaR grade) in 1 litre of water	Blood creatine estimation
Sodium sulphite (isotonic)	1.5 g of anhydrous salt in 100 cm³ of water	Blood glucose determination
Sodium tungstate	10 g in 100 cm³ of water	Blood glucose determination
Sucrose (1 M)	342 g in 1 litre water. Other molarities prepared by dilution as needed	Suction pressure and osmotic experiment
Sucrose (5–10%)	5–10 g in 100 cm³ of water	Pollen tube experiment
Stoke's reagent	2 g of tartaric acid and 3 g of iron(II) sulphate in 100 cm³ of water. Add 0.88 ammonia just before use until the precipitate formed just redissolves	
Sulphomolybdic acid	10 g of sodium molybdate (or molybdic acid) in 100 cm³ of concentrated sulphuric acid	Test for glucosides
Sulphosalicylic acid (sulpho-2-hydroxybenzoic acid)	25 g in 100 cm³ of water	Test for albumen and proteoses in urine. Bence-Jones protein test
Tannic acid	Dissolve 10 g in 10 cm³ of ethanol, then dilute to 100 cm³ with water	Test for gelatin and albumin
Tollen's reagent	Dissolve 3.29 g of silver nitrate in 100 cm³ of water. Just before use add 3 drops of 2 M NaOH to 5 cm³ of the silver nitrate solution. Dissolve the precipitate in a few drops of ammonium hydroxide	Reagent for reducing sugars and aldehydes
Topfer's reagent	0.5 g of dimethylaminoazobenzene in 100 cm³ of 95% ethanol	Test for free acid (HCl) in gastric contents
Tauber's reagent (**carcenogenic** – illegal in English schools)	1 g of benzidine in 25 cm³ of glacial ethanoic acid. Solution only keeps a few days	Specific test for pentoses and pentose compounds such as riboflavin and nucleic acid
Thymol in ethanol	5 g in 100 cm³ ethanol	
Toluene	—	Urine preservative

Trichloroethanoic acid	10 g in 100 cm³ of water	Protein precipitant
Trichloroethanoic acid	20 g in 100 cm³ of water	Plasma protein determination by colorimetric method
Tungstic acid	Add 10 g of sodium tungstate to 100 cm³ of water containing 1.9 cm³ of concentrated sulphuric acid	Blood protein precipitant
Ufflemann's reagent	Add iron(III) chloride to a 1% aqueous solution of phenol until an amethyst blue colour is produced. Add 5 cm³ of test solution to 5 cm³ of reagent	Lactic acid test
Uranium ethanoate	35 g in 1 litre of water	Estimation of phosphate in urine
Uric acid reagent (Folin's)	See *Folin's solution*	
Wij's solution	Dissolve 13 g of iodine in 1 litre glacial ethanoic acid. Pass dried chlorine gas through 975 cm³ of the solution until the colour changes to orange, then add the remaining 25 cm³ of solution to remove any excess chlorine	Determination of iodine number

ISOTONIC SALINE SOLUTIONS

A number of experiments require the bathing of tissues in a solution which is isotonic with the body fluid of the animal. For temporary microscopical preparations and short experiments, an isotonic solution of sodium chloride may be used. For more prolonged physiological investigations it is usually preferable to use one of the more comprehensive Ringer solutions. For simplicity of nomenclature, these more carefully balanced solutions are referred to as Mammal Ringer, Insect Ringer, etc., although the formulation given may be a more recent modification of the original Ringer's formula. In some experiments a trace of glucose is added to the Ringer solution to provide a respiratory substrate.

Saline solutions

These are sometimes referred to as 'normal salines' – a term which should never be used as it is easily confused with normal solutions in the chemical sense.

For invertebrate tissues
0.75 per cent NaCl made up in de-ionised water

For amphibian tissues
0.64 per cent NaCl made up in de-ionised water

For mammalian tissue
0.9 per cent NaCl made up in de-ionised water

For mammalian blood
0.6 per cent NaCl made up in de-ionised water

Ringer's solution

The quantities given are in grams per litre.

For frog

NaCl	6.5
$CaCl_2.6H_2O$	0.12
KCl	0.14
$NaHCO_3$	0.20

For mammal

NaCl	8.0
$CaCl_2.6H_2O$	0.2
KCl	0.2
$NaHCO_3$	1.0
$MgCl_2.6H_2O$	0.1
NaH_2PO_4	0.05

If required, 1 gram of glucose per litre may be added.

For locusts and other insects

NaCl	7.6
$CaCl_2.6H_2O$	0.22
KCl	0.75
$MgCl_2.6H_2O$	0.19
$NaHCO_3$	0.37
NaH_2PO_4	0.48

For earthworm

NaCl	6.0
$CaCl_2.6H_2O$	0.2
KCl	0.12
$NaHCO_3$	0.1

For 'Astacus'

NaCl	12.0
$CaCl_2.6H_2O$	1.5
KCl	0.4
$MgCl_2.6H_2O$	0.25
$NaHCO_3$	0.2

For marine crustaceans including 'Carcinus'

NaCl	31.0
$CaCl_2.6H_2O$	1.37
KCl	0.99
$MgCl_2.6H_2O$	2.35
$NaHCO_3$	0.22

For chick embryos

NaCl	7.0
$CaCl_2.6H_2O$	0.24
KCl	0.42

HISTOLOGICAL AND BIOLOGICAL SOLUTIONS

Biologists, because of the wide range of material with which they deal, require an enormous array of reagents. Many of those in use represent only slight modification of a basic recipe, and the author has, with some trepidation, undertaken some pruning of the usual lists in a field where empiricism sometimes seems to have taken over from science! Reagents are listed under the headings of

Fixatives and preservatives
Stains and mordants and differentiating solutions
Clearing fluids and mountants
Macerating fluids
Narcotising agents

Although a few stains and reagents must be freshly prepared where needed, the majority store well in cool, dry conditions if kept away from sunlight. Many stains rely upon a precise pH to function well, and it is wise therefore not to store them with acids or ammonia. Where this is not possible, some stains which are particularly sensitive to pH change can be made up in buffer solutions. G. T. Gurr produce a range of tablets for this purpose. In a few cases stains improve with prolonged oxidation and aging (e.g. some forms of haemotoxylin). Aqueous stains which are prone to mould growth may be preserved by the addition of an equal volume of chloroform (trichloromethane).

Table 7.24 Fixatives and preservatives

Solution	Preparation of working solution	Use and comments
Algal preservative	5 cm³ of formaldehyde (methanal), 2.5 cm³ of glycerol, 2.5 cm³ of glacial ethanoic acid, 1.5 g of uranyl ethanoate, and 1 g copper(II) chloride in 90 cm³ ethanol	
Bensley's fixative	2.5% mercury(II) chloride in water and add, prior to use, 2% osmic acid in the ratio 4 parts mercury(II) chloride to 1 part osmic acid	Mitochondria
Ammonium dichromate	10 g per litre of water	Fixing and hardening nervous tissue
Bouin fixative	Add 5 cm³ of glacial ethanoic acid and 25 cm³ of formaldehyde to 75 cm³ of saturated aqueous picric acid	Fix for 18 hours and wash in 50% ethanol, 70% ethanol and 70% ethanol containing lithium carbonate (saturated to remove picric acid). This is a good fixative for use before staining with Heidenhain's iron haematoxalyn, but is now superceded for most cytological work by La Cour's fixatives

Cajal's fixative	Add 1 g of cobalt nitrate and 15 cm³ of formaldehyde to 100 cm³ of water	Golgi bodies
Calberla's fluid	30 cm³ of ethanol, 30 cm³ of glycerol and 30 cm³ of water	Storage of fixed and partly dehydrated plant material, particularly fungi
Carnoy's fluid	10 cm³ of glacial ethanoic acid, 30 cm³ of trichloromethane and 60 cm³ of ethanol	Good fixative for meristems, root tips, and anthers
Champy's fluid	35 cm³ of 1% aqueous chromic acid, 35 cm³ of 3% aqueous potassium dichromate, 20 cm³ of 2% osmic acid	Fixation of mitochondria and animal tissues
Chromo-ethanoic	Widely used plant fixative, the formulation differing considerably according to material: (a) 1 g of chromic acid, 0.5 cm³ of glacial ethanoic acid, 400 cm³ of water (b) 2 g of chromic acid, 1 cm³ of glacial ethanoic acid, 300 cm³ of water (c) 1 g of chromic acid, 1 cm³ of glacial ethanoic acid, 150 cm³ of water (d) 4 g of chromic acid, 4 cm³ of glacial ethanoic acid, 400 cm³ of water (e) 5 g of chromic acid, 10 cm³ of glacial ethanoic acid, 145 cm³ of water (f) 7 g of chromic acid, 10 cm³ of glacial ethanoic acid, and 83 cm³ of water (g) 20 g of chromic acid, 20 cm³ of glacial ethanoic acid, and 160 cm³ of water	Fixative for: Marine algae Algae Filamentous algae General use for plant tissues Prothalli and fungi Meristems, ovaries, root tips Woody material and leaves
Chromic acid	0.5 g in 100 cm³ of water	Fixative for animal tissue for cytology
Clarke's fluid (ethanoic alcohol)	25 cm³ of glacial ethanoic acid added to 75 cm³ of ethanol	Fixative – particularly good before nuclear staining
Colchicine	0.2 g in 100 cm³ of water	Pre-fixitive for root tips; prevents clumping of chromosomes
Dubosq-brasil (alcoholic Bouin)	150 cm³ of 80% ethanol, 60 cm³ of formaldehyde, 15 cm³ of glacial ethanoic acid, and 1 g of picric acid	Fixitive for arthropods – particularly those containing parasites
Ethanoic (acetic) acid	1 cm³ (glacial) in 100 cm³ of water	Fixative for nuclei and chromosomes Connective tissue – white fibres disappear leaving yellow fibres distinct
Farmer's fluid	50 cm³ of glacial ethanoic acid, 100 cm³ ethanol	Fixative for nuclear material
Flemming's fluid (strong)	75 cm³ of 1% chromic acid, 20 cm³ of 2% osmic acid, 5 cm³ of glacial ethanoic acid	Cytological fixative – use small pieces of material as the rate of penetration is low
Flemming's fluid (weak)	0.1 cm³ of glacial ethanoic acid, 25 cm³ of 1% chromic acid, 5 cm³ of 2% osmic acid, and 70 cm³ of water	Fixative for fungi and rotifera
Flemming's fluid (Taylor's modification)	0.2 cm³ of 10% chromic acid, 2 cm³ of glacial ethanoic acid, 1.5 cm³ of 2% osmic acid, 0.15 maltose, and 8.3 cm³ of water	Excellent for root tips and smears
Flemming's without acetic (ethanoic) (F.W.A. fluid)	90 cm³ of 0.25% chromic acid, 10 cm³ of 1% osmic acid (or 5 cm³ of 2% plus 5 cm³ of water)	All Fleming fluids should be made up freshly as required. *They do not store*

Formaldehyde (methanal)	Sold commercially as a 40% solution under the trade names 'Formol' and 'Formalin'. Formaldehyde when used in this book refers to the 40% solution unless a lower concentration is specifically given	
Formaldehyde, 3% (3% Formalin)	75 cm³ of formaldehyde and 925 cm³ of water	Preservative
Formaldehyde, 4% (4% Formalin)	100 cm³ of formaldehyde and 900 cm³ of water	Fixative
Formalin alcohol	10 cm³ of formaldehyde and 90 cm³ of ethanol	Fixative
Formalin acetic alcohol (F.A.A. fixative)	5 cm³ of glacial ethanoic acid, 5 cm³ of formaldehyde, 90 cm³ of ethanol	Algal fixative
Formalin-calcium (Baker's formalin)	1 g calcium chloride, 25 cm³ of formaldehyde, 225 cm³ of water	Fixative
Formalin dichromate	0.5 g potassium dichromate, 10 cm³ of formaldehyde, and 90 cm³ of water	Fixative for hardening brain tissue
Formalin saline	8.5 g sodium chloride, 100 cm³ of formaldehyde, 90 cm³ of ethanol	Good general fixative
Formalin-mercury(II) sublimate fixative	90 cm³ of saturated aqueous mercury(II) chloride solution, 10 cm³ of formaldehyde	Fixative for animal histology
Gilson's fixative	5 cm³ of glacial ethanoic acid, 6 cm³ of concentrated nitric acid, 80 cm³ of ethanol, 20 g of dry zinc(II) chloride are added to 320 cm³ water	
Green algae preservative	(a) 1 g copper ethanoate and 5 cm³ of formaldehyde in 95 cm³ of water (b) 12 cm³ of formaldehyde, 88 cm³ of water	Soak in (a) for 24 hours and store in (b)
Kaiserling's fixative	50 g potassium ethanoate, 100 cm³ of glycerol and 500 cm³ of water	
La Cour's fixatives 2BX	60 cm³ 10% ethanoic acid 100 cm³ 2% chromic acid 120 cm³ 2% osmic acid 100 cm³ 2% potassium dichromate 10 cm³ 1% aqueous saponin 50 cm³ water	Bulk fixation
2BE	12 cm³ 10% ethanoic acid 100 cm³ 2% chromic acid 32 cm³ 2% osmic acid 100 cm³ 2% potassium dichromate 10 cm³ 1% saponin 90 cm³ water	Fixation of plant tissue
2BD	30 cm³ 10% ethanoic acid 100 cm³ 2% chromic acid 60 cm³ 2% osmic acid 100 cm³ 2% potassium dichromate 20 cm³ 1% sapomin 210 cm³ water	Fixative for general work
Merkel's fluid	25 cm³ 1% chromic acid 25 cm³ 1% platinum(IV) chloride 150 cm³ water	Alternative to chromo-ethanoic fixative – good for mitosis in flowering plants

Laboratory Solutions

Mercuric-ethanoic (acetic)	1 cm³ of glacial ethanoic acid in 100 cm³ of saturated mercury(II) chloride solution	Fixative for small animals and seminal vesicle smears from *Lumbricus*
Muller's fixative	25 g of potassium dichromate and 10 g of sodium sulphate in 1 litre of water	
Navashin's fluid	1 cm³ glacial ethanoic acid 10 cm³ 1% chromic acid 4 cm³ formaldehyde 5 cm³ water Make freshly when required	Modification of Flemming's solution giving better results for parasitic fungi in plant tissue
Osmic acid	0.25% aqueous	Fixing protozoa
Pampel's fluid	20 cm³ glacial ethanoic acid 150 cm³ ethanol 60 cm³ formaldehyde 300 cm³ water	Zoological preservative used particularly for insect material
Paradichlorbenzene	2 g in 100 cm³ of water	Pre-fixation bath to prevent clumping of chromosomes
Perenyi's fluid	18 cm³ ethanol 0.1 g chromic acid 4 cm³ concentrated nitric acid 40 cm³ water	Fixative and preservative for amphibian spawn from which it removes the jelly in a few weeks
Rossman's fluid	To 90 cm³ of a saturated solution of picric acid in ethanol, add 10 cm³ of formaldehyde just before use	Fixative for use in the periodic acid–Schiff method for glycogen in marrow and blood smears
Schaudinn's fixative	100 cm³ of saturated aqueous mercury(II) chloride and 50 cm³ of ethanol; add 3 drops of glacial ethanoic acid	Protozoan fixative
Sodium fluoride	2 g in 100 cm³ of water	Preservative of intestinal mucosa
Sodium thiosulphate (hypo)	0.75 g of sodium thiosulphate and 1 crystal of thymol in 100 cm³ of 10% ethanol	Removal of iodine from iodine treated tissue fixed with mercury(II) chloride
Susa fixative	4 cm³ glacial ethanoic acid 20 cm³ formaldehyde 4.5 g mercury(II) chloride 0.5 g sodium chloride 2 cm³ trichloroethanoic acid 80 cm³ water Avoid use of metal instruments with this solution	General fixative for animal tissue. Does not harden the tissue over much
Zenker's fixative	5 g mercury(II) chloride, 2.5 g potassium dichromate, 100 cm³ water Add 5 cm³ glacial ethanoic acid just before use	Used to fix fine detail in animal tissue, and also for mitosis
Zenker's formalin	5 g mercury(II) chloride 2.5 g potassium dichromate 1 g sodium sulphate 100 cm³ water Add 5 cm³ of formaldehyde just before use	Fixative in animal histology. Unsuitable for plant tissues Fix for 12 hours
Zirkle's fluid	81 g copper(II) sulphate 1.25 g ammonium dichromate 1.25 g potassium dichromate 100 cm³ water	Fixation of plant mitochondria

Mordants

In many staining techniques the dye becomes directly attached to the tissue which it stains. Some dyes, however, will not attach themselves directly to biological material, at least in a way which is effective for microscopic examination. These stains are used *indirectly* by a process called *mordanting*. Mordants penetrate or form a strong link with the surface of cytoplasmic structures. When the dye is added it is absorbed onto the mordant, thus giving satisfactory staining. In some cases the tissue is soaked in a mordant before staining, in others, such as Ehrlich's haematoxylin, the mordant is included in the stain mixture.

Table 7.25 Stains, mordants, and differentiating solutions

Solution	Preparation of working solution	Use and comments
Acetic aniline blue (Hoffmann's blue) (ethanoic phenylamine blue)	1 g of aniline blue in 98 cm³ of ethanol (50%) with 1 cm³ of glacial ethanoic acid	Stains protoplasm and sieve-plate callose
Acetic iodine green	Add 1 cm³ of glacial ethanoic acid to 99 cm³ of iodine green stain	Nuclear and plasma
Acetic lacmoid	Heat 1 g of lacmoid in 55 cm³ of water, add 45 cm³ of glacial ethanoic acid *or* 2.2 g lacmoid in 100 cm³ of glacial ethanoic acid; dilute 10 cm³ of this solution with 12 cm³ of water for use. (The dilute stain does not keep well.)	Stain-fixative for chromosomes
Acetic methylene blue	0.2 g of methylene blue in 99 cm³ of water and 1 cm³ of glacial ethanoic acid	Discharging the nematocysts of coelenterates
Acetic methyl green	1.5 cm³ of glacial ethanoic acid, 98.5 cm³ of water, and just sufficient methyl green to produce a pale blue-green colour	Killing and staining protozoa
Acetic orcein	2.2 g of orcein in 100 cm³ of glacial ethanoic acid; dilute 10 cm³ of this solution with 12 cm³ of water for use (dilute solution does not keep)	Stain fixative for chromosomes Also used for connective tissue
Acetic carmine (Acetocarmine)	Dissolve as much as possible of 1 g carmine in 45 cm³ of glacial ethanoic acid by heating. Dilute to 100 cm³ with water	Stain fixative for protozoa and nuclei
Acid alcohol	Add 3 cm³ of concentrated HCl to 100 cm³ of 70% ethanol	Differentiating reagent
Acid fuschin (Mallory) (Acid magenta)	1 g of acid fuschin in 100 cm³ of water *or* 1 g of acid fuschin in 100 cm³ of 50% ethanol	Excellent general stain. Used as a counterstain to iodine green, methyl green, malachite green, and aniline blue
Alcian blue	1 g in 100 cm³ of water with crystal of thymol as preservative	Permanent stain for mucin
Alkaline alcohol	Add 0.5 cm³ of ammonia (0.880) to 99.5 cm³ of 90% ethanol	—
Alizarin red S	0.1 g in 100 cm³ 1% KOH	Used to stain developing bone in whole mounts
Alum carmine (Mayer's)	Carmine 2 g, alum 5 g, water 100 cm³; boil for 1 hour	Stain for small terrestrial and freshwater animals

Ammoniacal fushin	5 g of basic fuschin in 100 cm³ of ethanol; add 0.88 ammonia to produce a permanent pale yellow colouration	Stain for lignin. Counterstain with fast green
Aniline blue (phenylamine blue)	1 g of aniline in 100 cm³ of 85% ethanol 1 g in 200 cm³ of water	Cellulose stain. Stain for algae and fungi
Aniline blue-lactophenol	1 g of aniline blue, in a solvent made up of 25 cm³ of each of phenol, water, lactic acid, and glycerol, added in that order	Excellent stain-fixative for fungi
Aniline blue-orange (Mallory)	5 g of orange G, 1.25 g of aniline blue, 2.5 g of phosphotunstic acid in 250 cm³ of water	Stain for collagen and connective tissue
Aniline gentian violet	Add 10 g of crystal violet and 5 cm³ of aniline to 175 cm³ of water at 40 °C. Add 20 cm³ of ethanol and filter	Mitotic stain. Gram stain for bacteria
Aniline red	Synonym for basic fuchsin	
Aniline sulphate	Make a saturated aqueous solution and add a few drops of concentrated sulphuric acid	Specific stain for lignin
Azan No. 2	0.5 g of aniline blue and 2 g of orange G in 100 cm³ of 8% ethanoic acid, heat, and filter	Connective tissue and collagen stain
Basic fuchsin	0.1 g of basic fuchsin in 160 cm³ of water with 1 cm³ of ethanol	Bacterial stain and nuclei
Best's carmine	2 g carmine 1 g potassium carbonate 5 g potassium chloride all dissolved in 60 cm³ of water and boiled gently. Do not filter	Stain for glycogen in wax embedded sections
Biebrich scarlet acetic	0.2 g Biebrich scarlet in 100 cm³ of 1% ethanoic acid	Plasma and connective tissue stain. Also used as a counterstain for haematal 8
Bismark brown Y	0.3 g in 100 cm³ of isotonic saline 0.3 g in 100 cm³ of ethanol	Nuclear stain which can be used with small living organisms or as an alcoholic stain for permanent preparations
Borax carmine (Grenacher's alcoholic)	Boil 2 g carmine and 8 g borax in 200 cm³ of water for $\frac{1}{2}$ hour. Cool and make up to 200 cm³ with water, then add 200 cm³ 70% ethanol. Stand for 48 hours and filter	General nuclear stain for bulk tissue and sections. Used for both plant and animal material. Small whole mounts and chick embryos
Borax methylene blue	Dissolve 5 g of methylene blue and 25 g of borax in 500 cm³ of hot water. Stain improves with age.	Negri bodies and connective tissue
Borrel's methylene blue	Precipitate silver hydroxide from 100 cm³ of 0.5% silver nitrate by adding excess NaOH (3%). Wash the precipitate until no longer alkaline and boil for 5 minutes in 100 cm³ of 1% aqueous methylene blue	Nerve tissue stain. Blood smears with 5% aqueous eosin
Breed's stain	Dissolve 0.3 g methylene blue in 30 cm³ of ethanol, and add 100 cm³ of $2\frac{1}{2}$% aqueous phenol	Stains bacteria in milk smears
Brilliant cresyl blue (aqueous)	1 g of stain in 100 cm³ of water with 0.9 g of sodium chloride	Blood stain and also used as supra vitam and intra vitam
Carbol fuchsin	To a solution of 5 g phenol in 100 cm³ of water, add 1 g of basic fuchsin dissolved in 10 cm³ of ethanol	Nuclear stain for bacteria, bacterial spores, yeast, and yeast spores. Also used for fungi

Carbol fuchsin (dilute for Gram stain)	To 10 cm³ of the above solution add 90 cm³ of water	Gram stain for micro-organisms
Carbol gentian violet	Dissolve 5 g of crystal violet and 10 g of phenol crystals in hot water, add 10 cm³ of ethanol, and filter	Gram stain (use Gram's iodine as counter stain) for bacteria, fungi to show nuclei, cell division, epithelia, and spermatozoa
Carbol methylene blue	Dissolve 1.5 g of methylene blue and 5 g of phenol crystals in 10 cm³ of ethanol, and add 100 cm³ of water	General stain for bacteria
Carbol thionin	1 g thionin and 5 g phenol crystals in 100 cm³ of water. Dilute 1 to 1 with water for use	Stain for bacteria and fungi. Good for showing fungi in host plant — fungi violet, lignin blue
Carmalum (Mayer's)	Dissolve 1 g carminic acid, 10 g alum in 200 cm³ of hot water with 0.1 g of salicylic acid or a crystal of thymol as a preservative. Filter	Cell contents, non-lignified tissue, and small nematodes
Chlorazol black E (Azo black)	0.5 g in 100 cm³ of ethanol (70%) *or* 0.5 g in 100 cm³ of water	General hystology and cytology Chromosomes
Congo red	1 g of stain in 100 cm³ of water, with a few drops of NaOH	Stains rusts but not host tissue after acetic — ethane fixation. Vital stain for protozoa. Stain for live yeast to be fed to ciliates
Corallin red	Saturated solution of aurine (Na salt) in 4% aqueous sodium carbonate; add preservative	Callose stain
Cotton blue lactophenol	See *aniline blue lactophenol*	
Crystal violet	0.5 g in 100 cm³ of water	Gram stain for bacteria, mitotic figures, and amyloid
Cyanine	Dissolve 1 g of cyanine blue in 100 cm³ of ethanol and add 100 cm³ of water	Lignified tissue with erythrosin counter stain
Ehrlich's tri-acid stain	Mix the following in a 500 cm³ flask in the order given, with shaking after each addition. Do not filter. 70 cm³ saturated aqueous orange G 35 cm³ 20% aqueous acid fuchsin 75 cm³ water 75 cm³ ethanol 63 cm³ 10% aqueous methyl green 50 cm³ ethanol 50 cm³ glycerol	Excellent blood stain
Eosin	1 g of water soluble eosin in 100 cm³ of water. Trichloromethane may be added as a preservative	Good cytoplasmic stain. Used as counterstain for Ehrlich's haematoxalin
Eosin alcoholic	1 g in 100 cm³ of 70% ethanol	Good cytoplasmic stain. Used as counterstain for Ehrlich's haematoxalin
Erythrosin	1 g of erythrosine in either 100 cm³ of water or 100 cm³ of 90% ethanol	Cytoplasmic stain. Counterstain for Delafield's haematoxalin
Fast green F.C.F.	Dissolve 0.5 g in 100 cm³ of ethanol, or in 100 cm³ of clove oil if the stain is to be used as a counterstain after dehydration	Because it does not fade in light but possesses the same qualities, it should replace light green SF in all formulations. *(contd.)*

		Use as a counterstain to safronin in plant sections. Feulgen reaction for chromatin which contains thymonucleic acid
Fast red 7B (Sudan red 7B)	Dissolve 1 g in 100 cm³ of hot propylene glycol	Fat stain
Feulgen stain (Schiff's reagent)	Dissolve 1 g of basic fuchsin in 200 cm³ of boiling water, cool to 50 °C, and add 2 g of potassium metabisulphite (disulphate(VI)). When cool, add 2 cm³ of AnalaR concentrated HCl. Stand in a closed flask overnight, and add 1 g of decolourising charcoal. Shake until water white and filter	Feulgen reaction for D.N.A.
Feulgen sulphurous acid solution	1 g of potassium metabisulphite (disulphate(VI)) dissolved in 200 cm³ of water with 10 cm³ of M HCl	Feulgen reaction
Giemsa's stain	3.8 g of Giemsa's stain powder heated in a flask in a waterbath for 1 hour with 250 cm³ of water and 250 cm³ glycerol	Blood and blood parasites
Gram's iodine	Dissolve 1 g of iodine and 2 g of potassium iodide in 25 cm³ of water. Add water when solids are fully dissolved to make 300 cm³.	Gram's stain for bacteria
Haemalium (Mayer's)	Dissolve 50 g alum, 0.2 g sodium iodate, 50 g chloral hydrate and 1 g of citric acid in 1 litre of water. Add 10 cm³ of ethanol containing 1 g of haematoxylin	Good stain for bulk use and sections of animal material

Haematoxylin
There is an enormous range of formulations of this stain, some of which are very specialised, whilst others would seem to be only one worker's variation on a theme. These are the main variants which should prove to be more than adequate in an educational laboratory:

alcoholic	1 g of haematoxylin in 70 cm³ of ethanol. When dissolved add 30 cm³ of water	
alum	Alum 5 g, haematoxylin 0.25 g, thymol 0.2 g dissolved in 100 cm³ of water. *Maximum shelf life 3 months*	Staining sections after Flemming's solution
Delafield's	Dissolve 4 g of haematoxylin in 25 cm³ of ethanol and add the solution to 400 cm³ of saturated aqueous ammonia alum. Stand in full light for 4 days, and then add 100 cm³ of glycerol and 100 cm³ of methanol. Add 1 cm³ of hydrogen peroxide and stand in a sunny place to ripen for several weeks. If after prolonged storage the stain has ripened too much to 'blue' with tap water, add a few drops of NaOH to the stock bottle	One of the best general stains for animal material
Ehrlich's	Dissolve 4 g of haematoxylin in 200 cm³ of ethanol and add 200 cm³ of water, 200 cm³ of glycerol, 20 cm³ of glacial ethanoic acid, and excess potassium alum. Place in a large flask loosely stoppered with cotton wool (allow plenty of air space above the liquid) and put in a warm, light place to ripen for several weeks	Excellent stain for animal histology
Harris's	Dissolve 50 g of ammonia alum in 500 cm³ of water and add to this 2.5 g of haematoxylin *(contd.)*	Good stain for general histology

dissolved in 25 cm³ of ethanol. Boil the mixture and add to it 1.25 g of mercury(II) oxide. Cool rapidly and filter

Harris's (Mallory's modification)	As above, but add 10 cm³ of glacial ethanoic acid	
Heidenhain's iron	*Solution A.* 3 g of iron alum in 100 cm³ of water *Solution B.* 0.5 g of haematoxylin in 10 cm³ of ethanol, diluted to 100 cm³ with water	Good stain for algae, skin, suprarenal bodies, and thymus. To use, mordant the section in solution A (warm) for 15 minutes. Overstain for 10 minutes in B, and differentiate in solution A under the microscope
Renaut's eosin	Mix together 30 cm³ of saturated aqueous eosin Y, 40 cm³ of saturated haematoxylin in ethanol, and 130 cm³ of saturated solution of potassium alum in glycerol. Leave in an open bottle until the ethanol has evaporated (3–4 weeks)	Stain for plant nuclei
Iodine	Make a saturated solution of potassium iodide and then saturate this solution with iodine. Filter and dilute with water until the solution is a pale golden brown	Temporary stain for plant material. Stains cellulose yellow and starch granules black
Iodine green	1 g iodine green in 70 cm³ of ethanol, diluted to 100 cm³ with water	Stain for lignin
Janus green B	Dissolve 0.5–0.1 g of Janus green B in 1 litre of isotonic saline	Vital stain for mitochondria and golgi bodies
Jenner's stain	0.5 g of stain in 100 cm³ of methanol	Blood smears
Lacto-fuchsin	0.25 g of acid fuchsin in 250 cm³ of lactic acid	Superb stain-mountant for fungal hyphae
Leishman's stain	Full preparation is very tedious; for educational purposes it is better to purchase ready-prepared Leishman stain in powder form. Dissolve 0.3 g of the powder in 200 cm³ of methanol	Very fine stain for blood and parasites. Also used for blood-forming tissues

Recipes for use of light green are given here as many institutions will have stocks of the powdered dye which they may wish to use up. *It is strongly recommended however that Fast Green F.W.F. is used in place of light green in all stains as it does not fade with age.*

Light green (Masson's)	4 g of light green and 4 cm³ of glacial ethanoic acid dissolved in 200 cm³ of water	Stain for plant tissues, best used with safranine or haematoxylin as a counterstain
Light green in clove oil	Dissolve 0.2 g of light green in 50 cm³ of 100% ethanol mixed with 50 cm³ of clove oil. Keep closely stoppered to prevent absorption of moisture from the air	Used as a counterstain to safranine or haematoxylin in plant sections after the dehydration stage
Loeffler's methylene blue (*see also* Barrel's methylene blue, and acetic (ethanoic) methylene blue)	Dissolve 1 g of methylene blue in 200 cm³ of water heated to 50 °C and then add 2 cm³ of 1% KOH and 60 cm³ of ethanol	Bacterial stain
Lugol's iodine	4 g of iodine and 6 g of potassium iodide in 100 cm³ of water	Modification of Gram's method for bacteria
Malachite green	1 g in 100 cm³ of water	Can be used as a vital stain for plant tissue

Laboratory Solutions

Methyl blue	1 g in 100 cm³ of water	Counterstain for carmine, eosin, and safranin. Also used as a supravital stain for aquatic animals
Mallory's triple stain	*Solution A*. 0.1 g of acid fuchsin in 100 cm³ of water *Solution B*. (Mordant) 1 g of phosphomolybdic acid in 100 cm³ of water *Solution C*. 0.5 g of watersoluble aniline blue, 2 g of orange G, and 2 g of oxalic acid in 100 cm³ of water	A fine stain for general histology. Material is best fixed in Zenker
Methylene blue	1 g in 100 cm³ of water, with 0.5 g of NaCl	Vital stain
Methyl green	1 g in 100 cm³ of 70% ethanol	Plant material, yeasts
	25 cm³ of the above solution added to 1.5 cm³ of glacial ethanoic acid and 100 cm³ of water	Kills and stains protozoa
Methyl violet 6B (Jensen's)	1 g of methyl violet in 200 cm³ of water	Jensen's modification of Gram's stain
Neutral red (Jensen's)	Dissolve 1 g of neutral red in 1 litre of water and add 1 cm³ of 2% ethanoic acid; filter	Jenson's modification of Gram's stain
Neutral red	0.1 g in 1 litre of isotonic saline	Vital stain for protozoa. Turns yellow in acid conditions – e.g. food vacuales
Neutral red – fast green	2 g of neutral red and 0.2 g of fast green in 1 litre of ethanol	Excellent bacterial stain
Nigrosine	5 g in 100 cm³ of water	Good stain for spores and capsules of bacteria
Phloroglucinol	5 g in 100 cm³ of 70% ethanol	Lignin – an excellent temporary stain. Soak section on slide and add 1 drop of concentrated HCl
Phloxine B	0.5 g in 100 cm³ of water	Filamentous algae
Safranine alcoholic	1 g in 100 cm³ of 50% ethanol	Good stain for both plant and animal tissue, particularly if counterstained with fast green
Safranine aqueous	Dissolve 1 g in 100 cm³ of water and filter if necessary	Bacterial spores
Schultze solution	30 g of zinc chloride, 5 g of potassium iodide, 1 g of iodine in 15 cm³ of water. Only keeps 2–3 weeks	Cellulose
Sudan black	Saturated solution in 70% ethanol *or* 1 g in 100 cm³ of hot propylene glycol	Suberin, cutin, and fat Fatty tissue
Sudan III	Saturated solution in a 50/50 mixture of 70% ethanol and propanone	Good for root endodermis. Superceded by Sudan black and Sudan IV for fat
Sudan IV	5 g in 100 cm³ of 70% ethanol	Fat and cutin
Thionin	Dissolve 0.5 g in 100 cm³ of 70% ethanol and add 1 cm³ of concentrated HCl	Cartilage
Van Gieson's stain	1 g of acid fuchsin in 100 cm³ of saturated aqueous picric acid solution	Good permanent stain for protozoa

Table 7.26 Clearing fluids and mountants

Fluid	*Preparation of working fluid*	*Uses and comments*
Acetone–xylene (propanone–1,2-dimethyl benzene)	20 cm³ of acetone (anhydrous) with 80 cm³ of xylene	Rapid clearing after 90% ethanol
Algal mountant	1.0% solution of copper(II) chloride in lactophenol	
Brun's glucose mountant	40 g of glucose, 10 cm³ of thymol (2% in ethanol) and 10 cm³ of glycerol in 140 cm³ of water	
Canada balsam	Dissolve sufficient Canada balsam resin in xylene to form a viscous solution	Popular mountant which discolours with age; when old it becomes acid and affects some stains. *Best replaced with a modern synthetic mountant such as Gerrard's 'Micrex'*
Cedarwood oil	As purchased	Clearing reagent
Celloidin	2, 4, 6, 8, and 10% solutions	Used as a series for embedding
Clove oil	As purchased	Clearing, particularly good for plant sections stained with safranine and light green
Dammar	Dissolve sufficient resin in xylene to form viscous solution	Mountant – less acid than balsam. Refractive index = 1.5191
Dioxan	Di-ethylene dioxide	Dehydrating and clearing agent with **toxic vapour**
Ethanol	30%, 50%, 70%, 95%, and 100% are all used. (10% used for narcotising)	Series used for dehydrating

Note 100% ethanol is very expensive and absorbs moisture from the air very rapidly once the seal is broken. For all normal uses it is best to make dilutions from industrial methylated spirit (74° of proof), rating it as 95%. For final dehydration, ethex (below) should be used.

'Ethex' (synonym 'Cellosolve')	Ethene glycol monoethyl ether (ethoxyethanol)	Miscible with water, alcohol, clove oil, and xylene. More rapid than alcohol as there is no need to 'grade up'. With delicate material add 'Ethex' drop by drop to the material in dilute alcohol and then transfer to 100% 'Ethex'
Glycerol	50 cm³ of glycerol and 1 cm³ of concentrated thymol solution made up to 100 cm³ with water	Temporary mountant
	20 cm³ of glycerol and 80 cm³ of water	Dehydration of delicate material (Miles' method) – place material in solution and allow water to evaporate for 7 days and transfer to 95% ethanol
Gum chloral (de Faure's mountant)	Dissolve 50 g of chloral hydrate in 50 cm³ of water; add 20 cm³ of glycerol followed by 40 g of crushed gum arabic. Stopper closely, and stand for some time with occasional shaking. Use without filtering	Mountant for insects, helminth eggs, and fungi

Gurr's water mounting medium	Use as purchased	Aqueous mountant suitable for fat stains. Refractive index = 1.4045
Lactic acid	70%–75% in water	Clearing agent for cellulose and fungal cellulose
Lactophenol	Dissolve 10 g of phenol in 10 cm³ of water without heating, then add 10 cm³ of glycerol and 10 cm³ of lactic acid	Mountant, particularly for fungi
Lacto-fuchsin	See *stains*	
Polyvinyl alcohol-lactophenol	Dissolve 22 g of phenol crystals in 25 cm³ of lactic acid. Add 56 cm³ of polyvinyl alcohol, stir, and heat in a waterbath until the solution is clear	Mountant for insect larvae and softening wood
Terpineol	As purchased	Good clearing agent for hard material
Toluol	As purchased	Clearing agent
Xylol (xylene) (1,2-dimethylbenzene)	As purchased	Clearing agent
Xylol-acetone	20 cm³ of anhydrous acetone added to 80 cm³ of xylol	Rapid clearing after 90% ethanol (dimethylbenzene-propanone)
Xylol-alcohol	50 cm³ of 100% alcohol and 50 cm³ of xylol	
Xylol-phenol	5 g of phenol in 100 cm³ of xylol	Helps to reduce the problem of 'milky' xylol due to the presence of water

Table 7.27 Macerating fluids

Fluid	Preparation of working fluid	Use
Cellulose ethanoate	12% cellulose ethanoate in acetone	Used to soften wood for sectioning
Chromic acid (Priestley)	5% solution in water	Used to macerate plant tissue
Chromic acid (Ranvier)	0.2% solution in water	Used to macerate animal tissue
Eau de Javelle	Dissolve 10 g of sodium *or* potassium carbonate in 100 cm³ of water. Dissolve 5 g of calcium hypochlorite in another 100 cm³ of water. Mix the two solutions, and decant the clear liquid when the precipitate has settled	Used to remove soft tissue which surrounds hard skeletal material
Ethanol (Ranvier)	30% ethanol	Macerating animal tissue
Jeffery's macerating fluid	Add equal quantities of 10% nitric acid and 10% chromic acid	
Schulze's macerating fluid	Dissolve 2 g of potassium chlorate in 100 cm³ of concentrated nitric acid	Maceration of woody tissues

Table 7.28 Narcotising agents

Substance	Working form	Uses
Carbon dioxide	From siphon	Coelenterates, Echinoderms, Annelids
Chloral hydrate (2,2,2-trichloroethomediol)	0.1 g of chloral hydrate in 100 cm³ of water	Narcotising small aquatic animals
Chloretone	0.01–0.1%	Invertebrates, vertebrate larvae, and fish
Cocaine	1% aqueous solution of cocaine hydrochloride *or* 30 cm³ of 2% aqueous cocaine hydrochloride added to 10 cm³ of ethanol and made up to 100 cm³ with water	Narcotising rotifers
Ethanol	10% aqueous	Used for fresh-water animals
Ether	As purchased. **Note** that explosions have occurred in refrigerators when ether from the lungs of dissection material has formed an explosive mixture with air and been ignited by a spark from the microswitch in the door	
Ethyl urethane	0.3%–1.0% aqueous	
Magnesium chloride	Saturated aqueous solution diluted with an equal volume of sea water	Used for marine animals
Menthol	Scatter crystals	All groups
Rouselet's solution (Baker)	See second formula for cocaine (above)	
Tobacco smoke		Ciliates and flagellates

NOMENCLATURE OF BIOLOGICAL DYES

Many of the dyes used in biological stains have more than one name; the name used here is that which is the preferred designation of the *U.S. Commission on Biological Stains*. The main divergence from the preferred designations has been in spelling; often the U.S. spelling omits the final 'e' of a word, but the spelling used here is that most likely to be found in the lists of an English supplier e.g. nigrosin/nigrosine; anilin/aniline. Prejudice has also led to the retention of 'ae' in preference to 'e' in such words as haematoxylin.

The lists below give the *preferred name* in italic, and the more common synonyms in biological use.

Aniline blue   China blue 22; China blue; Cotton blue; Marine blue V; Soluble blue M3; Soluble blue 2R

Aurantia   Imperial yellow

Biebrich scarlet   Acid red 66; Creoceine scarlet; Double scarlet B.S.F.; Ponceau B; Scarlet B; Scarlet E.C.

Bismark brown Y   Basic brown; Basic brown G.; Basic brown G.X.; Basic brown G.X.P.; Excelsior brown; Leather brown; Manchester brown; Phenylene brown; Vesuvin

Chlorazol black   Direct black 38; Direct black M.S.; Direct deep black E.W. extra; Erie black GX00; Pontamine black E; Renol black G. (Other letters of designation also occur after these names and all refer to Chlorazol black)

Cochineal   Natural red 4

Congo red   Direct red 28; Congo; Cotton red B; Cotton red C; Direct red C

Crystal violet   Basic violet 3; Gentian violet; Hexamethyl violet; Hexamethyl violet 10B; Violet C

Eosin Y   Acid red 87; Bromo acid J; Bromoflorescein; Bronze bromo ES; Eosin WS (eosin, water soluble)

Erythrosin bluish   Acid red 51; Dianthine B; Eosin J; Erythrosin B; Iodeosin B; Pyrosin B

Fast green F.C.F.   Food green 3

Fuchsin, acid   Acid violet 19; Acid magenta; Acid rubin; Fuchsin S

Fuchsin, basic   Aniline red; Basic rubin; Fuchsin RFN; Magenta

Basic fuchsin is made up of 3 dyes:

  Pararosanilin   Basic red 9; Basic rubin; Magenta O; Parafuchsin; Paramagenta

  Magenta II

  Rosanilin   Basic violet 14; Magenta I

A closely related compound to Magenta II is

  New Fuchsin   Basic Violet 2; Magenta III

Haematoxylin   Natural black I; Natural black II

Hoffman's violet   Dahlia: Iodine violet; Primula R water soluble; Red violet; Violet R

Iodine green   Hoffman's green

Janus green B   Diazin green S; Union green B

Light green SF. Yellowish   Acid green 5; Acid green; Fast green N; green 2 G

Magdala red   Basic red 6; Naphthalene pink; Naphthalene red; Naphthylamine pink; Sudan red

Malachite green   Basic green 4; Diamond green B; Light green N; Malachite green A, B, etc.; New victoria green extra O; Solid green O; Victoria green B

Martius yellow   Acid yellow; Manchester yellow; Naphthol yellow

Metanyl yellow   Acid yellow 36; Acid yellow R; Jaune metanyl; Orange MN; Orange MNO; Soluble yellow OL; Tropaeolin G; Yellow M

Methyl blue   Acid blue 93; Cotton blue; Helvetia blue

Methyl eosin   Solvent red 44; Alcohol soluble eosin; *this dye is not used in biological stains*

which require eosin Y *(or ethyl eosin if alcohol soluble eosin is needed)*

Methyl green   Basic blue 20; Double green SF; Light green

Methyl violet B   Basic violet I; Dahlia; Gentian violet; Methyl violet B; Methyl violet BO; Methyl violet R; Paris violet; Pyoktaninum coeruleum

Methylene blue   Basic blue 9; Methylene blue B; Methylene blue chloride; Swiss blue

Neutral red   Basic red 5; Toluylene red

New methylene blue   Basic blue 24; Methylene blue NN

Nigrosine   Acid black 2; Gray R; Indulin black; Nigrosine W

Nile blue sulphate   Basic blue 12; Nile blue A

Orange G   Acid orange 10; Crystal orange GG; Orange GG; Wool orange 2G

Orcein   Natural red 28

Phloxine B   Acid red 92; Cyanosine; Eosin 10B; Phloxine TA

Red corallin   Aurin R

Safranin O   Basic red 2; Cotton red; Gossypimine; Safranin A; Safranin Y

Sudan black B   Solvent black 3

Sudan III   Solvent red 23; Cerasil red; Fat ponceau R; Oil red AS; Scarlet B; Sudan G; Tony red

Sudan IV   Solvent red 24; Cerotin Ponceau 3B; Scarlet R (incorrectly); Scarlet red

Trypan blue   Direct blue 14; Azidine blue 3B; Benzamine blue 3B; Benzo blue 3B; Chlorazol blue 3B; Congo blue 3B; Dianil blue H3G; Naphthamine blue 3BX; Niagra blue 3B

Table 7.29 Manumetric solutions

Solution	Preparation of working solution
Eosin	Dissolve 1 g of eosin in 100 cm³ of ethanol, then dilute 1–2 cm³ of this solution to 1 litre with water
Fluorescein	As above
Mercury	Probably the most popular manumetric fluid – best avoided on account of its **toxicity** (unless the use of other fluids would necessitate a very long column because of the pressures involved in the experiment)

Table 7.30 Electrolytic solutions

Solution	Working solution
Sulphuric acid for accumulators	Add 220 cm³ of concentrated sulphuric acid (technical grade) to 750 cm³ of water *very carefully, stirring all the time* and make up to 1 litre. Check the specific gravity with a hydrometer, and adjust to 1.25 by adding more acid if the reading is low, or more distilled water if the reading is high
Daniell cell	(i) Add 80 cm³ of concentrated sulphuric acid to 900 cm³ of water and make up to 1 litre (ii) Make a saturated solution of copper(II) sulphate (approximately 400 g per litre) and add 2 cm³ of concentrated sulphuric acid)
Copper(II) sulphate solution (for electrolysis)	Dissolve 150 g of copper(II) sulphate in 1 litre and add 25 cm³ of concentrated sulphuric acid and 60 cm³ of ethanol
Leclanché cell	Dissolve 350 g of ammonium chloride and make up to 1 litre
Silver nitrate solution	Approximately 170 g per litre
Solution to platinize platinum electrodes	Add 10 cm³ of 5% platinum chloride solution and 1 cm³ of 0.1 M lead ethanoate solution to 25 cm³ of 4 M hydrochloric acid and make up to 50 cm³ with distilled water. Immerse the platinum electrodes in the solution for approximately 20 minutes, reversing the current every few minutes

Table 7.31 Making indicator papers

Many of the indicators listed in this chapter may be used to prepare papers. Sheets of filter or chromatographic paper are dipped in the required solution and hung up to dry

Paper	Preparation
Ferrox reagent	1 g of iron(III) chloride dissolved in 10 cm³ of ethanol, and 1 g of potassium thiocyanate in another 10 cm³ of methanol, are mixed and filtered. Dip the paper and dry — then dip and dry a second time. Used as a test for iron
Heat sensitive paper	1.5 g of cobalt chloride and 1.5 g of calcium chloride in 1 litre of water. Dip the paper and dry
Moisture sensitive paper	Dissolve 1 g of cobalt chloride in 100 cm³ of water. Dip the paper and dry
Potassium cobalticyanide	4 g of potassium cobalticyanide and 1 g of potassium chlorate in 100 cm³ of water. Dip and dry. Used as a test for zinc
Starch-potassium iodide	Mix 1 g of soluble starch to a thin paste with water, then add to 100 cm³ of boiling water. Add 1 g of potassium iodide. Dip the paper and dry
Turmeric	Soak ground turmeric root in water several times and discard the solutions. Dry the residue and soak for several days in an equal weight of ethanol in a stoppered flask. Dip the paper in the resultant solution and dry

MICROBIOLOGICAL CULTURE MEDIA

With only a very few exceptions all of the agars and broths likely to be required in the laboratory can be purchased in the form of tablets or granules. These commercially prepared media are to be preferred to home-made agars on grounds of both reliability and expense (home-made agars are costly in terms of technician or teacher time).

Preparation of Petri dishes

If a large number are required, the appropriate amount of agar should be made up from granules (which are more economical than tablets); allow 10 cm^3 of media per 90 mm dish. The granules are weighed out and shaken up in a flask with deionised water to make up the required volume. The mixture is then autoclaved for 15 minutes at 121 °C with a cotton wool plug in the neck of the flask, whilst the pre-sterilized, disposable Petri dishes are laid out in a row on the edge of the preparation bench. The sterilized medium should be allowed to cool to 50–55 °C before pouring to minimise the problem of condensation. If several flasks of agar are sterilized at once, they may be placed in a waterbath at 55 °C until needed. The lids of the Petri dishes should be raised just sufficiently to allow the neck of the flask to enter and then replaced as each dish is poured. If the flask is kept at the pouring angle there is only a minimal chance of airbourne spores falling into the neck. If all the medium is not poured, the mouth of the flask should be flamed and replugged.

If only a small number of plates of any type are required tablets may be most convenient. Place two tablets and 10 cm^3 of deionized water in a universal bottle, or a McCartney bottle, and autoclave with their lids *loosely* in place. Such bottles may be transferred from the autoclave to a waterbath at 55 °C ready for pouring by the student.

As soon as the agar has solidified, the dish should be inverted and stored in this position; this will prevent drops of condensation forming on the lid and then falling to the surface of the agar.

Preparation of agar slopes

For some experiments, and for the long term storage of cultures in the refrigerator, Universal bottles or McCartney bottles are to be preferred. If a large number are required the medium should be prepared in bulk as described above and 5 cm^3 poured into each bottle, which is then autoclaved. Alternatively one tablet and 5 cm^3 of deionized water are placed in each bottle, and then autoclaved with the cap loosely fitted. When the bottles are removed from the autoclave they should be laid in a row with their necks resting on a metal or glass rod about 1 cm in diameter – the agar will then set with the desired slope.

Preparation of broths

Nutrient broth may be prepared in bulk from granules and distributed to bottles before or after sterilization. Alternatively one tablet and 5 cm^3 of deionized water are placed in each bottle and autoclaved with the screw cap loose.

Despite the clear-cut advantages of using prepared media, some schools prefer to prepare their own; a few of the more popular recipes are given below. Potato slopes are to be discouraged as they may tempt the pupil to carry out microbiological experiments at home, with the subsequent risk of growing pathogens.

Potato dextrose agar (PDA)

Cut 250 g of washed potatoes into small pieces and steam or autoclave in a muslin bag for 1 hour. After steaming, allow all liquid to drain from the bag but do not squeeze. Dilute to 1 litre and add 20 g of glucose. Add 20 g agar and autoclave.

Malt agar

Dissolve 40 g of malt extract in 1 litre of water with 20 g of agar. Adjust the pH to 5.4 and autoclave.

Sabourard's agar

Dissolve 40 g of glucose, 10 g of peptone, and 20 g of agar in 1 litre of water. Adjust the pH to 5.4 and autoclave.

Dung agar

Soak 1 kg of dung (horse, cow, or rabbit) for 3 days. Decant and dilute the liquid until straw coloured. Dissolve 25 g of agar in 1 litre of dilute extract, and autoclave.

Nutrient agar for bacteria

Dissolve 10 g of 'Lemco' beef extract, 10 g of peptone, 5 g of sodium chloride, and 15 g of agar in 1 litre of water.

Ready to use agar

Several of the more popular agars are available made up in a 'sausage' which can be squeezed from its skin and cut into thin discs with a sterile knife; the discs are then placed in Petri dishes.

Tablet, granular, ready-prepared, and plain agars are available from Oxoid Ltd, Southwark Bridge Road, London S.E.1. Oxoid publish a very informative catalogue, which gives data and application techniques of all their products, called 'The Oxoid Manual' (free to bona fide customers), and also produce an introductory kit of microbiological material including Petri dishes, cultures, media, and antibiotic discs, complete with a well-written instruction book, to introduce schools to elementary microbiology.

Note The use of deionised water or glass distilled water is advisable for microbiological work as tap-water and metal distilled water may contain copper, zinc, or lead ions in sufficient quantity to inhibit microbial growth.

STERILISATION OF APPARATUS FOR MICROBIOLOGICAL WORK

Forceps points and inoculating loops Pass through a flame and allow a moment to cool before use. To save time, some technicians dip the hot needle or loop in ethanol to cool it, then touch the flame again with the loop which will burn off the alcohol without reheating.

Glass Petri dishes, flasks, test tubes, pipettes, instruments, and glass syringes Stopper flasks and tubes with cotton wool, wrap pipettes and instruments in brown paper or metal foil, and sterilize in a hot-air oven at 160 °C for 1 hour.

Mouths of culture tubes, cotton wool stoppers, slides, and cover slips Pass through a flame.

Cotton wool in bulk Autoclave, and then open the draincock to expel water before releasing the pressure. When all water and steam has been expelled, close the valve and allow to cool. The cotton wool will be vacuum dried.

Culture media and bottle caps containing rubber washers Use an autoclave or pressure cooker – 15 minutes at pressure.

Contaminated floors and benches Wash with 3 per cent Lysol, but do not allow the solution to contact skin. Cetrimide (available from Boots Chemists Ltd.) can also be used and is safe on the skin, but it is not effective against all bacterial spores.

Disposal of old cultures

There is always a risk that old cultures may become contaminated with pathogens. They should be destroyed by immersion in Lysol, or by autoclaving. If spore-forming bacteria are likely to be present, autoclaving is to be preferred.

Testing sterilisation equipment

If there is any reason to suspect the efficiency of ovens or autoclaves, they may be tested by the following methods. (The oven thermometer is not a good guide as it may well be in a 'warm spot' near the top of the oven.)

1 Spore strips – small pieces of paper soaked in a standardized suspension of Bacillus stearothermophilus spores. These spores are only killed by 12 minutes exposure to moist heat at 121 °C or by a suitable time and temperature combination of dry heat. Place in sterilizing equipment and after carrying out the sterilization procedure, place the strips in sterile dextrose tryptone broth and incubate for 7 days. An unsterilized strip must be incubated as a control and an uninoculated tube of broth incubated as a further control.

2 Browne's tubes – small tubes containing a red solution which turns green if exposed to the correct time/temperature combination. Store below 20 °C; use type 2 tubes for autoclaves and type 3 tubes for dry heat (available from Albert Browne Ltd., Chancery House, Abbey Gate, Leicester.)

Table 7.32 Composition of artificial seawater (from Barne's data)

Chlorinity = 19%
Salinity = 34.33%

Substance	Grams		
NaCl	23.991		
KCl	0.742		
$CaCl_2$	1.135	($CaCl_2.6H_2O$	2.240 g)
$MgCl_2$	5.102	($MgCl_2.6H_2O$	10.893 g)
Na_2SO_4	4.012	($Na_2SO_4.10H_2O$	9.100 g)
$NaHCO_3$	0.197		
NaBr	0.085	($NaBr.2H_2O$	0.115 g)
$SrCl_2$	0.011	($SrCl_2.6H_2O$	0.018 g)
H_3BC_3	0.027		

Dissolve the quantities of substances indicated in deionised water and make up to one litre.

8. Electronics

It is not possible in a book of this size and scope to give more than the simplest introduction to the circuitry encountered in teaching laboratories; the bibliography on page 144 will refer the reader to more detailed texts. This chapter will give basic information on electronic components, what they do in circuits, and an outline of some of the circuits which occur frequently in the laboratory either as complete units or as 'building blocks' in a larger unit.

At first sight, the inside of a piece of electronic equipment seems to be very complex, and to understand it one must break it down into a series of 'building blocks' or modules. Take for instance a simple sound level meter (a decibel meter) of the type which switches on a light or operates a meter whenever the sound level rises above a certain point. There will be a probe, in this case a microphone, which feeds into an attenuator circuit (a series of resistors which can be switched successively into the circuit); from the attenuator the signal passes to a switching circuit via an amplifier, and finally operates the bulb or a meter. Such a circuit may be represented in 'building block' form thus:

RESISTORS

These are components which offer a resistance to the passage of electrons; in so doing they consume some of the electrical energy which is dissipated as heat. The rate at which energy can be safely dissipated is stated on each component in watts. If this rate of dissipation is exceeded, the component will overheat and either fail completely or cause some degree of malfunction of the circuit. Before replacing a resistor which has 'gone high', decide whether it is some other circuit fault which has overloaded the resistance, or the resistance itself has been operating on its limits; if the latter, the resistance should be replaced by one of the same value but of higher wattage (e.g. a 100 ohm, $\frac{1}{8}$ W resistor could be replaced by a 100 ohm, $\frac{1}{4}$ W resistor).

Resistors are either of fixed, or variable value. *Fixed resistors* may either be made of special resistance wire wound on a ceramic former or they may be of the carbon rod type. These are made by mixing clay and powdered carbon, which is formed into a rod and baked. In recent years a new type of resistor,

This type of analysis enables the technician to predict what should happen to the signal at each stage in the circuit, and a suitable voltage measurement or an oscilloscope display at each point will quickly indicate which stage of the apparatus is faulty.

A number of basic circuits appear frequently as 'building blocks' in laboratory instruments. They are:

Probes – either pick-ups for an existing signal, or some type of transducer which converts the phenomenon to be measured into an electrical signal (e.g. a microphone converts sound into electrical waves or a selenium cell converts light energy into a voltage which can be measured). Some probe units, such as light- or temperature-dependent resistors, are used to modify a signal or voltage level in the measuring circuit.

Power supplies　　　　*Level switches*
Amplifiers　　　　　　*Meter or digital readout*
Oscillators　　　　　　*Potential dividers*
Attenuators　　　　　　*Voltage doublers*

which consists of a layer of metal oxide or a carbon film deposited on a glass rod, has been developed for miniature circuits with transistors. It is impossible to manufacture a batch of resistors which are of identical precise value, so manufacturers grade their products into different tolerances, e.g. a ∓ 10 per cent 100 ohm resistor may lie anywhere between 90 ohm and 110 ohm, whereas a ∓ 2 per cent component will be within the range 98–102 ohm.

For some applications resistors of precise values are needed and these may be purchased as wirewound precision resistors (provided that the value required is a stock item). Often when repairing instruments specially-manufactured, precision resistances have to be replaced but this is not such a difficult problem as it seems; suitable values can be built up by soldering two or more components in series (RS Components Ltd, supply wirewound resistors as low as 0.22 ohm), or careful checking of a batch of 5 per cent or 10 per cent resistors on an accurate bridge can locate a component of the precise value required.

Carbon and metal oxide film resistors are usually available in 0.125 W, 0.25 W, 0.5 W, 1 W, and 2 W ratings. If the application demands a higher rating wirewound resistors are made in 2.5 W, 5 W, and 10 W ratings.

The value of a resistor is usually indicated on the body in some way; wirewound resistors have their value printed in numbers on the side (some specially wound resistors have their value written on with black ink). Most carbon and film resistors are colour coded to show both their value and tolerance although one of the major suppliers has recently changed to printing the value on the component in accordance with the B.S.1852, which will eventually replace the colour code.

According to the B.S.1852 resistance code, the value is marked numerically, using the letter R to indicate the decimal place in resistances below 1000 ohm and the letter K to indicate the decimal place in resistances above 1000 ohm. These examples will show more clearly how the system works (the letters K and M are used to signify ' $\times 1000$ ' and ' $\times 1\,000\,000$ '; $1000\ \Omega = 1\ k\Omega$, $1\,000\,000\ \Omega = 1\ M\Omega$, meg-ohm):

0.22 Ω	would be marked	R22
1 Ω	,,	1RO
2.2 Ω	,,	2R2
22 Ω	,,	22R
100 Ω	,,	100R
1 kΩ	,,	1KO
1.2 kΩ	,,	1K2
10 kΩ	,,	10K
10 MΩ	,,	10M

In addition to the value, the tolerance is marked with a letter as follows:

$$F = \mp 1\%$$
$$G = \mp 2\%$$
$$J = \mp 5\%$$
$$K = \mp 10\%$$
$$M = \mp 20\%.$$

Thus a component marked

R33M is $0.33\ \Omega \mp 20\%$,
6K8G is $6.8\ k\Omega \mp 2\%$,
4M7M is $4.7\ M\Omega \mp 20\%$.

When soldering B.S.1852 components in place it is good practice to orientate them so that their values can all be read from the same angle. Although the colour code system is more difficult to read if one is not handling components daily, it has the advantage of readability in any position and is preferred by service engineers.

The colour code involves a series of rings around the resistor (see Figure 8.1). The first two rings represent digits, the third the number of noughts which follow these digits (sometimes called the multiplier). Usually a fourth ring is present which indicates the tolerance. In the case of grade 1, high stability resistors a fifth salmon pink ring or body colour may be present.

The colour series runs from black and brown through the colours of the rainbow to grey and white:

Black	0	Green	5
Brown	1	Blue	6
Red	2	Violet	7
Orange	3	Grey	8
Yellow	4	White	9

No tolerance band indicates ± 20 per cent, a silver band ± 10 per cent, a gold band ± 5 per cent, red ± 2 per cent, brown ± 1 per cent.

Some examples are:

Red, black, brown, and gold marks $200\ \Omega \mp 5\%$
Red, black, black, and silver marks $20\Omega \mp 10\%$
Green, blue, violet, and gold marks $560\ M\Omega \mp 5\%$

At first glance at a catalogue of resistors, the values seem to be strangely chosen. This is because preferred values in each tolerance range are designed to overlap, and obviously fewer components will be needed in the 20 per cent range than in the 10 per cent range. Preferred values are given below as whole numbers, but resistors can be obtained in all multiples of ten, and in some sub-multiples e.g. 12 is a preferred value and components are available in the preferred list at 1.2, 12, 120, 1.2 k, 12 k, 120 k, and 1.2 M.

Table 8.1 Preferred values of resistors

5%	10%	20%	5%	10%	20%
10	10	10	36		
11			39	39	
12	12		43		
13			47	47	47
15	15	15	51		
16			56	56	
18	18		62		
20			68	68	68
22	22	22	75		
24			82	82	
27	27		91		
30			100	100	100
33	33	33			

If the resistance of a circuit has to be delicately adjusted to produce the optimum performance, a *variable resistor* will have to be included in the

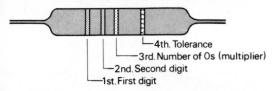

Figure 8.1 The colour code system for resistors

network. If frequent adjustment is necessary, the variable resistor will have a knob on the outside of the instrument. For applications of this type, there is a choice of carbon track controls (for heat dissipations up to 1 W) and wirewound controls (for higher wattage requirements). The relationship of the resistance track to the angle through which the knob is turned may be linear or logarithmic; the body of the control is usually marked with the value and type, e.g. 100 k Log. Tolerance is normally $\mp 20\%$. Controls are also made with built-in on/off switches.

If the circuit requires resistance balancing only when it is initially set up, a pre-set potentiometer is used (Figure 8.2); this may be of the skeleton type requiring a small screwdriver to turn it, or of the moulded track type with a knob. Straight sliding

track pre-sets are sometimes found, and in instruments of high accuracy rectilinear pre-sets (in which the slide is moved along the track by means of a screw) are used to give very fine control of the adjustment; where great precision is necessary, the track is lengthened to a ten-turn spiral, or helix. These controls may be found in long or short versions, but they are distinguished by a spindle which rotates through more than 360°. Some instruments use a tandem control, which consists of two resistance tracks mounted on the same spindle.

The symbols used to represent resistors in circuit diagrams are shown in Figure 8.3.

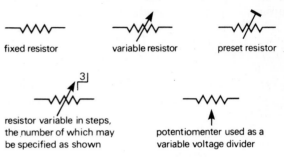

Figure 8.3 Symbols used to represent resistors

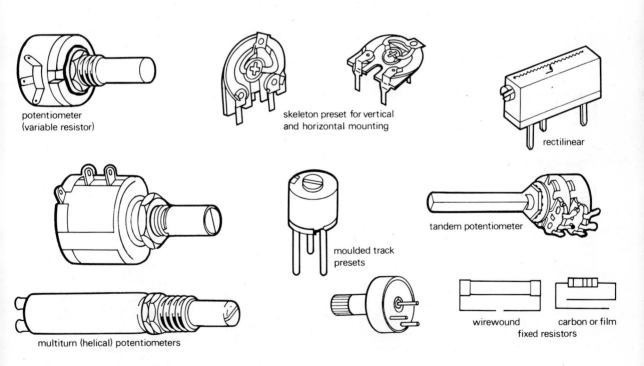

Figure 8.2 Types of potentiometer and resistor

CAPACITORS

These consist of two metal plates (sometimes more) separated by a non-conducting material called a dielectric. The metal plates may be flat or rolled up like a swiss roll. The dielectric may be air, paper, mica, plastic, glass, or ceramics depending upon the type of capacitor and its intended use. Air is a poor dielectric in that the plates must be well spaced to prevent sparking between them; it is used only for variable capacitors. Variable capacitors are of two types – large interleaved metal plates used to tune radio stations in a wireless, and small trimmer capacitors (sometimes called postage stamp capacitors) which are used as pre-sets in tuned circuits.

Capacitors, or condensers as they are occasionally called, have many uses. They will not permit the passage of d.c., but allow a.c. to pass acting only as a resistance. Capacitors can store an electric charge, which is built up at a rate dependant on a resistor in the circuit; this property allows them to be used in timing circuits.

A vast range of capacitor types is available and the development of new dielectric materials and encapsulating materials adds to the list every year. A telephone call to the technical service of any supplier will bring advice on the most suitable type for a given application. When making equipment for school use, only three types of fixed value capacitor are likely to be needed. For high capacitance *electrolytic* capacitors (Figure 8.4) are used – the dielectric is a polarised film of metal oxide (which can be destroyed if the component is placed in the circuit the wrong way round). The polarity of electrolytic capacitors may be shown by a red mark against the positive lead,

or by + or − signs. Some can types have only one lead which is positive, the case forming the negative connection. Double capacitor types also occur; the tags are marked yellow and red, with the negative tag plain. The values of all electrolytic capacitors are printed on the side.

Where high stability combined with accuracy is required, for instance in timing circuits, close tolerance *silvered mica capacitors* are used (see Figure 8.5). Their values are printed on the side.

Figure 8.5 Silvered mica capacitor

The third type of capacitor is known as a *polyester capacitor* because it uses a polyester plastic film as the dielectric, with a conducting film deposited on it. Polyester capacitors are suitable for use as general purpose capacitors; they may have their values printed on the side, or be colour coded.

The unit of capacitance is the *farad* which represents far too large a storage capacity for practical purposes, so much smaller units are used. These are the *microfarad*, which is one millionth of a farad, and the *picofarad* which is one million-millionth of a farad. The word microfarad is always abbreviated to μF, and the picofarad to pF.

Figure 8.4 Electrolytic capacitors

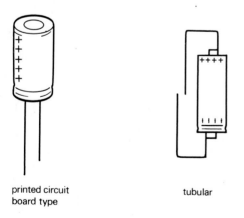

printed circuit
board type

tubular

can

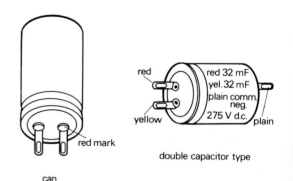

double capacitor type

When capacitors are colour coded they are marked with 3 or 4 spots or bands. The colours are the same as those used for resistors, the first two bands being digits and the third band the number of noughts. The unit used is always the picofarad. A tolerance mark is not always present and its value is different from the tolerance bands used in resistors.

Some manufacturers indicate by means of an arrow which way to read the marks and others place a large blob of paint to indicate which is the left hand end (always read from left to right or top to bottom). Some methods of coding are shown in Figure 8.6.

Some types of capacitor are commonly found in instruments; they are illustrated in Figure 8.7, which also shows the symbols used in circuit diagrams.

Table 8.2 Tolerance of capacitors

	Tolerance	
	10 pF or less	*over 10 pF*
Black	2 pF	20%
Brown	0.1 pF	1%
Red		2%
Orange		2.5%
Green	0.5 pF	5%
Grey	0.25 pF	
White	1 pF	10%

Figure 8.6   Some methods of coding capacitors

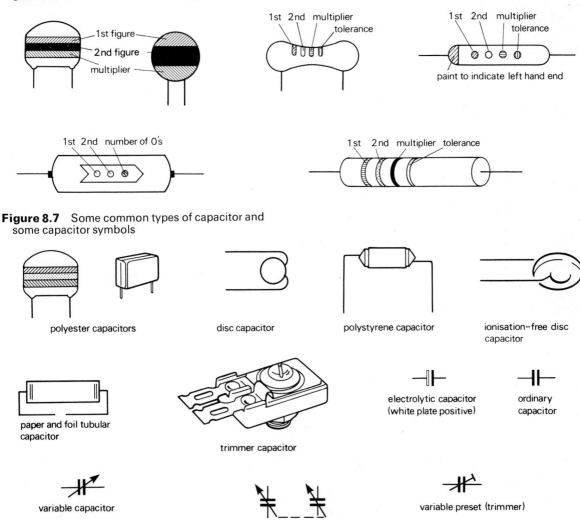

Figure 8.7   Some common types of capacitor and some capacitor symbols

When selecting components remember that electrolytic types can only be used with d.c., and that the voltage rating of a capacitor is the maximum d.c. voltage that it can withstand (in a.c. applications one must allow for the fact that the potential difference to be considered is the peak voltage of a sinusoidal wave and is equal to 1.4 times its root mean square value). In mains applications at 240 volts, the peak potential difference is about 336 volts and a 350 working voltage capacitor is the lowest practical rating which can be chosen.

As long as the storage value of a capacitor is correct, the voltage rating is unimportant as long as it exceeds any circuit voltage. The only penalty of using components of higher voltage rating than necessary is that they are more expensive, and bulkier, than the optimum component.

TRANSFORMERS

A number of different types of transformer are used in laboratory equipment, the most common being the mains transformer which converts 240 volts a.c. mains current to a higher or lower voltage, which can be passed through a rectifier to produce a direct current. Mains transformers are available with a wide range of outputs from 1 volt to 450 volts and some have three tappings, the centre tapping being 0 volts. In a centre-tapped 350–0–350 V transformer there is a potential difference of 350 V between the centre tapping and either of the outer tappings, but if only the outer tappings are used the potential difference is 700 V.

Multi-tapped transformers are available (Figure 8.8) which will give a number of different low voltages from the same transformer – one type gives 1–40 volts in steps of one volt.

For all educational purposes it is wise to choose a transformer with screened primary windings which will give the equipment user some protection from the mains. The type of transformers known as 'auto transformers' offer no protection, and should not be used in schools. Safety isolating transformers are available which give maximum protection from the mains.

In some audio equipment small transformers are found in the output stages. The function of these output transformers is to match the output impedance of the equipment with the input impedance of the loudspeaker.

In some equipment a specially designed transformer called a 'choke' is placed in the power supply circuit; this removes the ripple in the supply by preventing the ready flow of low frequencies.

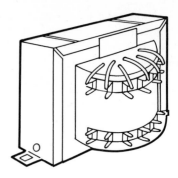

Figure 8.8   A multi-tapped transformer

British colour code for power transformers

Primary	
Common	Black
10 V	Black/Green
210 V	Black/Yellow
230 V	Black/Red
Electrostatic screen	Bare wire
Secondaries	
H.T.	Red
C.T.	Red/Yellow
Rectifier heater	Green
C.T.	Green/Yellow
Heater (1)	Brown
C.T.	Brown/Yellow
Heaters (2)	Blue
C.T.	Blue/Yellow

American colour code for power transformers

Primary	
Common	Black
Tap	Black/Yellow
Finish	Black/Red
Untapped primaries	Black
Secondaries	
H.T.	Red
C.T.	Red/Yellow
Rectifier	Yellow
C.T.	Yellow/Blue
Heater (1)	Green
C.T.	Green/Yellow
Heater (2)	Brown
C.T.	Brown/Yellow
Heater (3)	Grey
C.T.	Grey/Yellow

British colour code for output transformers

Inner primary	Brown
Outer primary	Green
Primary C.T.	Red
Inner secondary	Maroon
Outer secondary	White

British colour code for I.F. transformers

H.T. +	Red
Anode	Blue
Diode or grid	Green
Earth or	
A.V.C. return	Black

In a tapped secondary, the C.T. is black and the second diode or grid is black/green.

American colour code for A.F. transformers

Anode (OP)	Blue
H.T. + (plain or C.T.)	Red
Anode (IP) on C.T. primaries	Brown (blue if polarity is unimportant)
Grid (OS)	Green
Grid return (plain or C.T.)	Black
Grid (IS) on C.T. secondaries	Yellow (green if polarity unimportant)

Note The same code is used for valve-to-line and line-to-grid transformers.

Symbols for inductors including transformers

DIODES

Semiconductor diodes are used extensively in laboratory equipment. The basic characteristic of all diodes is that they will only conduct an electric current in one direction. This feature is used in many ways in the circuitry of laboratory instruments – the placing of a diode in front of any component which can be damaged by a change of polarity will prevent such damage. If a diode is placed in the path of an alternating current, it will allow only alternate half cycles to pass, blocking the others and therefore acting as a *half-wave rectifier* (see Figure 8.10).

Figure 8.10   Half-wave rectification

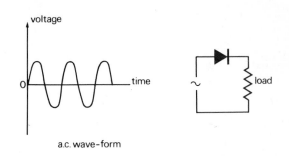

a.c. wave-form

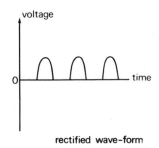

rectified wave-form

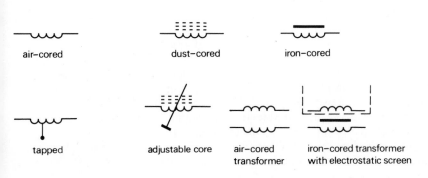

air–cored   dust–cored   iron–cored

tapped   adjustable core   air–cored transformer   iron–cored transformer with electrostatic screen

If four diodes are connected as a *bridge*, the negative part of the waveform is not just blocked but is also inverted as a positive current to give *full-wave rectification* (Figure 8.11); this principle is made use of in power supplies (see below). Bridge rectifiers may be made up from discrete components, or be in the form of a plastic capsule containing a complete bridge.

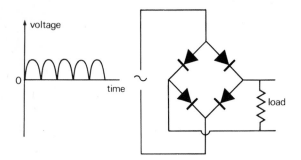

Figure 8.11   Full-wave rectification with four diodes

If a centre-tapped transformer is in use, full-wave rectification may be carried out with only two diodes (Figure 8.12).

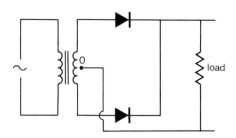

Figure 8.12   Full-wave rectification with two diodes

Diodes are used in several ways for temperature measurement as their characteristics undergo a small but linear change with temperature. If an operational amplifier is used to increase this change, temperature can be read on a meter. Although integrated circuit amplifiers such as the 741 series can be used, it is essential to include a feedback circuit as the diode change with temperature affects the linearity of the amplifier.

For extremely linear measurements with simple apparatus, the bridge circuit in Figure 8.13 using two diodes will be found useful if a suitable reference temperature, such as melting ice, can be provided for one of the diodes. The second diode is used as the probe in the environment to be measured.

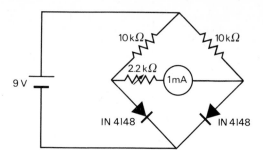

Figure 8.13   Differential diode circuit for temperature measurement

With an ordinary diode a current can be passed in one direction and the voltage increased until the diode breaks down irretrievably, an effect known as *the Zener effect*. Special diodes, called Zener diodes, are made which are able to withstand this effect – these diodes are of great importance because they act as ordinary diodes up to the Zener voltage at which point they allow large currents to pass but do not permit an increase in voltage (Figure 8.14). Diodes can be produced to show this Zener effect at a wide range of voltages, so that they can be used to act as voltage stabilizers in power supplies. Zener diodes are also used as protective devices across the terminals of meters – any current larger than that required to produce a full-scale deflection of the meter will be by-passed. One supplier lists 50 different Zener diodes to give voltage references from 2.7 V to 100 V.

TRANSISTORS

Transistors are used in electronic circuits to act as amplifiers or switches (see amplifiers and multi-vibrators below). In power supply circuits, a special *power transistor* is often used to stabilize the output by means of a feedback circuit. Transistors are made in two main types, p-n-p and n-p-n, and from the technician's point of view all that is necessary is to be able to identify the type from reference tables (see bibliography).

Another type of transistor has appeared on the market recently which is finding considerable application in laboratory equipment. This is the *field effect transistor (F.E.T.)*; this has a very high input impedance in the order of 10 MΩ which makes it very suitable for such applications as high impedance voltmeters (Figure 8.15).

Whenever possible choose silicon transistors rather than germanium as they are far less

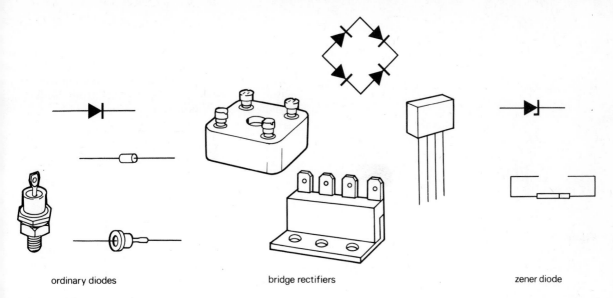

Figure 8.14 Diodes

heat-sensitive and therefore less likely to be damaged when soldering (it is good practice to use a heat shunt when soldering semiconductors into a circuit). Pliers can often be used, but a crocodile clip with two strips of copper soldered inside of the jaws is preferable.

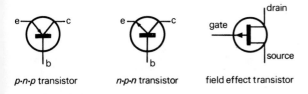

Figure 8.15 Types of transistor

Three wires lead out from most transistors and are known as the base, collector, and emitter. Some transistors used for very high frequency applications have a metal screen (usually the package which can be earthed) which gives rise to a fourth wire. Power transistors normally have two leadouts, the emitter and the base, which are marked E and B; the metal case acts as the collector connection. In some audio applications, matched pairs of transistors may be used; these can be etched on the same silicon chip, and a capsule with six wires arranged in two groups is likely to be of this type. F.E.T.s have three leadouts known as the drain, gate, and source.

The leadout wires are identified by their relationship to a spot of paint, or to a pip or flat, on the capsule. Figure 8.16 should help to identify the leadouts on most transistors.

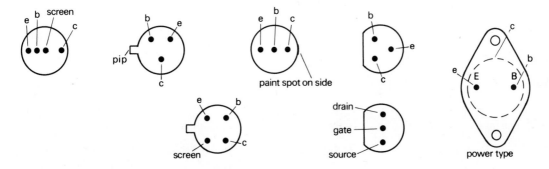

Figure 8.16 Types of transistor leadouts

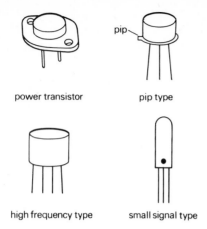

power transistor pip type

high frequency type small signal type

Figure 8.17 Styles of transistor

Transistors are wired into circuits in three different ways, named according to the leadout which is connected to the link between the input and the output (Figure 8.18).

The common emitter configuration is the most usual in amplifier circuits; if the transistor is to function well in a stable manner, it is necessary that some direct current should flow through the device, and that furthermore certain stable potentials should be established at the base, emitter, and collector. These conditions are called *bias conditions* and they

Figure 8.18 Ways of connecting transistors into circuits

are usually achieved by a technique known as *base voltage biasing*, the configuration of which is shown in Figure 8.19. C_1 is a coupling capacitor which prevents any d.c. potential from previous stages from affecting the base potential. C_2 is the emitter capacitor which prevents the emitter from following variations in base potential.

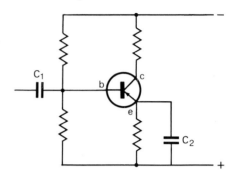

Figure 8.19 Configuration for base voltage biasing

If an unsuitable or wrongly biased transistor is used, the transistor junction tends to heat up and its resistance is reduced; more current flows, and the heating effect is further increased, until the transistor is damaged. This condition is known as *thermal runaway*.

Power transistors often dissipate a great deal of heat and for this reason they are usually mounted on metal heat sinks which conduct the heat away from the transistor, and disperse it by radiation to the air.

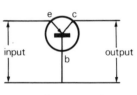

common base connection

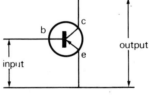

common emmitter connection

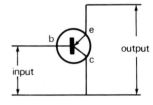

common collector connection

Table 8.3 Characteristics of transistor arrangements

	Common base	Common emitter	Common collector
Current gain	1	High	High
Voltage gain	High	High	1
Input impedance	Low	Medium	Fairly high
Output impedance	High	Medium	Low
Power gain	Medium	High	Low
Phase inversion relative to input	Nil	180°	Nil

THERMISTORS

Thermistors are semiconductors with a high resistance which decreases as temperature rises. They can be manufactured to extremely small sizes which makes them invaluable as temperature sensors in scientific work. When selecting thermistors for temperature measurement, look at the temperature/resistance graph and choose a device which is virtually linear over the temperature range to be measured. The best results are obtained using a bridge circuit, an example of which (based on the Standard Telephone & Cables Ltd. device STC F23) is illustrated in Figure 8.20. The cell RM1R is a hearing aid battery of 1.4 V, and the two variable resistors are used to adjust zero and full scale deflection.

Thermistors are used in many circuits as a pro-tective device to prevent thermal runaway. They are of two types – *miniature glass-beaded*, which are used for measurement, amplitude control, and timing circuits, and *rod types* which are used primarily in protective circuits (Figure 8.21).

PHOTO DEPENDANT RESISTORS

Photo dependant resistors, such as the Mullard ORP 12, contain a semiconductor material, cadmium sulphide, which changes its conductivity according to the light intensity. Dark resistance is in the order of 20 kΩ, falling to a few hundred ohms in very bright sunlight. The characteristic curve of these devices has two linear portions, that covering the higher light intensities being rather insensitive (Figure 8.22). For measurement of bright light it is best to cover the cell with a neutral density filter ($\times 5$), or half a table tennis ball, to bring the device onto its more sensitive linear range.

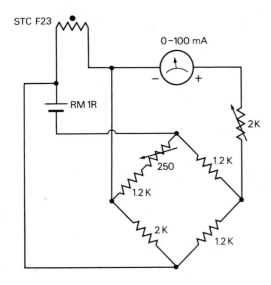

Figure 8.20   A 0–50 °C bridge

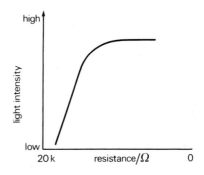

Figure 8.22   Response curve of a CdS cell

PHOTOVOLTAIC CELLS (SOLAR CELLS)

Photovoltaic cells (solar cells) generate a voltage dependant on the light intensity. They are sometimes known as selenium cells but are used less in measurement now than they used to be, although they still have some important applications. They are useful to the biologist for total light energy comparisons as they can be wired direct to an amp/hour meter such as the 150S, available from Electrosil Ltd. (Figure 8.23).

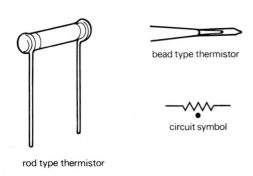

Figure 8.21   Thermistor types and symbol

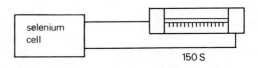

Figure 8.23   Total light energy meter

PHOTODIODES

These are the latest type of light-sensitive device and can be built into linear light meters with the FET operational amplifier, FET-MOPA, sold by R. S. Components Ltd.

One recent application of photodiodes has been in a new type of colorimeter, with a probe unit containing a photodiode and a light-emitting diode at a fixed distance (Figure 8.24). The probe is used by immersing it in the sample.

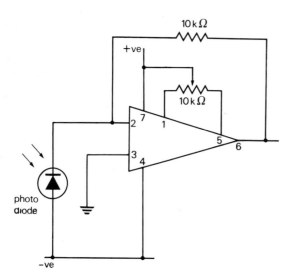

Figure 8.24 A linear photometer

Many other devices are coming onto the market and will appear from time to time in equipment. The *thyristor* group, which includes silicon controlled rectifiers (SCRs), diacs, and triacs, are being used increasingly for power control circuits and rectification. *Tunnel diodes* are used to produce oscillations, and the *Gunn diode* is used to produce microwave frequencies.

INTEGRATED CIRCUITS

One of the important developments from transistor technology has been the growth of integrated circuitry, which provides us with a wide range of complete circuits etched onto a minute silicon chip (Figure 8.25). These integrated circuits, or ICs, are of the utmost importance in laboratory instrument design as they can be modified with a few discrete external components for a wide range of functions. For example, the 741 operational amplifier can be used as an inverting amplifier, non-inverting amplifier, differential amplifier, level switch, or as an integrator. The 741 could therefore be used as the basis for a conductivity meter, pH meter, decibel meter, diode thermometer, lightmeter, colorimeter, and many other instruments. The potential of such a device can easily be appreciated as it is available in a dual in-line package which can be plugged into a socket and is of low cost, thus revolutionizing the servicing problems of school instruments. A fault can be cured by plugging in a new 741.

The nomenclature of ICs often causes difficulty, each manufacturer helping to confuse the picture by adding his own code letters. Most devices have a number which identifies the silicon chip used; this should be the same for all manufacturers, and is the only part of the name which really matters. Letters in front of the number indicate the manufacturer, or sometimes the function, and a letter after the number usually indicates a modification to the original chip design to give slightly improved characteristics. For example, BP 741, 741 OPA, 741 C, and LM 741 CN are all similar devices from different manufacturers, and for most purposes are interchangable.

ICs may be very specialized, or relatively simple, building blocks ranging from simple operational amplifiers and flip-flops, through logic circuits, to complete digital voltmeter circuits on a single chip. Some ICs are used in a wide range of applications; for example, the CA3035 IC contains 10 transistors, 1 diode, and 15 resistors etched onto the same silicon chip and is an 'Ultra-high-gain wide-band amplifier array' housing three separate amplifiers which can be used individually or cascaded to give a total gain of 129 dB (i.e. $\times 2.8$ million). R. M. Marston, in his book *110 Integrated Circuit Projects for the Home Constructor* (Newnes–Butterworth), describes 30 circuits based on this device, many of which are useful laboratory instruments. This book is recommended to anyone wishing to gain an appreciation of the versatility of ICs.

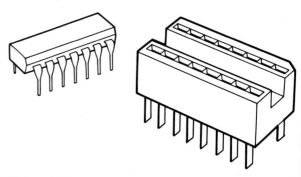

Figure 8.25 Integrated circuits

METERS

For most school instruments the *moving-coil galvano-meter* is chosen as the readout device. With suitable circuit layout, it can be used as a voltmeter, ammeter, frequency meter, a.c. voltmeter, ohmeter, and power meter (watt or dB). Whatever scale is printed on the moving coil galvanometer dial, it is a device to *measure current*, usually in milliamps (milliammeter) or microamps (microammeter); its range can be altered by wiring a resistor (or shunt, as it is usually called) across the meter terminals (Figure 8.26). For example, a milliammeter with a 1 mA full scale deflection and an internal meter resistance of 75 ohms is connected across a battery of 1.5 V. If we ignore the small internal resistance of the battery, the current flowing will be

$$I = \frac{V}{R} \text{ i.e. } \frac{1.5}{75} = 0.02 \text{ amp, or 20 milliamps;}$$

as this is outside the range of our meter, we must attach a suitable resistance in parallel with the meter to 'shunt' a proportion of the current past the meter. In this instance it would be convenient to measure

1/10 th of the current and shunt the remaining 9/10 ths. This could be achieved by placing a resistance of exactly 8.33 ohms across the meter so that full scale deflection now indicated 10 milliamps, and our meter needle will move 2/10 ths of the way across the scale if reconnected to the battery. If 99/100 ths of the current is shunted with a resistance of 0.74 ohms, our meter will read up to 100 milliamps, and so on.

Milliameters intended for *voltage measurements* are usually made with a higher resistance, e.g. 1250 ohms, and voltage ranges are increased by wiring resistors *in series* (see Figure 8.27) with the meter (note that when used for d.c. voltage measurement the meter still measures current flowing through the circuit resistance).

The moving-coil meter cannot respond directly to an a.c. supply, so any current measured must be rectified to d.c. (Figure 8.28). The meter now receives pulsating d.c. and the pointer registers the 'root mean square' or effective value, which is what we usually need to know. ('Effective value' is that which has the same heating effect or power as an equivalent direct current.)

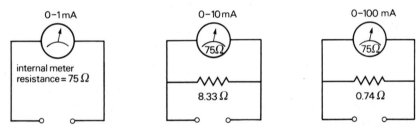

Figure 8.26   Use of a 1 mA (1000μA) f.s.d. meter to read different current ranges

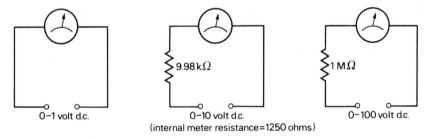

Figure 8.27   A 1 mA f.s.d. meter used as a d.c. voltmeter

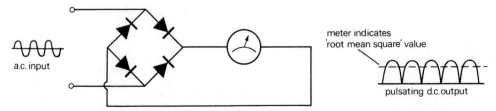

Figure 8.28   A 1 mA f.s.d. meter set up for a.c. voltage measurement

The milliameter can be used in two ways to *measure resistance*; the simple set-up shown in Figure 8.29 is often used for approximate resistance measurements. If the test leads from the sockets are shorted, the variable resistor can be used to adjust the pointer to full scale deflection. This represents zero ohms. The leads are now separated, and the meter reads zero current, which indicates an infinite resistance. Any resistance placed between the test leads will give an intermediate current depending on the value of the resistor, and the scale can be calibrated by Ohm's Law.

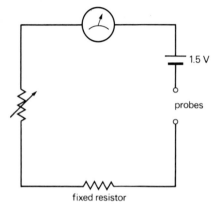

Figure 8.29   Simple Ohmeter

If the value of the unknown resistance is to be determined with a high degree of accuracy, a bridge circuit is used. In the circuit in Figure 8.30, R_A and R_B are of known value (or ratio), and R_C is an accurately calibrated variable resistor or a decade box selected by switches. The value of the unknown resistor, R_X, is found by varying R_C until the meter shows no current flowing in the bridge.

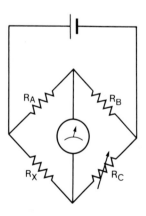

Figure 8.30   The bridge method for determining resistance

Capacitor values can be measured by the resistance of the capacitor to a.c., or by capacitance bridges.

Frequency can be measured by means of a *Wein bridge* (Figure 8.31), the value of C_1 and R_1 being varied until a nul point is obtained. At the nul point, $C_1 \times R_1$ is equal to the frequency.

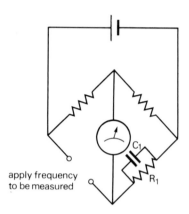

Figure 8.31   Wein bridge

With suitable circuitry, and a multi-range switch, it is possible to build a general-purpose instrument which will measure resistance, a.c., and d.c. voltages, a.c. and d.c. currents, frequency, capacitance, and power output. Such instruments are called 'multimeters', the best known of which is probably the 'Avo testmeter'.

When choosing a testmeter, the higher the impedence the better.

For the measurement of very small voltages some amplification may be needed, and for this reason 'millivolt' meters are often provided with valve, or FET, amplifier stages.

Digital voltmeters are becoming increasingly popular, but at present are expensive in relation to moving coil meters.

Whenever an instrument based on a moving coil galvanometer is used, remember that the accuracy, at best, is likely to be 1 per cent of full scale deflection, and with cheaper types it may only be 3 per cent. If a 3 per cent meter is used for a full range pH meter, accuracy will only be in the order of 0.4 of a pH unit even though the scale may be divided into 0.1 pH unit divisions.

Ampere-hour meters

The ampere-hour meter, or Curtis meter (Figure 8.32), distributed by Electrosil Ltd. is a low cost component which is very versatile as an energy integrator. It consists of a capillary tube filled with mercury, with a break in the column filled with electrolyte. When a potential is applied across the meter, mercury is 'plated' across the electrolyte, causing the bubble of electrolyte to move along. The distance moved is directly proportional to the time for which the circuit is switched on, and to the current applied. If the current is 5 mA (the maximum which the instrument can handle), the meter will record for up to 2 hours. If the current is reduced by means of a ballast resistor the meter will record for anything up to 10 000 hours. Used in conjunction with light level, temperature level, or sound level switches, the device can record the number of hours exposure in a given period. Used in a series circuit with a photoresistor and a battery, total light energy can be integrated; with two such instruments comparisons between sites can be made. Cheap plug-in sockets are available for the meter so that one meter can be used for several instruments if they are not required simultaneously. *Note that if a current much in excess of 5 mA is applied the two parts of the mercury column are likely to join up, thereby destroying the meter. If the meter is allowed to run off its scale it cannot be reset* – it is essential therefore to include a ballast resistor in the circuit to give a running time greater than the duration of the experiment. Reversing the polarity will drive the bubble back to the start.

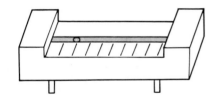

Figure 8.32   A Curtis amp/hour meter

RELAYS

In instruments which require a switching circuit to operate a secondary circuit, it is often necessary to use relays to handle the current of the secondary circuit if it is too large to be switched directly by the transistors. The circuit in Figure 8.33, which utilises an R.S. Components LAS 15 integrated circuit light-operated switch for counting purposes, illustrates the point. The level at which the switch operates is determined by the values of R_1 and C_1. The output from the switch is amplified by the external transistor to operate a 12 volt relay, which in turn switches a 12 volt circuit to operate the electromagnetic counter.

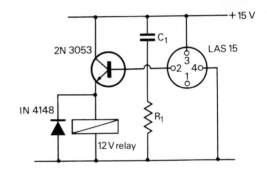

Figure 8.33   An example of a relay used in an integrated circuit

The relay is a device which is operated by an electromagnetic field, and consists of a reed switch surrounded by an induction coil. It is represented in circuit diagrams by the symbols in Figure 8.34, symbol A being used if both primary and secondary circuits are represented in the diagram.

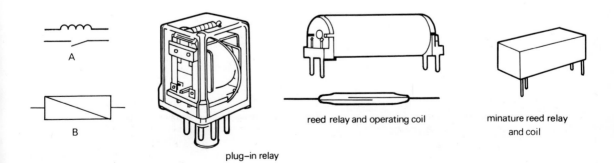

Figure 8.34   Relay types and symbols

131

CIRCUIT CONSTRUCTION

It is possible to construct simple circuits by the 'bird's nest' method, but this should be rejected except for the roughest of prototypes. Circuits can be built on printed circuit boards, veroboard (which is a laminated board with strips of copper drilled for component wires at regular intervals), group boards, tag strips, insulated matrix boards with turret tags inserted where required, and sometimes on the tags on a switch wafer. Figure 8.35 shows some of the types available.

When making up a circuit for the first time, component layout and modification of the circuit can be most easily carried out on a 'T-Dec', which is a matrix board with push-in contacts (Figure 8.36).

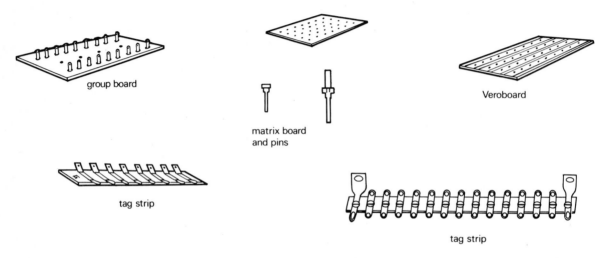

group board

Veroboard

matrix board
and pins

tag strip

tag strip

Figure 8.35   Circuit boards

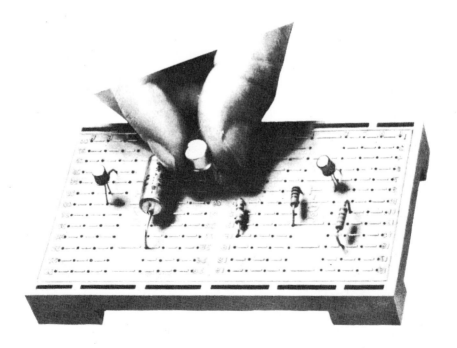

Figure 8.36

Other electronic symbols used in circuit diagrams

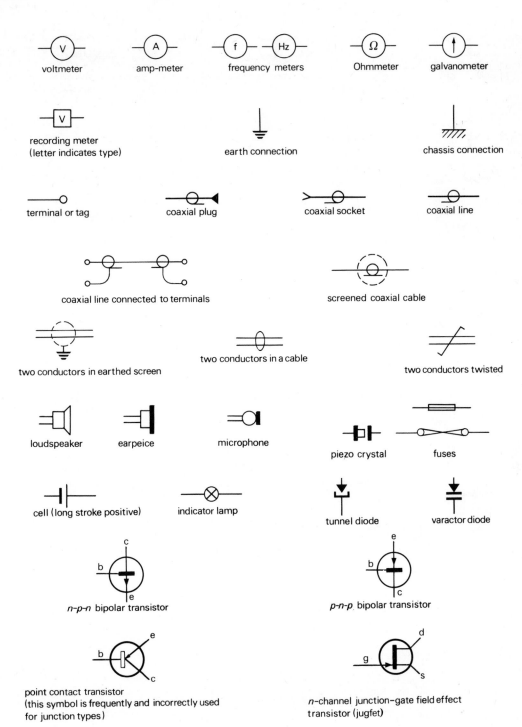

voltmeter

amp-meter

frequency meters

Ohmmeter

galvanometer

recording meter
(letter indicates type)

earth connection

chassis connection

terminal or tag

coaxial plug

coaxial socket

coaxial line

coaxial line connected to terminals

screened coaxial cable

two conductors in earthed screen

two conductors in a cable

two conductors twisted

loudspeaker   earpeice   microphone

piezo crystal   fuses

cell (long stroke positive)

indicator lamp

tunnel diode

varactor diode

n-p-n bipolar transistor

p-n-p bipolar transistor

point contact transistor
(this symbol is frequently and incorrectly used
for junction types)

n-channel junction–gate field effect
transistor (jugfet)

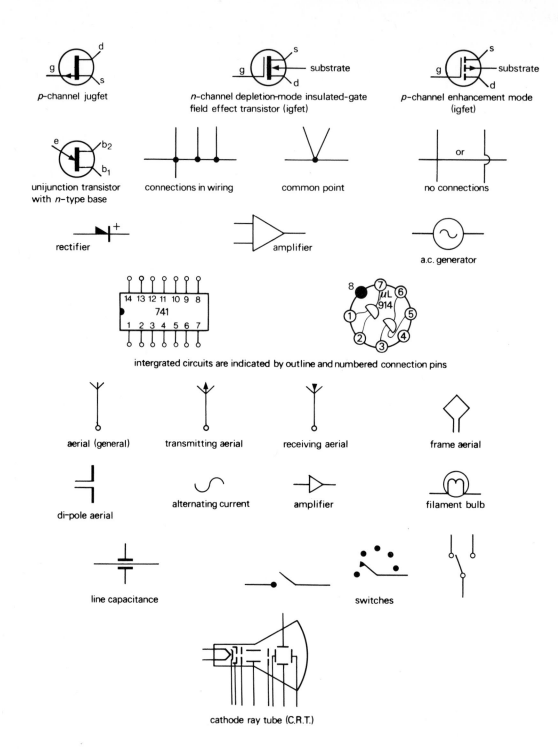

Figure 8.37 Some other symbols used in circuit
diagrams (U.K. and U.S.A.)

Some European symbols differ from those used in the U.K. and U.S.A., and a few selected examples are given in Figure 8.38 to enable the circuit diagrams of imported equipment to be understood.

Figure 8.38 Some European circuit symbols

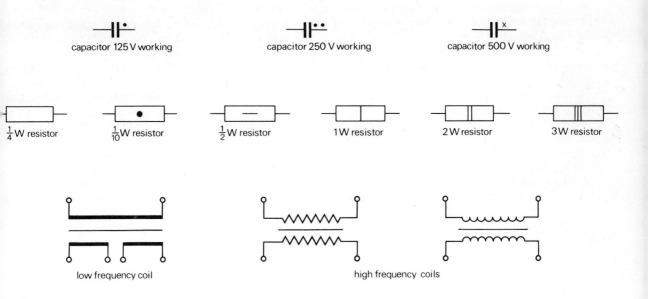

capacitor 125 V working   capacitor 250 V working   capacitor 500 V working

$\frac{1}{4}$ W resistor $\frac{1}{10}$ W resistor $\frac{1}{2}$ W resistor   1 W resistor   2 W resistor   3 W resistor

low frequency coil   high frequency coils

Symbols of some common valves

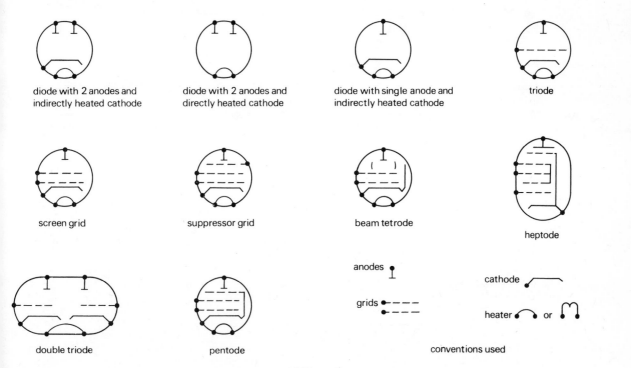

diode with 2 anodes and indirectly heated cathode

diode with 2 anodes and directly heated cathode

diode with single anode and indirectly heated cathode

triode

screen grid   suppressor grid   beam tetrode   heptode

double triode   pentode

anodes

grids

cathode

heater   or

conventions used

COMMON CIRCUIT 'BUILDING BLOCKS'
Power supplies

For many transistor circuits the only power supply may be a battery, which may be connected to the equipment directly or through a pair of diodes to prevent damage should the battery be connected with the wrong polarity.

If the equipment operates from the mains, the alternating current will probably be fed into a transformer to reduce the voltage, which is still a.c. The transformed supply is then rectified by means of a rectifier, diodes, or a diode rectifier bridge. The supply is now d.c., but will not be stable as there is inevitably some ripple – the placing of a large capacitor across the rectifier output will 'smooth' the supply. Very precise stabilization may be achieved by placing a power transistor in the circuit (Figure 8.40). The voltage level may be controlled by means of a Zener diode.

When designing power supplies, or replacing components of existing power supplies, the information in Table 8.4 (published by R.S. Components Ltd) will enable components of the correct rating to be selected for a range of rectification applications.

Amplifiers

Amplifiers are basic building blocks in many laboratory circuits, but it is possible to give only a simple outline of the more common types here. There are two basic types, *current amplifiers* and *voltage amplifiers*; the section above on transistors deals with the current and voltage gain characteristics of different transistor configurations. Often the choice of configuration depends upon the input and output impedances which have to be matched. In *d.c. amplifiers* (Figure 8.41) the input is passed direct to the transistor base, and if several amplifier stages are cascaded, the collector output of the first is fed direct to the base of the second. In *a.c. amplifiers* (Figure 8.42) the configuration is similar, except that the input to each stage is fed through a capacitor to block any d.c. present in the signal from the previous stage. For voltage amplifiers, a load resistor is connected to the collector.

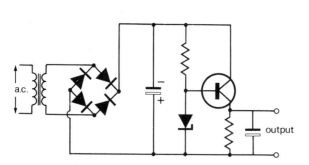

Figure 8.40   Stabilized transistorized power supply

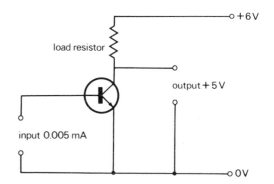

Figure 8.41   A d.c. amplifier

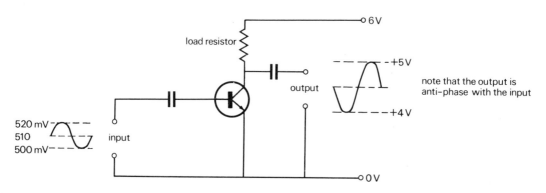

Figure 8.42   An a.c. amplifier

Table 8.4 Basic single phase rectifier circuit relationships
(Reproduced by permission of R.S. Components Ltd)

Full-wave bridge – capacitive input filter	*Full-wave bridge – choke input filter*	*Full-wave bridge – resistive load*
Circuit factor: 0.62	**Circuit factor: 0.94**	**Circuit factor: 0.9**

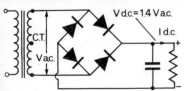 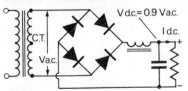 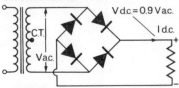

Rating per rectifier
   Peak inverse voltage = 1.4 V a.c.
   Average current = 0.5 A d.c.
   Peak current: See Note 1

Rating per rectifier
   Peak inverse voltage = 1.4 V a.c.
   Average current = 0.5 A d.c.
   Peak current = 1.0 A d.c.

Rating per rectifier
   Peak inverse voltage = 1.4 V a.c.
   Average current = 0.5 A d.c.
   Peak current = 1.57 A d.c.

Full-wave – capacitive input filter	*Full-wave – choke input filter*	*Full-wave – resistive load*
Circuit factor: 1.0	**Circuit factor: 1.54**	**Circuit factor: 1.39**

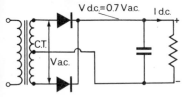 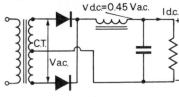 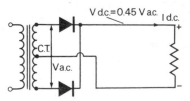

Rating per rectifier
   Peak inverse voltage = 1.4 V a.c.
   Average current = 0.5 A d.c.
   Peak current: See Note 1

Rating per rectifier
   Peak inverse voltage = 1.4 V a.c.
   Average current = 0.5 A d.c.
   Peak current = 1.0 A d.c.

Rating per rectifier
   Peak inverse voltage = 1.4 V a.c.
   Average current = 0.5 A d.c.
   Peak current = 1.57 A d.c.

Half-wave – capacitive input filter	*Resistive–load*	*Half-wave – resistive load*
Circuit factor: 0.27 (0.43)	**Circuit factor: 1.15**	**Circuit factor: 0.64**
See Note 2		

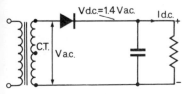 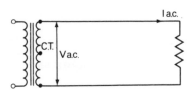 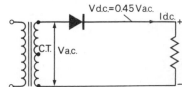

Rating per rectifier
   Peak inverse voltage = 2.8 V a.c.
   Average current = 1.0 A d.c.
   Peak current: See Note 1

Rating per rectifier
   Peak inverse voltage = 1.4 V a.c.
   Average current = 1.0 A d.c.
   Peak current = 3.14 A d.c.

Note 1
Peak current is only limited by the effective series
resistance of the transformer and associated components.

Note 2
The 0.43 circuit factor should only be used when the
transformer has been specially designed for half-wave
capacitive input filter circuits.

   To estimate ratings for conditions not included in a transformer specification:

$$\frac{\text{Load current}}{\text{in new circuit}} = \frac{\text{Load current}}{\text{in specified circuit}} \times \frac{\textbf{Circuit factor shown above for new circuit}}{\textbf{Circuit factor shown above for specified circuit}}$$

Reference to the diagram in Table 8.4 for the new circuit will then show the relationship between transformer
secondary volts and output volts and also rectifier ratings.

Relationships shown assume no circuit losses, and should only be used as a guide.

In practical circuits, a bias current is provided by potential divider resistors R_1 and R_2. R_4 is added to give temperature stabilization, as well as some negative feedback for wanted a.c. signals. If the feedback is not required, a bypass capacitor C_1 is added. Figure 8.43 gives a practical circuit for a.c. voltage amplification using n-p-n transistors (with p-n-p transistors the polarity of the power supply, and the electrolytic capacitor, must be reversed).

In some instruments a type of amplifier called a *long-tail-pair amplifier* (Figure 8.44) is used. This is a *differential amplifier* in which the output is a measure of the difference between two inputs.

In a *push-pull amplifier* (Figure 8.45) each transistor conducts during alternate half cycles. This type of amplifier is often used in audio circuits to feed a loudspeaker.

When a large gain is required in a.c. applications, an *R.C. (resistive-capacitive) coupled two-stage amplifier* (Figure 8.46) is used.

Operational amplifiers derive their name from their use in carrying out mathematical operations. Although they can be built from discrete components, this is scarcely worthwhile as they are available in integrated circuit form at low prices. Operational amplifiers form the basis of many laboratory instruments. A versatile device, such as the 741, can be connected to a few discrete components to give an *integrating amplifier* (Figure 8.47), where the output is the equal to the product of the two inputs, a *differential amplifier* (Figure 8.48), where the output is the difference between the two inputs, a *level detector* or *voltage comparator*, which will operate a relay, and *inverting* or *non-inverting a.c. or d.c. amplifiers* (Figure 8.49). Sometimes an amplifier is used, not for its amplification, but for its high input impedance; e.g. an FET operational amplifier may be used with a gain of unity to give a millivolt meter a very high input impedence.

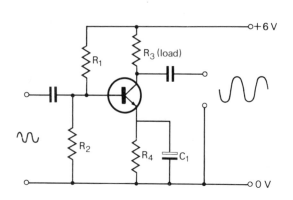

Figure 8.43   A practical a.c. amplifier

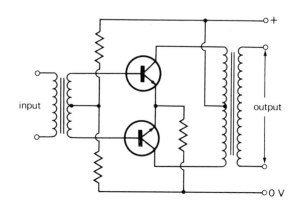

Figure 8.45   A push-pull amplifier

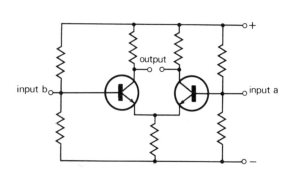

Figure 8.44   A long-tailed pair amplifier

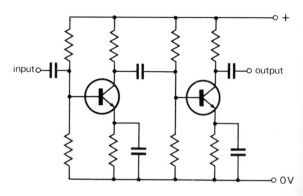

Figure 8.46   R.C. coupled two-stage amplifier

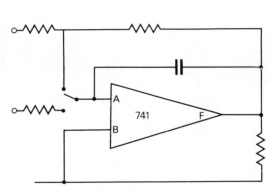

Figure 8.47 The 741 as an integrating amplifier

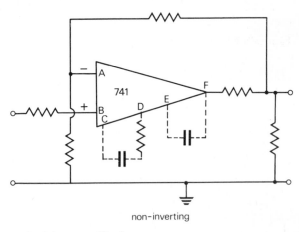

Figure 8.48 The 741 as a differential amplifier

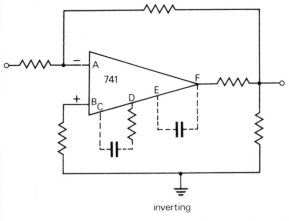

inverting

non-inverting

(The links CD and EF are used only in a.c. amplifiers)

Figure 8.49 The 741 as an inverting or non-inverting amplifier

Level switches

These are used to detect a signal of given strength. The most frequent applications in laboratory instruments are for operating a relay or time recording device whenever their threshold value is exceeded. Any form of energy can be transduced by an appropriate probe to provide a potential which is compared with a reference potential.

The circuit shown in Figure 8.50 will operate a relay or similar device at any predetermined level of the d.c. input voltage. Adjustment of R_2 will give the level of operation where the output V_0 switches over from -0.5 V to $+3.2$ V or vice versa. The operating level can be noted on the meter.

The d.c. input voltage can be in excess of 1 V provided resistor R_3 is incorporated to limit V to a maximum of 1 V, e.g. for an input of 100 V d.c. R_3 must be 99 kΩ.

Oscillators

Oscillators are used to produce sine waves. A *square wave generator* is an oscillator designed to clip the positive and negative peaks of a sine wave and is often used for timing purposes. Although they are drawn rather differently in circuit diagrams, oscillators are in fact amplifiers in which part of the output is fed back to the input. All amplifiers can be forced into oscillation. One of the most useful oscillators in instrumentation is the astable multivibrator, or flip-flop, which is in effect a two transistor amplifier with the transistors coupled input to output. The rate of oscillation is dependant upon the resistors and capacitors R_1 R_3 and C_1 C_2. Figure 8.51 shows a multivibrator as it is normally drawn, and Figure 8.52 shows it as an amplifier with feedback.

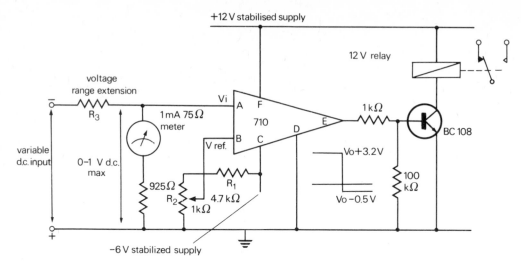

Figure 8.50 A level switch using a 710 OPA operational amplifier

Attenuators

These are circuit blocks which may take several forms, the simplest being a resistance which reduces the input signal. Often such attenuators are variable or switched decade resistors, and they attenuate the signal as a whole. Capacitors and resistors are sometimes built into an attenuation network to reduce part of the signal – often an unwanted frequency. In very sophisticated circuits, transistorised attenuators are used.

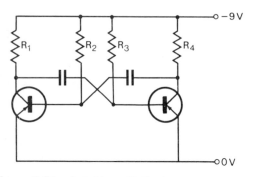

Figure 8.51 A stable multivibrator

Potential divider networks

When the circuit 'building blocks' described above are assembled to produce a complete instrument, it is likely that different parts of the circuit will require to be energized by different supply voltages. This does not mean that we require a separate battery for each part of the circuit – we can use a series of resistors to attenuate the voltage. The voltage developed across a resistance in a series circuit is proportional to the value of the resistance. Thus if the resistor is 10 per cent of the total resistance in the circuit then the voltage developed across it will be 10 per cent of the total voltage applied to the whole circuit. Consider the simple potential divider network in Figure 8.53; four 10 kilohm resistors are connected in series across a 100 V potential. The circuit may be tapped between A and B to give 25 V, between A and C to give 50 V, between A and D to give 75 V, and finally across A and E to obtain the full 100 V potential difference.

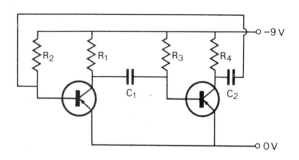

Figure 8.52 A multivibrator circuit redrawn as an amplifier with feedback

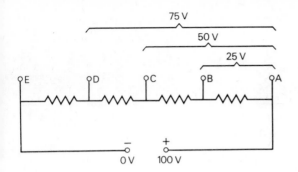

Figure 8.53 A potential divider

Voltage doubler circuits

These can be used to provide very high voltages from small voltages. The circuit consists of two half-wave rectifier circuits joined together in such a way that the output from one is added to the output from the other (Figure 8.54). It is possible to connect a number of such circuits in series to produce very high voltages. Another method of increasing voltage is to use a step-up transformer.

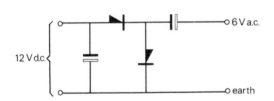

Figure 8.54 Voltage doubler circuit

Logic circuits

A number of very useful circuits used for instrumentation are available in integrated circuit form. These include binary counters, logic gates, and divide by ten circuits; a series of this last circuit is sometimes used to reduce a crystal oscillation to the value needed for a timing circuit.

TESTING FAULTY EQUIPMENT

Always start by checking the obvious such as power supply, input and output contacts, and the condition of any batteries or cells used.

If a check of these obvious points does not reveal the fault, the circuit must be examined logically. Consider the whole instrument as a series of building blocks, and try to think out which block or blocks could be responsible for the particular type of malfunction. If the fault can be attributed to a particular block, such as an attenuator or an amplifier, the subsequent search for the faulty component will be minimised.

A visual examination of the circuit should be made, looking for any signs of components which have overheated. If any component shows signs of heat damage, it should be de-soldered and replaced. In a circuit which is correctly designed, it is most unusual for integrated circuits of semiconductors to break down; capacitors breaking down, or resistors changing their value, should always be the prime suspects.

If visual examination and/or logic alone do not indicate the faulty component, it will be necessary to check through the circuit step by step. Often a service diagram is available which gives the voltage/current, or the shape of the waveform, at each point in the equipment. Simple measurements with a multimeter, electronic voltmeter, or an oscilloscope may reveal the cause of the problem.

If a service diagram is not available, it may be necessary to draw out the circuit diagram so that its functioning can be understood and an intelligent guess, or better still a calculation, made of the circuit values which can then be checked.

Mains voltage equipment and very complex apparatus should not be tested unless the technician concerned is trained and competent to do so, and has the necessary test gear available. A trained electronic technician is often available in technical colleges and many education authorities have arrangements for the college to carry out the more complex servicing of school equipment.

It is beyond the bounds of this volume to describe the use of an oscilloscope in the tracing of equipment faults – so versatile is the oscilloscope that it requires a whole book to describe its uses in servicing; the reader is strongly recommended to obtain a copy of *Servicing with the Oscilloscope* by G. J. King (Newnes-Butterworth).

BASIC CIRCUIT CALCULATIONS

The majority of calculations needed in electronics stem from Ohm's law, which can be stated simply as 'For any circuit, the current × the resistance is equal to the voltage'. This is true as long as a coherent system of units is used, i.e. volts, amps, and ohms. Provided that two of these values are known, the other can be calculated by a rearrangement of the formula. Thus,

Voltage $=$ Current $\times$ Resistance

Current $= \dfrac{\text{Voltage}}{\text{Resistance}}$

Resistance $= \dfrac{\text{Voltage}}{\text{Current}}$

A convenient way of memorising the formulae is to use the triangle in Figure 8.55 – by covering up the value to be found, the correct formula is left.

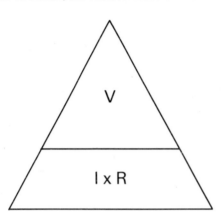

For example, if a current of 1 mA flows through a 100 kΩ resistance, what is the voltage, or potential difference?

$$V = I \times R$$

By substitution,

$$V = 0.001 \times 100\,000 = 100 \text{ volts}$$

Power calculations

The power consumption of a piece of equipment is stated in watts, or if the consumption is small, milliwatts

Power in watts $=$ Voltage $\times$ Current in amps.

For example, a 48 watt car headlamp running from a 12 volt battery consumes Y amps.
By substitution, $48 = Y \times 12$

$\therefore \qquad Y = \dfrac{48}{12} = 4 \text{ amps}$

Resistances in series

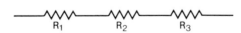

Total resistance $= R_1 + R_2 + R_3$

Resistances in parallel

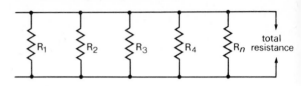

Total resistance of the circuit

$$= \cfrac{1}{\dfrac{1}{R_1}+\dfrac{1}{R_2}+\dfrac{1}{R_3}+\dfrac{1}{R_4}\cdots\dfrac{1}{R_n}}$$

For 2 parallel resistances

$$\text{Total resistance} = \frac{[R_1 \times R_2]}{[R_1 + R_2]}$$

For 3 parallel resistances

Total resistance

$$= \frac{[R_1 \times R_2 \times R_3]}{[R_1 \times R_2]+[R_2 \times R_3]+[R_3 \times R_1]}$$

For 4 parallel resistances

Total resistance

$$= \frac{[R_1 \times R_2 \times R_3 \times R_4]}{[R_1 R_2 R_3]+[R_2 R_3 R_4]+[R_3 R_4 R_1]+[R_4 R_1 R_2]}$$

Resistances in series-parallel circuits

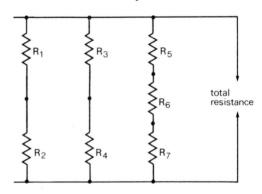

Total resistance

$$= \cfrac{1}{\left[\dfrac{1}{R_1+R_2}\right]+\left[\dfrac{1}{R_3+R_4}\right]+\left[\dfrac{1}{R_5+R_6+R_7}\right]}$$

Calculations in a.c. circuits

Ohm's law can be restated for application in a.c. circuits as follows:

$$E = P \div (I \cos X) \text{ or } \sqrt{PZ \div \cos X} \text{ or } IZ$$

$$I = P \div (E \cos X) \text{ or } \frac{E}{Z} \qquad \text{or } P \div (Z \cos X)$$

$$Z = \frac{E}{I} \qquad \text{or } \frac{(E^2 \cos X)}{P} \qquad \text{or } \frac{P}{(I \cos X)}$$

$$P = IE \cos X \qquad \text{or } \frac{(E^2 \cos X)}{Z} \qquad \text{or } I^2 Z \cos X$$

where I = current in amps
Z = impedance in ohms
P = wattage
E = voltage across the impedance
X = degrees of the phase angle

Kirchhoff's Laws

(a) *The algebraic sum of the currents at a junction in a network is zero* – if this were not true a charge of electricity would accumulate at the point in the network. Any positive currents flowing to the point must be balanced by a negative current flowing from it.

(b) *The algebraic sum of the voltage drops around any closed path in a circuit is zero* – otherwise some point would have more than one potential at the same time, which is clearly impossible.

These laws are invaluable in solving electrical network problems. Consider the circuit in Figure 8.59.

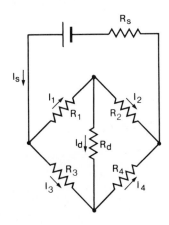

Figure 8.59

By Kirchhoff's laws,

$$I_s - I_1 - I_3 = 0$$
$$I_1 R_1 + I_d R_d - I_3 R_3 = 0$$

The four junctions and three loops give seven simultaneous equations which make it possible to solve for the six unknown currents.

Capacitors in series

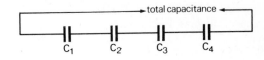

$$\text{Total capacitance} = \cfrac{1}{\cfrac{1}{C_1} + \cfrac{1}{C_2} + \cfrac{1}{C_3} + \cfrac{1}{C_4}}$$

Capacitors in parallel

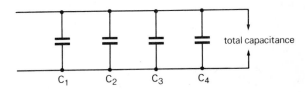

$$\text{Total capacitance} = C_1 + C_2 + C_3 + C_4$$

Series-parallel capacitors

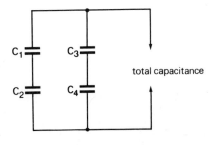

$$\text{Total capacitance} = \cfrac{1}{\left[\cfrac{1}{C_1} + \cfrac{1}{C_2}\right]} + \cfrac{1}{\left[\cfrac{1}{C_3} + \cfrac{1}{C_4}\right]}$$

BIBLIOGRAPHY

BALL A. M. (ed.) (1970), *Radio valve and Transistor Data*, London, Iliffe (distributed by Newnes-Butterworth)

BROPHY, J. J. (1966), *Semiconductor Devices*, London, Allen and Unwin

CRANE P. W. (1971), *Electronics for Technicians*, Oxford, Pergamon

DANCE J. B. (1969), *Photoelectronic Devices*, London, Iliffe (distributed by Newnes-Butterworth)

Electronics and Technology Series – covering a wide range of subjects. Published by Holt, Rinehart and Winston (Toronto, Canada, distributed by Holt-Blond Ltd, 120 Golden Lane, London EC1)

HIBBERD R. G. (1968), *Transistor Pocket Book*, London, Newnes-Butterworth

HUGHES J. and JOHNSTON T. N. (1970), *Using Semiconductors*, London, Heinemann Educational

JENKINS J. and JARVIS W. H. (1966), *Basic Principles of Electronics, Volume 1, Thermionics*, Oxford, Pergamon

JENKINS J. and JARVIS W. H. (1971), *Basic Principles of Electronics, Volume 2, Semiconductors*, Oxford, Pergamon

KING G. J. (1969), *Servicing with the Oscilloscope*, London, Newnes-Butterworth

KING G. J. (1971), *Radio, Television and Audio Test Instruments*, London, Newnes-Butterworth

KING G. J. (1973), *Rapid Servicing of Transistor Equipment*, London, Newnes-Butterworth

MARSTON R. M. (1969), *110 Semiconductor Projects for the Home Constructor*, London, Iliffe (distributed by Newnes-Butterworth)

MARSTON R. M. (1971), *110 Integrated Circuit Projects for the Home Constructor*, London, Iliffe (distributed by Newnes-Butterworth)

MOORSHEAD H. W. (1972), *Radio Engineer's Pocket Book*, London, Newnes-Butterworth

MULLARD DATA BOOK (annual)

PRICE L. W. (1969), *Electronic Laboratory Technique*, Edinburgh, Churchill Livingstone

PROJECT TECHNOLOGY HANDBOOK No 13 (1974), *Basic Electrical and Electronic Construction Methods*, London, Heinemann Educational

SCROGGIE M. G. (1971), *Radio and Electronic Laboratory Handbook*, London, Iliffe (distributed by Newnes-Butterworth)

WARE G. C. (1967), *Basic Electronics for Biologists*, Edinburgh, Churchill Livingstone

9. Laboratory Workshop Materials

GLASS

The technology of glassworking could well be the subject of a book on its own. Very few schools or colleges will employ a technician who is skilled in glassworking, and it is doubtful whether there is any economic sense in a semi-skilled technician attempting any but the simplest operations. The cost in time and wasted material will almost certainly outweigh the cost of buying the finished article. Laboratory suppliers list 'T' joints, 'Y' joints, and 'U's at only a few pence each and a stock of these will save much valuable time. For the construction of more elaborate glass apparatus, a stock of ready-made components with ground-glass joints will be found invaluable (an illustrated catalogue is available from Quickfit & Quartz).

The two main types of glass which will be needed are 'soda glass' and 'Pyrex'. Soda glass softens at a lower temperature when being worked and requires very careful annealing. 'Pyrex' glass is more expensive, requires longer and stronger heating to bring it to a workable state, and is very resistant to breakage from thermal shock. It is a mistake to hold large stocks of glass, as glass becomes brittle with age; this makes clean cutting difficult – it is almost impossible to cut intricate shapes from old window panes.

Cutting

For *sheetglass*, a wheel cutter or diamond cutter is used to score a line of weakness along which the glass should break when suitable pressure is applied. Wheel cutters have a limited life; the wheels should be changed as soon as it becomes difficult to make a continuous scratch without going over the line a second time. When cutting across a sheet, make sure that the line goes right to the edge – scoring in from the edge, rather than out to the edge will reduce the likelihood of chipped corners. If shapes are to be cut, tangential scores should be made in the waste from the shape to the edge so that the waste can fracture away in clean pieces. For long, straight cuts the pane can be placed on the bench with a metre rule edge acting as a fulcrum when pressure is applied.

Tubing is best cut by scoring with a glasscutting knife; roll the tube on the bench while scoring around it. Tubing cutters and triangular files are often sold to schools, but these are a waste of money. With small diameter tubing it is not necessary to score all the way round – a nick on one side will suffice. The scored tube can then be broken by applying pressure behind the nick *on the opposite side of the tubing* with the thumbs, at the same time exerting an outward pull (see Figure 9.1).

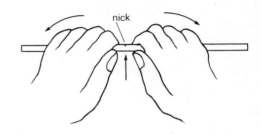

Figure 9.1 Cutting glass tube

If long lengths are to be cut, make sure that the free ends will not knock over apparatus or hit someone. When a short piece is to be cut from a long length it is safest to tuck the long end under the armpit for greater control as the two pieces separate.

Glass tubing of over one centimetre in diameter is not easy to cut by scoring and pulling. Large diameter glass tube (including bottles and jam jars) is best cut by scoring all around with the glass knife and then applying a hot wire to the score-mark with the homemade device illustrated in Figure 9.2.

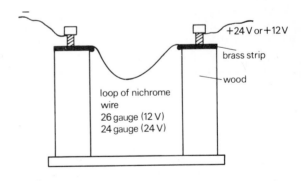

Figure 9.2 Hot-wire device for cutting glass tube

Bending

If glass rod or tubing is to be bent successfully, it is most important that it is heated evenly in a flame. A spreader or fishtail burner should be fitted to the bunsen so that a length of tube can be uniformly heated. The tube is held in both hands and rolled between forefingers and thumbs – *it is essential that the tube moves through just over 180° on each rolling* to ensure that the glass is heated evenly.

When the tube has softened to a workable state, remove it from the flame and bend to the desired angle. If the bend is to be greater than a right angle there will be a tendency for the tubing to flatten at the vertex of the bend – the secret of preventing this is to roll the bend along the heated part of the tube several times before the glass cools. Practice this by making a 'U' tube, allowing the 'U' to hang down vertically and rolling the bend by raising and lowering each arm alternately (Figure 9.3).

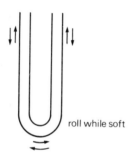

roll while soft

Figure 9.3   Forming a U-tube

Other simple glassworking techniques

Cutting large diameter tube

This is carried out by scoring a line all the way around the tube and applying heat accurately with a hot glass rod; this should be sufficient to initiate a local crack, which will run around the scored line. Another method is to place the scored line on a piece of red hot nichrome wire held in a support as shown in Figure 9.2 (use 26 or 24 gauge wire with 12 or 14 volts supply respectively).

Drawing out tube

Rotate the glass tube in a flame spreader until the area to be drawn is uniformly soft. Remove from the flame and, still rotating the tubing, draw the ends apart – failure to rotate at this stage will result in the drawn-out portion of the tube being asymmetrical.

Sealing tubes

Draw the tubing out very slowly until the waist is of small diameter, and then reheat the narrow portion

very strongly whilst rotating the tube in the fingers. This will separate the two portions of tubing and seal the ends at the same time. With a little practice and skill it is possible to remove any thick lump by blowing gently into the tubing whilst it is still soft. If the tube shows any tendency to bend at the sealed end, it should be softened, and then rotated slowly in a vertical position away from the flame until it hardens.

Making bulbs in tubing

If the bulb is to be in the middle of a length of tubing, the tube must be either plugged or sealed (the seal being cut off later). If the bulb is to be at the end of the tube, seal the tube as above and soften the end two centimetres; remove the tubing from the flame and blow the bulb with a quick but gentle puff. If a larger bulb is required, heat the next two centimetres of tubing and puff it out in the same way then, still rotating the work, reheat the waist, blow gently with the tubing held vertically, and rotate until a spherical bulb is formed (Figure 9.4).

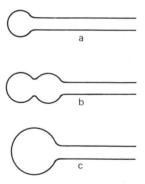

Figure 9.4   Forming a bulb in glass tubing

Making a hole in glass tubing

Use a blowpipe with the bunsen flame (or a needle flame from a propane brazing torch) to heat a small spot on the wall of a tube which has previously been plugged or sealed. When the glass is soft, blow firmly to produce a small, thin-walled bubble which can be broken off with a tool. Reheat the edges to smooth the hole.

Making a T joint

Proceed as for hole making, then seal or plug the other end of the tube. Heat the two pieces of glass to be joined until both are soft, rotating continuously, bring them together, and gently blow into the open tube to straighten and clear the joint. Anneal in a smokey flame.

Sealing wires into tubing

If nichrome or platinum wire is just placed in a length of capillary tubing and the end heated, the joint will prove to be very fragile. The correct method is to draw out a piece of tubing slightly and then to insert the wire with a small collar of tubing. This can be fused to form a bead, which joins the wire to the glass (Figure 9.5). Pyrex tubing should always be used.

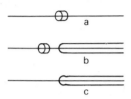

Figure 9.5 Sealing a wire into tubing

To make glass stirring rods

Cut suitable lengths of glass rod and heat one end strongly. Press this end vertically onto a flat metal surface or a tile to produce a head. Now heat about one centimetre of the other end until very soft, lay on tile, and press the end flat to form the stirring blade with an old file; anneal.

Annealing

The glass must be allowed to cool slowly to remove tensions set up by the processes of heating and bending. Ideally this is carried out in an annealing oven, but in practice with the limited range of work which will be undertaken in a school workshop, it will suffice to turn the air supply to the flame down until the flame is luminous and move the bend, and about four centimetres of tubing either side of the bend, through the flame until carbon begins to deposit on the tube. The component may then be set aside on a sheet of asbestos.

Adhesives

There are times when apparatus is being constructed from glass when the use of an epoxy-resin adhesive such as 'Araldite' may be considered. A joint made with this type of adhesive will be at least as strong as the glass. This technique is particularly useful for the construction of smoke cells and pond-life cells for microprojectors.

WOOD

This familiar material needs little comment beyond stressing its advantages and disadvantages as a construction material. It is easily obtained and worked and has a very good strength to weight ratio. The chief disadvantages for laboratory use are combustability and the tendency to shrink, expand or warp with slight changes in moisture content. Wood must always be finished with wax, varnish, button polish, or paint to seal the surface pores against dirt. The tendency to shrink and warp is most pronounced in boards which are cut from the tree tangentially, and least in boards which are radially or 'quarter' sawn (Figure 9.6).

Figure 9.6 Tangential and radial sawing

If solid wood is chosen for a particular construction an appropriate joint will have to be made. Where appearance is not the prime consideration, a dowel joint requires the least time and woodworking skill, and if much work is contemplated, a dowelling jig should be purchased or made. This simple device acts as a guide for the drill and enables accurate work to be carried out quickly by an unskilled person. The hardwood dowels can be purchased in 1–2 metre lengths from D.I.Y. shops and timber merchants. The joint is, of course, glued with a suitable modern adhesive.

Plywood

Plywood is usually available from timber merchants in a range of thicknesses from 3–18 mm. Sheets are normally 8 ft × 4 ft (244 cm × 122 cm) but 4 mm plywood is usually available in mahogany finish in 7 ft × 3 ft (213 cm × 91.4 cm) door panels. Small timber yards will usually cut to size if a whole panel would not be used. Offcuts in useful sizes for labwork are often available cheaply in D.I.Y. shops. Provided that a suitable thickness for the job is chosen, plywood is relatively free from shrinking and warping troubles, and is very strong. The chief disadvantages are that the end grain will not take screws or nails and that the material may delaminate if it gets wet. For this reason alone, plywood in the laboratory should always be varnished. If the plywood is likely to be in contact with water, British Standard 1188 marine ply should be specified as this will stand prolonged immersion in boiling water without de-lamination. If exceptional stress is likely there is a specially made plywood developed by *Thamesply* for sailing dinghy centre-plates.

Blockboard

This material is available in 8 ft × 4 ft (244 cm × 122 cm) panels, 12–24 mm thick with birch or decorative hardwood veneers, the latter being more expensive. Blockboard is manufactured by gluing square-sectioned strips of wood edge to edge, and then facing the reconstructed board with a thin three-ply on each side (occasionally a thick single veneer is used). The advantages are that a wide, easily worked board is obtained which has little or no tendency to warping because the strips of wood used in construction are laid so that alternate strips pull in opposite directions as they shrink or expand (Figure 9.7).

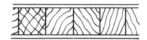

Figure 9.7   Blockboard

Chipboard

Chipboards or particle boards are made by coating small particles of wood with glue, and then compressing the mixture into sheets, which are usually 8 ft × 4 ft (244 cm × 122 cm) and 18 mm thick although 12 mm and 24 mm can sometimes be obtained. Plain chipboard is relatively cheap but the veneered versions are little cheaper than solid wood. The advantages are wide warp-free boards; the disadvantages are the difficulty of making strong joints without woodworking machinery and the low holding power of screws in this material.

If a strong joint is required, it is best to design the job with countersunk and clearance holes in the chipboard and to use a hardwood fillet to hold the screws (Figure 9.8). Glue all contact surfaces.

If a circular saw is available for accurate grooving, the joint shown in Figure 9.9 can be used for carcase work in plywood, blockboard, and chipboard. If the material is veneered, a space is left for a small fillet of matching hardwood to be glued into place. When the glue is set, this fillet is planed flush with the veneer surface.

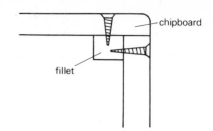

Figure 9.8   Using a fillet with chipboard

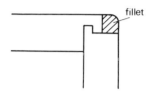

Figure 9.9

METALS

The metals used in the laboratory workshop fall into two main categories, *ferrous* and *non-ferrous* metals. Ferrous metals are alloys of iron, and the majority of them are prone to rusting. Although pure metals are sometimes used in the laboratory, the majority of metals which the technician will handle will be mixtures, or 'alloys'. The metallurgist can manipulate the differing physical and chemical properties of pure metals by alloying them to produce the most advantageous compromise for a particular job. For present purposes, it will be sufficient to consider what the different properties of metals are and then to look at the characteristics of the common alloys and pure metals which may have to be worked.

All metals, with the exception of mercury, could be said to possess the following mechanical properties, although the degree to which they exhibit these properties will vary very considerably.

Strength – the ability to resist compression, shearing, torsion, and tension, or a combination of these forces without breaking.

Toughness – the ability to resist a sudden application of force in the form of bending, twisting, or blows without fracturing. Toughness could be regarded as a combination of strength and ductility.

Ductility – the ability of a metal to withstand mechanical deformation without cracking. It is this property which enables copper, silver, and gold to be drawn out into fine wires.

Brittleness – the characteristic of being easily broken with little or no deformation at the point of fracture.

Elasticity – this is related to strength and is the ability of the metal, after deformation by force, to return to its former shape when the force is removed.

Hardness – the ability to resist scratching and penetration by 'hard' materials.

Creep – the tendency of some metals to yield progressively (i.e. as a function of time) under a steady load; this is most common at high temperatures, but some soft metals such as lead and zinc will creep at ambient temperatures. This is why lead roofs on churches have to be renewed occasionally – the sheet lead will thicken at the eaves and become thin in other places.

Fatigue – the tendency of some metals to fracture under repeated application of a cycle of stress, which may be well below their ultimate tensile stress.

Malleability – the ability of a metal to withstand hammering or rolling into thin sheets or bars without cracking. Gold is the most malleable metal and can be beaten into sheets of gold leaf only 0.000 000 25 in. thick. Malleability is not related to strength.

Plasticity – the ability of a metal to be shaped and take a permanent form without rupture. This characteristic is important in forging metals. Steels become plastic at bright red heat – if they are hammered to shape when they have cooled below bright red heat the metal tends to 'tear'.

All metals are good conductors of heat and electricity; some possess a very high degree of conductivity – a property which is utilized in a number of laboratory applications.

Ferrous metals

After iron has been extracted from its ores in a blast furnace, it is cast into blocks or 'pigs'. This pig iron may then be used for casting with little further purification or, after some of its carbon content has been removed, it may be used as wrought iron. By careful adjustment of the carbon content, and the addition of the correct proportions of alloying material, steels with a wide range of properties may be produced.

Cast iron

So named because it is cast to the required shape in a special mould, cast iron has the following properties:

brittle – easily fractured by sudden blows, very little elasticity in tension or torsion;

very strong under compression;

cannot be forged;

fairly easily machined and cheap to produce.

The fracture surface of cast iron has a grey or white colour and a sandy, granular appearance. It is used for producing intricate shapes and for the beds of machine tools.

Malleable iron castings

If cast iron components are packed in a furnace with an oxidizing material and heated to a red heat for several days, the carbon content of the cast iron is reduced. Malleable iron castings are tougher, more easily machined, and more resistant to shock than ordinary cast iron.

Wrought iron

This is purified pig iron. The carbon content is reduced, but a certain amount of fibrous slag remains evenly distributed throughout the metal, giving it a characteristic fibrous structure with the following properties:

does not fracture easily and can be readily bent, either hot or cold;

blue–black in colour, resistant to atmospheric corrosion;

very malleable and ductile;

resistant to fatigue and shock, therefore used for chains and crane hooks; iron rivets are made from wrought iron.

Mild steel

Mild steel is the most commonly used ferrous metal today (it has superseded wrought iron for most workshop purposes). It contains about 0.25 per cent carbon and a small amount of manganese. It is blue–grey in colour, and is easily worked by machine or hand tools. Mild steel is available in a wide range of sections, wire, rod, bar, tube, and sheet. It cannot be hardened by heat treatment, only by case-hardening.

Free cutting steel

Free cutting steel is a mild steel with a trace of lead added to improve its machining qualities.

Stainless steel

This alloy is very resistant to rusting. The most common type of stainless steel, 18/8 stainless steel, contains 18 per cent chromium and 8 per cent nickel with small amounts of copper, molybdenum, and titanium. 18/8 stainless steel cannot be hardened by heat treatment, but can be work-hardened. Special stainless steels which can be hardened and tempered in a similar manner to carbon steels are used in the production of stainless steel tools and knives.

Silicon steel

The addition of silicon and manganese to steel greatly increases its elasticity. Silicon steel is used for car springs.

Tool steels

Various alloying materials are used to improve the properties of steel for different purposes. The addition of *cobalt* or *tungsten* enables steels to retain their hardness at high temperatures; tungsten steels are in fact 'self-hardening'. 0.5 per cent *vanadium* added to chromium steel improves the grain structure, making the steel easier to forge and stamp as well as more resistant to shock.

Some common tool steels are:

Silver steel

The name refers to the appearance of the alloy; it does not contain any silver, but small proportions of silica, manganese, and chromium, and 1.25 per cent carbon. It is usually supplied in rods or bars 13 in. long to give convenient short lengths when cut to make small tools. The silver steel rods and bars are strong, have a bright ground finish, are accurately sized, can be hardened and tempered, and are used for tools and components.

Cast steel (carbon steel)

This steel has a high carbon content, is very brittle and very hard. It is usually annealed by heating to a cherry red and cooled slowly before use as punches, scribers, vice-jaws, files, and hammer heads.

High speed steel (HSS)

High speed steel contains tungsten and manganese, is generally a dark blue colour, is self hardening, and sometimes fractures if locally overheated when grinding. It is used for cutting tools and drills.

Although they are not steels, three other materials used as cutting tips on machine tools should be mentioned here.

Tungsten carbide

Particles of tungsten carbide in a matrix of cobalt is often used to tip lathe tools and drills. Cutting tools tipped with tungsten carbide will retain their cutting edge for very long periods, but when sharpening is necessary, this cannot be done on an ordinary grindstone – most good toolshops will undertake the work. Tungsten carbide tools are very good for machining abrasive materials such as iron and bronze.

Titanium tungsten carbide

Although less hard than tungsten carbide, this material is used to machine steel as it is resistant to the tendency of 'chips' to become welded to the cutting edge.

Both tungsten carbide and titanium tungsten carbide cutting tips are brazed onto carbon steel shanks.

Stellite

Stellite is an alloy of cobalt, chromium, and tungsten which has to be cast into shape as it cannot be forged. It is used in the rapid machining of very hard materials as it retains its hardness and cutting edge at red heat.

Nickel steel

When 36–50 per cent of nickel is alloyed with iron, metals are produced which expand and contract very little at room temperatures. Such alloys are used for measuring tapes, length standards, and pendulum rods.

Nickel chrome steel

Nickel chrome steel contains 4.4 per cent nickel, 1.2 per cent chromium, 0.5 per cent manganese, 0.3 per cent carbon and 0.2 per cent molybdenum and is used where very high tensile strength is needed (up to 100 tons in.$^{-2}$).

Non-ferrous metals

These are non-magnetic and non-rusting (although they may be subject to some other form of superficial corrosion). Copper, tin, lead, zinc, and aluminium will all be used at times in the laboratory workshop and, if electrical conductivity is critical, gold, silver, nickel, and platinum may be encountered on occasion. Sometimes when a highly conductive, nonarcing switch is needed, mercury is used as one of the contacts (the other contact dipping into the mercury in the 'make' position).

Non-ferrous metals are frequently alloyed to improve their working properties. The main characteristics of some non-ferrous metals and alloys are given below.

Copper

Copper:
  is malleable and ductile,
  workhardens when hammered or bent (may have to be annealed),
  can be drawn into fine wire,
  is a good conductor of heat and electricity (used for electrical work because it is the cheapest of the good conductors),
  can be joined by soft or hard soldering and can be welded commercially,
  is used where good conduction of electricity or the rapid conduction of heat is important.

Tin

A soft (expensive) metal with a shiny, silvery appearance, tin does not corrode and is used for tin plating steel sheet to produce a rust-free finish, and in low melting point alloys such as pewter and solders.

Zinc

Zinc:
  is a soft metal, easily worked into intricate shapes,
  is very resistant to corrosion in air and water,
  work-hardens when hammered,
  is used for 'galvanizing' steel and alloying with other metals.

Lead

Lead:
  is very heavy, with a low melting point (can easily be cast to a required shape),
  is malleable (very easily worked and shaped),
  is resistant to corrosion,
  exhibits little elasticity,
  is resistant to acids (used as a lining material for troughs in which winchesters are stored),
  is used as shielding against ionizing radiations, and in alloys (e.g. pewter, leadbased solders, and white metal bearings).

Aluminium

Aluminium:
  is the most abundant metallic element,
  is very light in weight,
  has good corrosion resistance due to the formation of a very tenacious oxide film which acts as a barrier to further attack,
  is a good conductor of heat and electricity,
  can be cast, wrought, spun, and extruded,
  can be welded, brazed with a special aluminium/silicon alloy, but cannot be soldered in the usual way,
  scrap is easily recovered and re-used (some schools melt down old motorcycle pistons for recasting),
Because of its strength/weight ratio and corrison resistance, aluminium is an important engineering material, but it is usually alloyed to improve its properties.

Brass

An alloy of copper and tin, resistant to corrosion and malleable, machines well, can be joined by soft or hard soldering and by brazing, and can be given a satin finish with fine steel wool or by sandblasting. It will take a high polish.

Bronze

An alloy of copper and tin, resistant to corrosion and with good wearing properties, bronze is used for bearings and for castings which can be machined.

Phosphor bronze

0.1–0.5 per cent phosphorus added to bronze alloys increases the strength and wearing qualities of bearings. Phosphor bronze is less ductile than ordinary bronze.

Lead bronze

Lead bronze contains up to 30 per cent lead which acts as a metallic lubricant in bearings if the oil film breaks down. It is used extensively in aero engines. Ordinary bronze may contain up to 1 per cent lead to improve its machinability.

Aluminium bronze

Contains 5–12 per cent aluminium, and is used as a substitute for steel when strength and anti-magnetic properties are required. It is used for boat fittings on account of its corrosion resistance.

Gunmetal

An alloy of copper, tin, and zinc, gunmetal is tough and strong and is a very good material for casting; the name derives from the fact that it was originally used for casting cannon. It is very resistant to corrosion, and its modern uses include marine fittings, gears, and housings for bearings.

White metal (Babbit's metal)

An alloy of 88 per cent tin, 4 per cent copper, and 8 per cent antimony which is used for bearings.

Duralumin

An alloy containing aluminium, copper, magnesium, and manganese which is stronger than aluminium but less resistant to corrosion, duralumin has the unusual quality of age-hardening (after heating and quenching in water, it hardens and gains strength for several days).

PLASTICS

These are man-made materials which at some stage in their processing are in a 'plastic' state which makes them eminently suitable for moulding. Some of these materials regain their plasticity whenever they are heated to a critical temperature and are called *thermoplastics*; another large group of plastics set into an irreversible state and are known as *thermosetting* plastics.

Although the earliest plastics were regarded as cheap substitutes for other matrials, their modern counterparts have a highly sophisticated technology of their own and are used on an enormous scale in manufacturing and engineering industries. In a work of this type it is only possible to give outline information on the handling of plastics, but their special and often unique properties should always be considered whenever materials are being selected for laboratory use.

Thermoplastics

Cellulose acetate

The first plastic to be injection moulded (1930), but now almost completely replaced by high impact polystyrene except in the form of clear acetate sheet, cellulose acetate has low strength for functional moulding but has poor dimensional stability when moulded. Ethylcellulose and cellulose acetate propionate are more recent developments, and are often used for blister packaging. All cellulose-based materials can be repaired or joined to wood with cellulose cement (Balsa cement).

Polystyrene

Polystyrene has high dimensional stability and is used very extensively for mouldings. 'Straight' polystyrene is hard, brittle, and transparent but its optical qualities are not as good as those of acrylics; addition of rubber reduces brittleness to produce high impact polystyrene, also called toughened polystyrene (TPS). Polystyrene is very easily jointed with cements made by dissolving polystyrene in ethylene dichloride or carbon tetrachloride, and these solvents can also be used for 'solvent welding' polystyrene. The material is softened and 'surface crazed' by a number of organic materials and is easily coloured by additives.

Expanded polystyrene

Used considerably for packaging and ceiling tiles. Density as low as 16 kg m^{-3} (1 lb ft^{-3}): expanded polystyrene is available in sheet, block, or as pellets which can be expanded in a mould. It is an excellent material for low temperature thermal insulation.

Expanded polystyrene tends to crumble if cut with a knife or a saw. With care and very light pressure, sheet may be cut with a razor blade or fretsaw. Accurate cutting is best carried out with a hot nichrome wire connected to a low voltage supply; the temperature should be just above that needed to melt the plastic. Polystyrene cement can be used as an adhesive, but only in the minutest quantities as the foam readily dissolves in the solvent vapour. PVA glue is the most suitable adhesive.

Expanded polystyrene is invaluable in the workshop for casting in aluminium and zinc. Patterns can be quickly constructed from tiles and thick packaging, embedded in moulding sand in an ordinary wooden or metal box, and poured. The molten metal vaporises the polystyrene and replaces it in the mould. *Note* that this type of casting must always be carried out with good ventilation as poisonous styrene fumes are released.

Acrylic plastics
('Perspex' in UK, 'Plexiglass' in U.S.A.)

These have a glasslike clarity which is not affected by ageing or discoloured by ultra violet light; they can be used for moulded optical systems and lenses, but are easily scratched. Acrylics are available in sheet, rod, and block form, clear or opaque in a wide range of colours; offcuts may be available cheaply or even free to schools from shop sign manufacturers.

They can be jointed by use of acrylic cements which are of two basic types – acrylic material in a solvent such as chloroform, or special solvents which bond Perspex to give very clean joints for museum jars, but which need UV light for hardening. A home-made cement can be made by placing Perspex scraps, cut into small pieces, in a screw-top jar and covering with chloroform; leave for one week to dissolve. Avoid inhalation and working near naked lights. Acrylics can be glued to other materials with epoxy resins.

If mechanical fastening are needed, use cheese-headed or cap screws. Avoid the use of self-tapping and countersink head screws as these usually cause stress cracking. Perspex can be tapped to take machine screws.

Perspex can be moulded to simple shapes by the application of heat. Warm the sheet in an oven controlled to 160–170 °C for about 30 minutes for a 6 mm sheet, or 20 minutes for a 3 mm sheet. The Perspex must be supported on a flat surface until it is softened. On removal from the oven the Perspex is held over a metal or wooden mould and allowed to

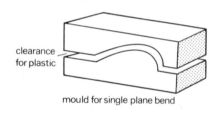

mould for single plane bend

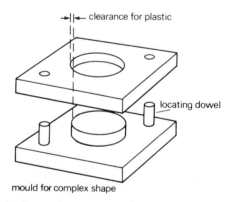

mould for complex shape

Figure 9.10 Perspex moulds

cool (Figure 9.10). Simple bends in a sheet or rod may be made by using a strip heater. This is made by mounting a glass-carcased electric heating element 2.5 cm below a 5 mm gap in a box made of 'Sindanyo' (TAC Construction Co Ltd, Ashburton Rd, Trafford Park, Manchester M17 1RU). Do not use the cement sheet used by builders as it is unsuitable. Lay the part of the sheet to be bent over the gap until it has softened, turning the sheet several times. Remove from heat and bend to the required angle, preferably with the aid of a simple jig. *Note* that when cutting asbestos always keep the sawcut wet to control dust which can be injurious to health.

Nylon

'Nylon' is a trade name for a group of polyamide plastics of which about five are frequently used in production, the most common being Nylon '6' and Nylon '6.6'. These are both general-purpose Nylons, 6.6 being the more rigid.

Nylon mouldings sometimes warp after manufacture because of internal stresses. A warped moulding can often be successfully treated by soaking in water at 70 °C for a period, and then allowing to cool slowly.

Nylon has sufficient strength and wear resistance to be used as an engineering material, and a recent development has been the inclusion of glass and carbon fibres to produce an even tougher material for gears and bearings. When considering the use of Nylon as an engineering material it is important to remember that it has a fairly high thermal expansion coefficient and that it can absorb water, giving rise to some dimensional instability.

Nylon can be sawn, drilled, filed, and turned in a lathe quite easily. When machining, use a high speed with a slow feed with plenty of coolant to reduce heat.

Nylon is used extensively for bearings. When designing bearings in Nylon, generous clearances are needed as the bearings would otherwise tend to bind because of thermal expansion. The recommended clearance is 0.006 inches per inch of bore diameter (0.01 mm per 1 cm), with a minimum clearance on shafts of small diameter of 0.002 in. (0.05 mm).

Note Nylon bearings should not be lubricated as this will only very slightly affect the friction, but will considerably reduce the bearing capacity. As far as possible bearing length should be equal to the bore – overlength bearings give no working advantage other than axial support and only increase the likelihood of binding occurring.

Suitable adhesives are epoxy resin, phenolic rubber, rubber base, and resourcinol adhesives.

Polyvinyl chloride (PVC)

Most commonly met in the laboratory as an electrical insulator, particularly on wires, PVC is flexible, but may harden with age if left in contact with other plastics which tend to draw out the plasticiser. PVC can also be made in a rigid form by polymerisation.

Flexible film and rigid film are available in thicknesses from 0.002 in. to 0.02 in. Thicker rigid sheets can also be obtained.

PVC has very good chemical resistance, but is softened or dissolved by chlorinated solvents such as ethylene dichloride, carbon tetrachloride, and trichlorethylene. It is joined by heat welding or by PVC cement (PVC dissolved in a chlorinated solvent). Some synthetic rubber solutions which have a chlorinated solvent may be used, especially when PVC has to be joined to other materials.

Polythene

There are two types: *low density polythene* which is used for polythene film and many moulded containers such as buckets and bowls, and *high density polythene* which is used for stiffer and more rigid containers.

Polythene is available in a wide range of coloured, clear, and opaque films and in powder form which can be used, with suitable equipment, for rotational and slush moulding. Polythene powder is often used in school craftrooms for coating metal by the fluid bed technique.

Polythene hardens and deteriorates on exposure to sunlight or UV light. Although polythene is chemically very inert and can be used for storage bottles for most chemicals, it is hardened and discoloured by mineral oils. The deterioration is limited to hardening and discolouration, the plastic otherwise remaining sound.

Polythene cannot be cemented, as no suitable solvent is available; it is best joined by heat welding and can be attached to other materials with rubber based adhesives.

Polypropylene

Introduced commercially in 1964 as an engineering plastic almost as hard as Nylon, although not as strong, polypropylene exhibits very low water absorption, hence it is more dimensionally stable than Nylon, and is used as an alternative to Nylon in wet applications.

Its chemical resistance is generally good, but it is affected by mineral oils and fuels and is very prone to building up static charges.

It will withstand continuous temperatures around 100°C – it is used in the laboratory for animal cages which can be autoclaved.

Polystyrene cannot be cemented, but with care cracks may be sealed in polypropylene equipment by fusing the plastic with a hot wire.

Polycarbonate

Polycarbonate is a high strength thermoplastic which is very tough, has low water absorption and high dimensional stability. It is used as an alternative to Nylon, but it is not as strong and its abrasion resistance is much less. Its strength can be improved by the addition of glass fibre and carbon fibres.

Polycarbonate is used for gears and tool handles which require high impact strength. Commercially it is usually injection moulded for components such as gears to avoid the high stresses set up by machining. For one-off jobs it can be machined as for Nylon, the internal stresses being relieved after machining by lengthy soaking in an oilbath at 120–130 °C.

Polycarbonate can be used at a temperature of 115 °C continuously, and can be autoclaved; distortion begins at 135 °C. It can be cemented with a cement made by dissolving polycarbonate in methylene chloride, and can be glued to other materials using rubber based adhesives and epoxy resin.

Polyacetal

One of the most recent plastics, polyacetal is similar in strength to Nylon but has better dimensional stability and low water absorption. Because of its toughness, strength, and fatigue resistance, it is better than Nylon or metals for gears provided that running temperatures do not exceed 100 °C. (Strength to weight ratio and fatigue resistance are better than metals.)

Polyacetal should not be used in stress situations above 100°C, or 80°C if immersed in water. It is inert to most chemicals, therefore cementing is impossible.

Polyacetal components are usually fixed with metal screws, bolts, self tapping screws, and rivets. It can be bonded to other materials with epoxy resin, provided the polyacetal surface is first acid-etched. It can only be painted if the surface is first treated with a special etching primer.

Acrylobutyl styrene (ABS) and Styrene acrynitrile polymer (SAN)

These are new plastics not yet used extensively in UK, but they may be regarded as very superior alternatives to polystyrene, having far greater impact strength and other improved properties.

Cement joints can be made with a cement prepared by dissolving the material in methyl ethyl ketone or ethyl isobutyl ketone. Solvent welded joints can be made if pressure is maintained for 48 hours, but solvent welded joints are not as strong as cemented joints.

They may be joined to other materials using epoxy resins. Some rubber-based adhesives may be used, provided the solvent is not a chlorinated hydrocarbon or a ketone.

Silicon rubbers

These are flexible and remeltable plastics which can be used for making re-usable moulds which will stand temperatures up to 200 °C without loss of properties. They can also be used once only for casting zinc alloys of low melting point. Silicone rubbers are available in different degrees of flexibility, and can be intermixed to give intermediate properties.

If a pattern is available, a mould can be formed on it by dip-coating in molten silicone rubber in a beaker or saucepan. When cold, the mould can be pealed off and used for casting plastic, plaster, wax, or low melting point zinc alloy.

'Vinamold' is the trade name of a silicone rubber available in the UK and sold by toy shops.

Thermosetting plastics

Thermosetting plastics tend to be harder, stronger, and much more rigid than thermoplastic materials. They are resins which undergo a permanent change of state when 'cured'. The curing process may be the application of heat or the addition of a chemical hardener. If a chemical hardener is used, the resins are described as 'cold setting' (often misleading as the resin may become very hot as a result of the chemical setting process).

Thermosetting resins tend to be strong but brittle and their properties are often improved by the addition of reinforcing fibres.

Although a wide range of resins are used industrially, there are only two types of resin which are likely to be used in the school workshop or laboratory – the remainder require heating equipment for curing.

Polyester resins

Unsaturated polyester resins are suitable for cold curing by the addition of a peroxide catalyst. Cold curing polyester resins are highly exothermic – temperatures may rise to 150 °C during the setting reaction. The heat of reaction can lead to cracking of the plastic, but this problem can be controlled by good design, i.e. a fairly high surface area to volume ratio, and by the addition of a suitable quantity of inert filler to the resin.

When polyester resins are used for casting it is important to remember that they contract between 7 and 10 per cent on curing.

An excellent range of free technical leaflets is available to schools from Trylon Ltd, Wollaston, Northants. These include leaflets on embedding in clear resin, casting, and laying up glass fibre in plastic resin.

Always use special barrier cream when handling resins and avoid skin contact as far as possible. A special cream is available to remove resin from the skin.

Polyester resin should never be used under damp conditions as it may not set properly and delamination may result.

Polyester resins may be used to lock screws and nuts by coating the thread before tightening up. Gears which are a light push fit may be secured to their shafts with resin.

Polyesters are used for glass fibre work in most school workshops, and detailed advice should be available if a major moulding job is to be undertaken.

Epoxy resins(Epoxides)

Epoxy resins can be used in much the same way as polyester resins in laying up glass fibre reinforced plastics. They produce superior quality laminates if heat cured, but cost twice as much as polyesters. If cold cured, they are little better than polyesters.

Epoxy resins have superior electrical properties to polyester resins and are used in the manufacture of laminates for printed circuit boards. They do not produce volatile substances on curing, and shrinkage is minimal.

Epoxy resins such as 'Araldite' form one of the most useful types of adhesive as they will form a strong bond on many smooth materials such as glass, metal, and some plastics.

Phenolic laminates

Heat cured phenolic resins are available in sheet, rod, tube, and block form. They are often reinforced to give properties suitable for specific use, e.g. 'Paxolin' with paper reinforcement for electrical work, 'Tufnol' with cotton fabric reinforcement for engineering applications, and asbestos wool reinforcement where high service temperature is required. For machining purposes, treat them as metals.

Melamine laminates

Similar to but more costly than phenolic laminates. Main advantages are that the surface is harder and more scratch resistant, and that the electrical anti-tracking properties are better when used in a damp atmosphere.

Glass Reinforced Plastics (GRP) and Carbon Fibre Reinforced Plastics (CFRP)

Resins may be reinforced by the inclusion of glass fibre in the form of chopped strand mat where considerable strength is needed, or by woven cloth where a thinner, strong laminate is required. Surface tissue is used to strengthen the final layer, and woven glass ribbon is used to seal joints and seams.

BIBLIOGRAPHY

ALMEIDA O. (1971), *Metalworking for Schools, Colleges and Home Craftsmen*, London, Mills and Boon

BARBOUR R. (1968), *Glassblowing for Laboratory Technicians*, Oxford, Pergamon

BRADLEY I. (1971), *The Amateur's Workshop*, Hemel Hempstead, Model and Allied Publications

BROUGH L. S. (1973), *Plastics*, London, Hutchinson Educational

CLARKE P. J. (1973), *Plastics for Schools*, London, Mills and Boon

PAIN F. (1957), *The Practical Woodturner*, London, Evans Brothers

SPAREY L. H. (1972), *The Amateur's Lathe*, Hemel Hempstead, Model and Allied Publications

Table 9.1 Some properties of plastics

Material	Specific gravity	Tensile strength (kN/m^2)	Maximum service temperature $(^\circ C)$
ABS	1.07	34–55	76
CAB	1.2	20–35	93–121
Cellulose acetate	–	–	60–82
Nylon '6'	1.14	68–82	93–121
Nylon '6.6'	1.14	68–82	135–148
Polyacetal	1.42	62–68	–
Polycarbonate	1.2	55–70	93
Polyethylene (low density)	0.92	6–17	49
Polyethylene (high density)	0.95	20–31	62
Polypropylene	0.90	34	93–121
Polystyrene	1.05	34–55	65–76
Polystyrene, high impact	1.06	24–46	52–74
Rigid PVC	1.4	48–55	65
Rigid PVC, high impact	1.35	34	52–74
Polyvinyl butyral	1.07–1.2	–	–
Polyether (chlorinated)	1.4	41	54–65
PTFE	2.2	10–20	260
PTCFE	2.1	31–37	193
Phenolic (paper base)	1.3–1.4	44–96	93–148
Phenolic (fabric base)	1.3–1.4	48–86	93–176
GRP (polyester)	1.6–2.0	172–380	121–204
GRP (epoxy)	1.6–2.0	275–586	121–204

10. Workshop Processes

JOINING MATERIALS

Soft soldering

Soft soldering is a method of joining metals by using a low melting point alloy which will form a bond between them. It is a fairly simple job if the following points are understood:

The surfaces to be joined must fit well together.
The strongest joint is obtained by capillary action.
The surfaces must be really clean – even grease from fingers can spoil the work.
The correct flux must be used to prevent the heated surfaces from oxidizing, and to assist the flow of molten solder.
The work and the soldering iron must be at the correct temperature. If the work is too cool the solder will 'ball' on the surface without bonding to the work.
The correct solder should be used.

Soldering irons

For the work which is likely to be encountered in the normal laboratory workshop, two electrical soldering irons will be needed. A small iron with a fine point will be used for electrical work and a larger iron with a heavy copper bit for sheet metal work (the iron must be large enough to heat the metal surrounding the joint to the melting point of the solder). Irons are available with interchangeable bits.

Fluxes

These may be *active* or *safe*. Active fluxes are the easiest to use as they have an etching action on the surface of the metals to be joined; after soldering is complete, they must be washed off with hot water to remove any corrosive residue. For this reason *active fluxes must never be used on electrical work* – a special, low melting point solder, containing a core of safe resin flux, is available for electrical work (Multicore Savbit).

Some *active fluxes* in use are:
Paste fluxes containing zinc or ammonium chloride in petroleum jelly.
Ammonium chloride.
Zinc chloride (killed spirits); add scraps of zinc to hydrochloric acid until all bubbling ceases, filter, and add 1 part water to 2 parts of this solution.

Safe fluxes do not have an etching action and hence do not have to be washed off. Those in use are:
Resin
Oleic acid
Tallow
Proprietary fluxes; Bakers Soldering Fluid and paste fluxes are available from every ironmonger.
Solder paint consists of small particles of solder distributed in its own flux.

Most fluxes are to some degree poisonous and should be applied with a brush, stick, or piece of card.

Soft solders

Solders are alloys which melt at relatively low temperatures. The basic ingredients are tin (melting point 232 °C) and lead (melting point 327 °C). By varying the proportions of metals in the alloy, solders with widely varying characteristics can be produced; some of the basic possibilities may be understood by reference to the phase diagram in Figure 10.1.

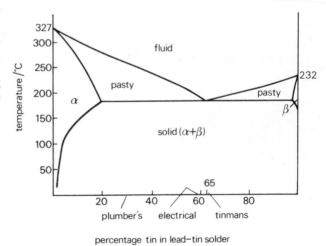

Figure 10.1 Phase diagram for lead–tin solder

As solder cools it goes through a 'pasty' stage when it is no longer fluid but remains workable; this is an essential characteristic of plumber's solder, which is used for 'wiped' joints. Tinman's solder has a lower lead content and passes rapidly from liquid to solid. Electrical solder has a relatively high tin content and melts at approximately 183 °C.

157

Table 10.1 Fluxes and solders for soft soldering various metals

Metal to be soldered	Flux	Type of solder
Copper Brass Gilding metal Gun metal	Zinc chloride, paste flux, resin, solder paint	Tinman's solder, a solder containing 65% tin, 35% lead
Electrical work	Use Multicore electrical solder, which has a built-in resin flux	Electrical solder; usual composition is tin 60%, lead 39.5%, antimony 0.5%
Lead	Tallow	Plumber's solder; tin 30%, lead 70%
Zinc Galvanised iron	Zinc chloride or dilute hydrochloric acid	Tinman's solder, or any solder containing more than 40% tin
Tin plate	Any flux. Care must be taken not to destroy the thin coat of tin by overheating	Tinman's solder, or any solder containing 50–65% tin
Cast iron	Any active flux; tinning is difficult	50% tin, 50% lead
Pewter Britannia metal	Tallow, resin, or glycerine which has been acidified with a few drops of hydrochloric acid	Pewter wire, Tinman's solder (65% tin, 35% lead), or Bismuth low melting point solder (lead 25%, tin 25%, bismuth 50%)

Low melting point solder is available which melts just above the boiling point of water; this solder contains a high proportion of bismuth and is often called 'Bismuth solder' – it is used mainly for soldering pewter and Britannia metal.

The soft soldering process

1 File the copper bit of the soldering iron lightly to clean it.
2 Heat the soldering iron.
3 Dip it in the flux.
4 Dip the iron in solder; this 'tinning' coats the soldering iron with solder – it can be kept clean by dipping in flux occasionally.
5 Flux the work, using a brush, stick, or piece of card.
6 Place the hot iron on the work, and touch solder onto the iron.
7 When the surrounding metal becomes sufficiently hot for the solder to 'run', draw the iron slowly along the joint.
8 Take off any surplus solder from the end of the joint with the tip of the iron.

Hard soldering (Silver soldering)

The hard soldering process is used when a strong joint is required. Carefully executed hard solder joints are very neat and will withstand higher temperatures than soft solder joints; for these reasons they are often preferred.

Hard solders are available with a wide range of melting points; this is not for use with different metals, but to allow very complex soldering jobs to be carried out in stages. If one piece of metal is soldered to another using a silver solder melting at 800 °C, a third piece can be added by using a solder which melts at a lower temperature without risk of disturbing the joint between the first two pieces.

For most purposes, three grades of solder will be sufficient. Hard soldering alloys are named according to the ease with which they melt, but as nomenclature varies from one supplier to another, it is as well to check the melting point when purchasing. Table 10.2 gives the names and melting points of the more readily available alloys as well as the approximate composition of the three most useful silver solders.

Table 10.2 Hard solders

Name	Melting point	Composition
Easy-flo	630 °C	
Easy solder	700 °C	silver 50%, copper 40%, zinc 10%
Medium	750 °C	silver 20%, copper 40%, zinc 40%
Hard	800 °C	silver 10%, copper 50%, zinc 40%

Flux

Powdered borax is mixed with water to form a paste and then brushed onto the work. Proprietary liquid fluxes such as 'Auflux' are available and can be strongly recommended.

The hard soldering process

1 The joint must be very carefully fitted as the technique depends upon the silver solder being drawn into the joint by capillary action.
2 Apply flux with a brush to the joint after fixing the components in place with bindings of iron wire. Keep the flux only on the joint as the solder will follow the flux.
3 Place small pieces of solder along the joint at intervals.
4 Heat the work with a blowlamp, blowtorch, or propane torch until the solder runs. (It is the work, not the solder, which must be heated.)
5 Allow the joint to cool and then clean it up with a fine file if necessary.

Hard soldering is most useful when joining copper, brass, gilding metal, or precious metals. For joining iron, steel, brass, or a combination of two of these metals, a very similar technique is used; this is called *brazing* because the joint is made by melting brass to join the metals.

Brazing

Brazing is a very useful method for joining metals in the laboratory workshop – it provides a stronger joint than hard soldering and is an ideal way of joining ferrous metals. Brazing can be used to make large constructions, but in the interests of safety this is best carried out in the school or college metalwork shop.

It is not generally realised that almost all of the jobs done by welding in school workshops could equally well be brazed, provided that the fitting of the parts is accurate.

Flux

Use powder or paste borax.

Spelter

The material used as a solder in brazing is known as 'spelter'; it is available in the form of rods and wires, or as a granular material. Spelter consists of brass and is made with varying compositions so that different melting points may be employed in planning

Table 10.3 Grades of spelter

Grade	Melting point	Composition
Soft	870 °C	50% copper, 50% zinc
Medium	880 °C	54% copper, 46% zinc
Hard	900 °C	60% copper, 40% zinc

complex constructions. The most commonly used grades are:

The brazing method

1 Prepare a well-fitting, clean joint as for hard soldering.
2 Place the joint on the brazing hearth and pack around carefully with firebrick – this is essential as success depends on getting the joint really hot. The greatest problem with school equipment is heat loss.
3 Place borax powder over the joint and heat strongly with a blowlamp, blowtorch, or propane torch.
4 When the joint is red hot and the borax molten, touch the brazing rod against the joint. *It is the heat from the workpiece which must melt the brass.* Once the joint has been heated, the torch should be kept on it until the job is complete – if the flame is removed excessive oxidation may prevent a good joint from being made.
5 If granular spelter is being used, it is placed on the heated joint with a spatula.
6 Brazing is complete when the joint 'flashes', i.e. the spelter ceases to be a bead on the surface and runs into the joint.
7 Cool and clean with a file.

In the laboratory workshop brazing and hard soldering should be carried out on an asbestos bench. (*Do not use thin sheet asbestos cement as this can shatter dangerously if heated locally.*) Firebricks, which can be purchased from builders' merchants, should be used as floor, back, and sides for a temporary hearth. After the job is completed the firebricks must be left on the asbestos bench until completely cool. (A clear notice should warn of the possible presence of hot objects.)

For laboratory soldering and brazing, a Sievert torch with interchangeable jets operating from a Calor gas (propane) cylinder is excellent (see Figure 10.2). A brazing torch is normally available in school metalwork shops.

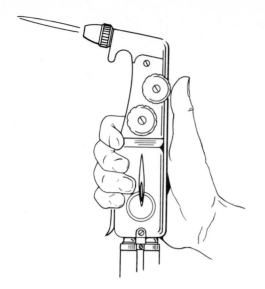

Figure 10.2   A brazing torch

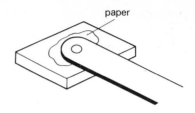

Figure 10.3   Riveting joints

Riveting

This is a simple way of joining metal components together. When joining sheet metal, it is usual to drill all the holes required in one sheet, but only one hole in the second sheet. The first rivet is then set, the remaining holes are drilled, and then the rest of the rivets set.

Rivets can be used to make pivot or hinge joints with a piece of paper between the components (Figure 10.3). When the rivet is set, the paper is burnt out.

For good riveting, it is essential to use rivets of the right size and correct material for the job in hand:

Soft iron for general engineering.

Soft aluminium for aluminium and duralumin.

Soft aluminium or copper for one-metalic materials.

Length – the rivet should project beyond the plate to be riveted by a distance equal to the diameter of the rivet.

Diameter of a rivet should be $1.2 \times$ the thickness of the thinnest plate to be riveted.

Spacing – the centre of the rivet hole to the edge of the plate should equal twice the diameter of the rivet. Rivets should be three times the diameter of a rivet apart.

Holes should be just large enough for the rivet to pass through.

Some common rivets are shown in Figure 10.4.

When riveting wood, it is necessary to use a close-fitting washer beneath both the head and tail of the rivet.

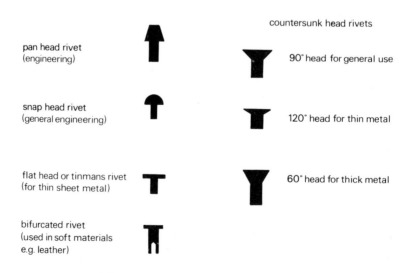

Figure 10.4   Some common types of rivets

The tail of a rivet may be spread by striking with a hammer and then working around the edge with the ball of the hammer. To obtain a neat finish, use a rivet set and rivet snap; the set presses the plates together and the snap domes the tail of the rivet neatly (Figure 10.5). Sets and snaps may be purchased or made by drilling a four inch (100 mm) length of silver steel in the lathe.

Pop riveting

Where the rivet is needed for light fastening only and no great shearing forces are involved, hollow pop rivets are ideal. They are probably the most useful type of rivets in the laboratory workshop.

Pop rivets are tubular rivets on a steel pin. The end of the pin at the tail of the rivet has a cone-shaped head just above a constriction. The rivet is placed in the hole drilled in the material to be joined and the pin is gripped in special pop rivet pliers. Then the pliers are squeezed, the pin is drawn towards the head of the rivet and the cone-shaped pin head spreads the tail. When this has been fully spread, a further squeeze of the pliers snaps the pin at the constriction, allowing the pin head to fall away and the pin shaft to be withdrawn (Figure 10.6).

The method is rapid, requires no skill, and is ideal for chassis and instrument box construction. One great advantage over other forms of riveting is that *all the work can be done from one side*, thus making it possible to rivet to tubing.

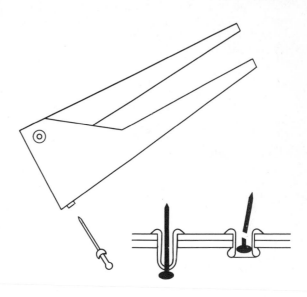

Figure 10.6   The pop rivet

Figure 10.5   Use of a rivet set and snap

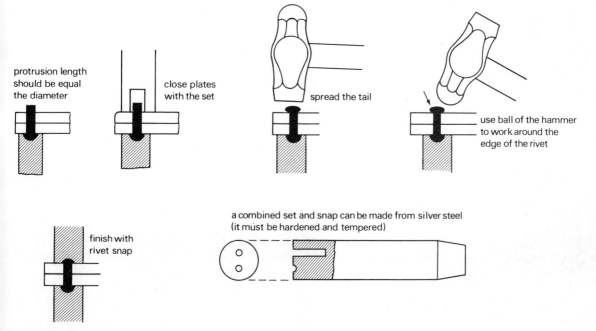

protrusion length should be equal the diameter

close plates with the set

spread the tail

use ball of the hammer to work around the edge of the rivet

finish with rivet snap

a combined set and snap can be made from silver steel (it must be hardened and tempered)

Screws

This term is used to include all threaded fastenings – bolts, set screws, studs, grub screws, thumb screws, Phillips screws (or the more recent Posidrive screws), cap screws, carriage bolts, coach screws, and wood screws. Woodscrews and coach screws have tapering threads; all the other types have parallel screw threads and are known as machine screws. The following notes and Figure 10.7 will enable most types to be identified.

Bolts

Used as a temporary fastening which can be removed, they can withstand tensile stress (pulls). The length of a bolt is measured from the end of the shank to the underside of the head. The thread does not occupy all of the shank. Always used with a nut, and usually with a washer. The head is usually hexagonal or square.

Set screws

Used as a temporary fastening, similar to a bolt, but the shank is threaded right up to the head. The head may be hexagonal, square, or any of the types used on machine screws.

Machine screws

These have parallel threads and various heads, which may be slotted for a screwdriver, Phillips headed for a Phillips screwdriver, or socketed for a hexagonal socket key.

A new type of tamper proof head has come onto the market and should be valuable in school laboratories. It has two shoulders against which the screwdriver acts when tightening the screw, and two slopes which give no purchase to a screwdriver when an attempt is made to loosen the screw. If, however, the screw has to be removed, this can only be done by drilling a pilot hole and inserting a screw remover.

Stud

Stud is threaded either at both ends or continuously (Figure 10.8); one end is screwed into a metal casting, the other end takes a nut. When placing a stud in position, use a nut and locknut with the spanner on the locknut; when removing, place the spanner on the lower nut. Studding rods (12 in. (30.5 cm) long in a range of diameters and threads) are available from good tool shops, or from K. R. Whistons, in steel or brass.

It is most important when ordering the above screws that a full description is given – diameter, type of thread, length, type of head, and material, e.g. $\frac{3}{8}$ in Whitworth, 2 in. countersunk Phillips head, brass.

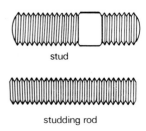

stud

studding rod

Figure 10.7   Types of screws

Figure 10.8

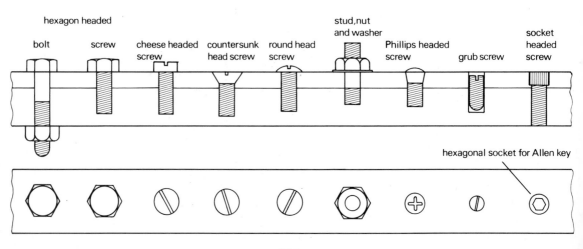

Nuts

These are usually square or hexagonal, but occasionally round nuts with a milled or knurled outer edge are used when hand tightening is desired, or, if nuts have to be removed frequently, wing nuts are used. Slotted nuts (castle nuts) are used where vibration is a problem, the tightened nut being secured through a hole in the bolt with a split pin (Figure 10.9).

If one nut is used to keep another in position (Figure 10.9), it is termed a locknut. Any nut can be used as a locknut, but special locknuts are made which are only half the thickness of a standard nut (standard nut thickness is equal to the diameter of the bolt). Self-locking nylon nuts, such as 'Nylock' nuts, are being used increasingly.

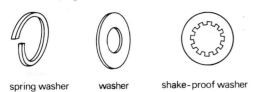

Figure 10.9

Washers

These fittings are available in steel and in brass. Steel washers may be bright, black, or plated. Plain and bevelled washers are available. Washers are usually ordered by hole size. Thickness varies; the thinner washers may be used in combination for accurate packing and spacing – they are sometimes called *shims*.

Spring steel washers are available in various patterns, and are used to prevent nuts from being vibrated free (Figure 10.10).

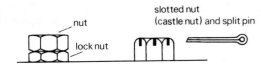

spring washer   washer   shake-proof washer

Figure 10.10

Carriage bolt and nut

A temporary fastening, made in a wide range of lengths and diameters, used to attach wood to metal. They have a domed head and square shank neck which is driven into the wood to prevent shank rotating as the nut is tightened (Figure 10.11). Carriage bolts can be used to join two pieces of wood if a washer is used below the nut.

Coach screw

Coach screws are used to fasten metal to thick wood. They are available in many sizes, with square or hexagonal heads, and a tapered thread of the wood screw type (Figure 10.11).

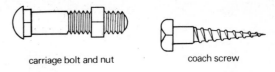

carriage bolt and nut   coach screw

Figure 10.11

Wood screws (Nettlefolds)

Fortunately, there is only one standard of wood screw in common use, although these are now made with a variety of heads – countersink, round, Posidrive, or Phillips headed (Figure 10.12). In addition to the head types, screws are made in steel, stainless steel, brass, and bronze; steel screws are often rendered rust-resistant by chromium or cadmium plating, galvanising, or japaning (painting).

It is very easy to snap brass screws before they are driven fully home, particularly in hardwoods – even when the correct clearance and counterbore holes have been made. It is good practice when using brass screws to drive home a steel screw of the same length and gauge first, then replace it with the permanent brass screw. Although this can be time consuming, it is advisable wherever there is a risk of the brass screw shearing off.

If wood screws are to provide a firm anchorage without splitting the wood, a clearance hole for the unthreaded shank of the screw must be drilled and a tapping hole or counterbore for the threaded portion is required. In the case of countersink headed screws it will also be necessary to use a countersink bit (or large twistbit). D.I.Y. shops sometimes stock combination bits which will countersink, clear, and counterbore for the popular sizes of woodscrew; a number 8 size, which is the most useful, will save a great deal of time.

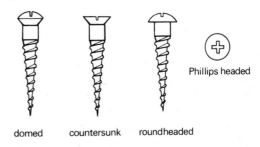

domed   countersunk   roundheaded

Figure 10.12   Woodscrews

Table 10.4 Woodscrews

Gauge	Counterdrill		Clearance drill	
	Imperial	Metric	Imperial	Metric
2	No. 56	1.20 mm	No. 44	2.20 mm
4	No. 54	1.40 mm	No. 35	2.80 mm
6	No. 50	1.80 mm	No. 28	3.60 mm
8	No. 44	2.20 mm	No. 17	4.40 mm
10	No. 38	2.60 mm	No. 9	5.00 mm
12	No. 31	3.00 mm	No. 1	5.80 mm
14	No. 30	3.20 mm	Letter F	6.40 mm

Woodscrew gauges run from No. 1 (0.066 in. shank diameter) to No. 32 (0.500 in. shank diameter). It is very unlikely, however, that the technician will meet any gauges other than 2, 4, 6, 8, 10, 12, and 14.

Screw thread systems

Different countries have, in the past, agreed on standardisation of thread forms but threads designed for use in one country cross the borders into other countries. The competant technician will therefore need to be aware of the existance of the most commonly found threads and to have some idea of the purposes to which they are suited.

The most common thread systems which the technician is likely to meet are the British Standard Whitworth (B.S.W.) and its derivatives the British Standard Fine (B.S.F.) and British Standard Pipe (B.S.P.); the British Association (B.A.); and the Unified Fine (U.F.) and the Unified Coarse (U.C.).

The position is further complicated in that the B.S.W. thread is due to be phased out and replaced by the I.S.O. metric thread, as the process of metrication proceeds.

Two other thread systems which are occasionally met in the laboratory workshop, but which will not be discussed in detail, are the British Standard Cycle (B.S.Cy.) and the British Standard Conduit (B.S.Con.).

The I.S.O. Metric Thread

Full details of the I.S.O. metric thread are published in BS 3643; the basic form of the thread is shown in Figure 10.13. Tapping drill sizes can be obtained by subtracting the pitch from the major diameter; for example, for the 6 mm size which has a pitch of 1.0 mm, the tapping drill will be 5 mm.

Table 10.5 Tapping and clearance sizes for I.S.O. Metric Coarse Thread (all sizes in mm)

Diameter	Pitch	Tapping drill	Clearance drill
2	0.4	1.6	2.2
2.5	0.45	2.05	2.6
3	0.5	2.5	3.2
4	0.7	3.3	4.2
5	0.8	4.2	5.2
6	1.0	5.0	6.2
8	1.25	6.75	8.2
10	1.5	8.5	10.2
11	1.5	9.5	11.2
12	1.75	10.25	12.2

Note The use of the 11 mm size may be preferable to the 12 mm size in school workshops as the pitch of the 12 mm size is difficult to cut using the leadscrew of many lathes.

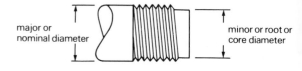

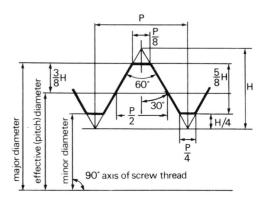

P = pitch
H = height of fundamental triangle = 0.86600 P
H/8 = 0.10825 P
H/4 = 0.21651 P
3H/8 = 0.32476 P
5H/8 = 0.54127 P

Figure 10.13 Basic form of the I.S.O. metric thread

The British Standard Whitworth (B.S.W.) Thread

Introduced in 1841 by Sir Joseph Whitworth, this has been the basic British engineering thread ever since

(Figure 10.14). The principal variations are the British Standard Fine (B.S.F.) and the British Standard Pipe (B.S.P.), both of which use the Whitworth form but have a finer pitch. The B.S.F. thread gives a screw a larger diameter core which has greater strength than the corresponding B.S.W. screw, but with a slight loss of thread strength. One advantage of the B.S.F. is that the nut is less likely to be loosened by vibration than that of the B.S.W. thread. The B.S.P. thread has a fine pitch, giving a shallow thread, which is essential so that it does not break through the wall of the pipe.

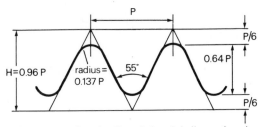

BSF has same form but finer pitch and shallower thread
BSP has same form but much finer pitch and thread depth

Figure 10.14 Basic form of the British Standard Whitworth thread

British Association Thread (B.A.)

This is often used on threads below $\frac{1}{4}$ in. diameter. The B.A. thread has a fine pitch (which is metric). B.A. threads are used for delicate work, and the B.A. series is very finely graded (see Table 10.8). The form of the thread is shown in Figure 10.15.

Table 10.6 Tapping and clearance sizes for British Standard Whitworth Thread

Screw diameter (inches)	Tapping drill Imperial	Metric	Clearance drill Imperial	Metric
$\frac{1}{8}$	No. 40	2.50 mm	$\frac{9}{64}$ in.	3.30 mm
$\frac{3}{16}$	No. 27	3.70 mm	$\frac{13}{64}$ in.	4.80 mm
$\frac{1}{4}$	No. 10	5.40 mm	$\frac{17}{64}$ in.	6.60 mm
$\frac{5}{16}$	Letter F	6.80 mm	$\frac{21}{64}$ in.	8.20 mm
$\frac{3}{8}$	Letter N	8.40 mm	$\frac{25}{64}$ in.	9.80 mm
$\frac{7}{16}$	$\frac{23}{64}$ in.	9.20 mm	$\frac{25}{64}$ in.	11.50 mm
$\frac{1}{2}$	$\frac{13}{32}$ in.	11.20 mm	$\frac{33}{64}$ in.	13.00 mm

Table 10.7 Tapping and clearance sizes for British Standard Fine Threads

Screw diameter (inches)	Tapping drill Imperial	Metric	Clearance drill Imperial	Metric
$\frac{1}{4}$	$\frac{7}{32}$ in.	5.60 mm	Letter F	6.50 mm
$\frac{5}{16}$	$\frac{17}{64}$ in.	6.60 mm	Letter P	8.20 mm
$\frac{3}{8}$	$\frac{21}{64}$ in.	8.20 mm	Letter W	9.80 mm
$\frac{17}{16}$	$\frac{3}{8}$ in.	9.80 mm	$\frac{29}{64}$ in.	11.50 mm
$\frac{1}{2}$	$\frac{27}{64}$ in.	10.80 mm	$\frac{33}{64}$ in.	13.00 mm
$\frac{9}{16}$	$\frac{31}{64}$ in.	12.20 mm	$\frac{37}{64}$ in.	14.00 mm
$\frac{3}{8}$	$\frac{35}{64}$ in.	13.20 mm	$\frac{41}{64}$ in.	15.70 mm

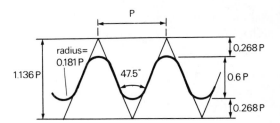

Figure 10.15 Basic form of the British Association thread

Table 10.8 Tapping and clearance sizes for British Association (B.A.) Threads

B.A. gauge no.	Diameter (inches)	Tapping drill Imperial	Metric	Clearing drill Imperial	Metric
0	0.236	No. 10	5.20 mm	1/4 in.	6.20 mm
1	0.209	No. 17	4.40 mm	7/32 in.	5.60 mm
2	0.185	No. 24	4.00 mm	3/16 in.	4.80 mm
3	0.161	No. 28	3.70 mm	11/64 in.	4.20 mm
4	0.142	No. 32	2.95 mm	5/32 in.	3.70 mm
5	0.126	No. 37	2.65 mm	9/64 in.	3.50 mm
6	0.110	No. 43	2.25 mm	1/8 in.	2.95 mm
7	0.098	No. 46	2.05 mm	7/64 in.	2.60 mm
9	0.075	No. 53	1.50 mm	No. 48	1.95 mm
10	0.067	No. 55	1.30 mm	No. 50	1.80 mm
11	0.059	No. 56	–	No. 52	1.60 mm

Square Thread

This is sometimes used when a strong thread is needed, capable of exerting pressure in either direction (Figure 10.16). It offers less friction than a vee thread and is found in vices, load-carrying presses, car jacks, and the cross slides of machine tools.

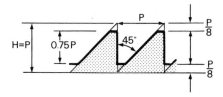

Figure 10.16 A square thread

Buttress Thread

The buttress thread (Figure 10.17) combines the strength of the vee thread with the easy transmission quality of the square thread. It is used when thrust is required in only one direction, e.g. some vices and presses.

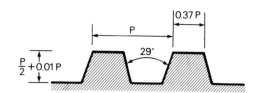

Figure 10.17 The Buttress thread

Acme Thread

The acme thread (Figure 10.18) is often used for the lead screws of lathes because it is easier to engage the half nuts on this very slightly vee'd thread than on a square thread. It is capable of transmitting motion in either direction.

The Unified Screw Thread System

The earliest thread to be used in American engineering was the 60° sharp V thread. This was later modified with a slight flat at the top and bottom of the thread and became known as the American National Form Thread. When the motor industry required an international standard a compromise was agreed between the A.N.F. and the B.S.W. threads and the new system became known as the Unified Screw Thread System (Figure 10.19). The Unified system is in fact completely interchangeable with the A.N.F. thread system.

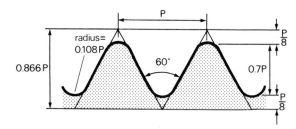

Figure 10.18 The Acme thread

Figure 10.19 Basic form of the Unified Screw Thread (U.N.C. and U.N.F.)

Table 10.9 Unified Screw Thread tapping and clearance drills (all distances in inches)

Major diameter (in.)	Unified National Coarse (U.N.C.)			Unified National Fine (U.N.F.)		
	Threads per inch	Tap drill (in.)	Clearance drill (in.)	Threads per inch	Tap drill (in.)	Clearance drill (in.)
$\frac{1}{4}$	20	No 7, $\frac{13}{64}$	$\frac{17}{64}$	28	No. 3, $\frac{7}{32}$	$\frac{17}{64}$
$\frac{5}{16}$	18	F, $\frac{17}{64}$	$\frac{21}{64}$	24	I, $\frac{9}{32}$	$\frac{17}{64}$
$\frac{3}{8}$	16	$\frac{5}{16}$	$\frac{25}{64}$	24	$\frac{21}{64}$	$\frac{25}{64}$
$\frac{7}{16}$	14	U, $\frac{23}{64}$	$\frac{29}{64}$	20	$\frac{25}{64}$	$\frac{29}{64}$
$\frac{1}{2}$	13	$\frac{27}{64}$	$\frac{33}{64}$	20	$\frac{29}{64}$	$\frac{33}{64}$
$\frac{5}{16}$	12	$\frac{31}{64}$	$\frac{37}{64}$	18	$\frac{1}{2}$	$\frac{37}{64}$
$\frac{5}{8}$	11	$\frac{17}{32}$	$\frac{41}{64}$	18	$\frac{9}{16}$	$\frac{41}{64}$
$\frac{3}{4}$	10	$\frac{21}{32}$	$\frac{45}{64}$	16	$\frac{11}{16}$	$\frac{49}{64}$
$\frac{7}{8}$	9	$\frac{45}{64}$	$\frac{57}{64}$	14	$\frac{13}{16}$	$\frac{57}{64}$
1	8	$\frac{7}{8}$	$1\frac{1}{32}$	14	$\frac{15}{16}$	$1\frac{1}{32}$

Table 10.10 Number of screw threads per inch (t.p.i.)

	B.A.		B.S.W.	B.S.F.
No. 10	72.6	$\frac{1}{8}$ in.	40	–
9	65.1	$\frac{3}{16}$ in.	24	32
8	59.1	$\frac{1}{4}$ in.	20	26
7	52.9	$\frac{5}{16}$ in.	18	22
6	47.9	$\frac{3}{8}$ in.	16	20
5	43.0	$\frac{7}{16}$ in.	14	18
4	38.5	$\frac{1}{2}$ in.	12	16
3	34.8	$\frac{9}{16}$ in.	12	16
2	31.4	$\frac{5}{8}$ in.	11	14
1	28.2	$\frac{3}{4}$ in.	10	12
0	25.4	1 in.	8	10

Wherever possible schools should use the B.S.F. and B.A. series in the sizes listed in Table 10.11 which approximate to, but are not interchangeable with, the I.S.O. metric series which will replace them. (Industrially, the changeover to metric has almost been completed, but the school technician will meet the various Imperial threads for many years to come.)

Table 10.11 Interim set of screw threads

	B.A.					B.S.F.				
Imperial	8	6	4	2	0	$\frac{1}{4}$	$\frac{5}{16}$	$\frac{3}{8}$	$\frac{7}{16}$	$\frac{1}{2}$
Metric	2	2.5	3	4	5	6	8	10	(11)	12

The 11 mm I.S.O. metric thread is included as it may be preferred to the 12 mm thread which cannot be cut on all English screwcutting lathes.

Table 10.12 Tapping and clearance sizes for the interim set

	Tapping drill		Clearance drill	
	Metric	Imperial	Metric	Imperial
8	1.8	No. 50	2.4	$\frac{3}{32}$ in.
6	2.4	$\frac{3}{32}$ in.	3.0	No. 31
4	3.0	No. 31	3.8	No. 35
2	4.0	No. 22	4.8	No. 12, $\frac{3}{16}$ in.
0	5.2	No. 6 or $\frac{13}{64}$ in.	6.2	D
$\frac{1}{4}$ in.	5.4	No. 3	6.6	G
$\frac{5}{16}$ in.	6.8	H	8.2	P
$\frac{3}{8}$ in.	8.4	$\frac{21}{64}$ in.	9.8	W
$\frac{7}{16}$ in.	9.8	$\frac{3}{8}$ in.	11.5	$\frac{29}{64}$ in.
$\frac{1}{2}$ in.	11.2	$\frac{7}{16}$ in.	13.0	$\frac{33}{64}$ in.

Adhesives and cements

From time to time a wide range of materials have to be joined together in the laboratory workshop. Bonded construction using *adhesives or cements* can often give a neater and stronger joint than mechanical fixing. *Glues or adhesives* may be of animal, vegetable, or mineral origin or they may belong to the modern generation of man-made resins which are so efficient that they have made glued joints the preferred method of construction in many industrial processes. *Cements* are usually made by dissolving some of the material to be joined in a suitable solvent; to all intents and purposes, the materials are welded together when the solvent has evaporated. In a glued joint however there is always a thin film of adhesive between the joined surfaces.

The technology of adhesives is very complex and some of the more difficult selection problems may best be discussed with a manufacturer's technical laboratory. It is possible here only to give general advice which will make the selection of suitable adhesives for the majority of applications fairly easy.

The factors which affect choice are:

1 Adhesion to both surfaces must be good.
2 Is the joint a rigid structural one, or is it flexible? Modern resins are very strong but often brittle; rubber-based adhesives are flexible but have a tendency to 'creep' if subjected to continuous stress from the same direction.
3 Will the joint need to be taken apart at any time in the future? If so, a water-soluble or a peelable adhesive may be the correct choice.
4 Will the joint be subjected to considerable heat?
5 Will the joint be subject to sudden mechanical shock or vibration. If so, 'straight' resins may be too brittle, but the slightly weaker resins which have been modified by the addition of rubber may be suitable.
6 The time available for setting. For example, the urea-formaldehyde resin sold as 'Aerolite' with its normal hardener sets in about $2\frac{1}{2}$ hours at 20 °C, but if the rapid hardener which can be purchased for this adhesive is used a set is obtained in about 3 minutes, which enables permanent repairs to wooden apparatus to be effected during the lesson in which the apparatus was damaged.

Although not strictly adhesive, organic solvents are used to bond many plastic materials – both surfaces of the joint are wetted with the solvent and then pressed firmly together. This method, which is called 'solvent welding', produces very strong, clean joints but it is as well to remember that the joint will not reach full strength until all the solvent has evaporated from it.

How adhesives work

Adhesives provide a bond between two surfaces and the strength of this bond will depend upon the strength of the adhesive film, and the adhesion between the film and the surfaces being joined. This implies that there is a film of adhesive present; if a joint fits too perfectly and is of non-porous material, *all* the adhesive may have been squeezed out of the joint and no bond will occur. With well-fitting joints, it is usually good practice to roughen the surfaces with a file or course sandpaper to give both 'key' and space for the glue film.

Adhesives stick by either *mechanical* or *specific* adhesion. In mechanical adhesion the glue forms an interlocking bond of a mechanical nature by penetrating the porous surface of the material. In specific adhesion there is a molecular attraction between the surface of the adhesive film and the substrate. With a suitable specific adhesive, smooth, non-porous surfaces can be bonded with a molecular force as strong and cohesive as that of a single substance.

When using mechanical adhesives on hard smooth materials, it is good practice to roughen the surface and 'double coat' the joint. Double coating gives the surface more time to absorb the adhesive and reduces the chance of a joint being squeezed dry under pressure.

With modern adhesives, the bond is often so strong that when a joint fails it breaks not at the glue line but in the solid material.

General classification of adhesives

Cements are basically a plastic dissolved in a suitable solvent. The majority of cements are made to bond the material from which they are derived, but a few such as 'balsa cement' are useful general-purpose adhesives for porous materials such as wood and paper.

Polystyrene cement is perhaps the most commonly used of the true cements. (It is unsuitable for use on expanded polystyrene as the vapour dissolves the foam. Special adhesives are now available for polystyrene foam from modelmakers' shops, or PVA adhesive can be used.)

Animal glues – albumen, isinglass, fish glue, gelatine, and casein; casein is the only one much used today and the type used is synthetic.

Vegetable glues – cellulose, dextrins, starches, gum resins, and flours; dextrins are still extensively used as 'office' glues for papers, but most of the others have been superceded by synthetics, although wallpaper adhesives such as 'Polycell' are useful for papier mâché work when making models.

Synthetic resins – may be *thermosetting* or *thermoplastic* types. The thermoplastic type, e.g. PVA, are dissolved in a solvent and adhesion occurs when the solvent evaporates. Thermosetting resins change from a plastic to a solid state irreversibly on heating. If a chemical hardener is used with thermosetting resins instead of heat to bring about the desired change of state, the adhesive is called *cold-curing* or *cold-setting*. The more important modern wood adhesives (UF resin, PF resin, and Resorcinol resin) all belong to this cold-curing group. Some other thermosetting resins have very useful specific adhesion properties; epoxy resins such as 'Araldite' bond to non-porous surfaces, such as metal and glass, and polyester resins are used as the bonding agent for glass fibre work (glass reinforced plastic, or GRP).

For many *woodwork* joints, a *synthetic casein glue* is the best choice as it is easy to use and has a longer 'pot life' than the synthetic resins once mixed. All types of glue which are temperature-dependant should be mixed with cold water; their pot life can be extended considerably if they are placed in a refrigerator between applications on a run of small jobs. Used in this way, the pot life of 'Cascomite' can be to extend from its usual 30 minutes to about 3 hours.

Urea-formaldehyde (UF) resins are available as 'one shot' powders activated by water, or as two part mixtures. UF resins are very strong and are cold-water proof; the two-part types can be mixed immediately before application, or the resin may be applied to one side of the joint and the hardener to the other. UF resins can also be used for bonding phenolic plastics, melamine, urea plastics, and laminates such as 'Formica'.

Phenol-formaldehyde (PF) resins are frequently used in marine plywood, as they are even stronger than UF resins and will withstand boiling water. They are basically industrial resins and are not as readily available as UF resins. PF resins are also modified industrially with rubber, nylon, or epoxy resins to make adhesives for bonding metals. Heat treatment is required to set PF resins.

Resorcinol-formaldehyde resins are cold-setting, and their water resistance is superior to both UF and PF types.

When using UF and RF resins, the temperature must be kept above 50 °F (10 °C) or the hardening process will be delayed. A typical setting time at 10 °C is about 20 hours, and this could be reduced to about $2\frac{1}{2}$ hours at 70 °F (21.1 °C). Most laboratory workshops will provide the higher temperature, but it then becomes advantageous to put the mix in a refrigerator between joints. Some hardeners will stain wood black if they come into contact with iron. Hardener can be applied with a small piece of sponge or rag tied to the end of a stick. *Never use bare hands to apply the hardener*, or skin fissures may be caused.

PVA glue is a white emulsion giving good bond strength with metal, wood, paper, ceramics, glass, leather, and some plastics. It is not waterproof, but the glue line remains elastic and is impervious to oil. PVA joints are damaged by heat above 50 °C, and by freezing. For structural joints in balsa wood, PVA is now the preferred adhesive but it does take much longer to set than balsa cement. (PVA joints can usually be handled after an hour, but full strength is not developed for at least 24 hours.)

Epoxy resins will stick almost anything to anything that is dry and grease-free. They are virtually immune to solvents, but because of their high cost in relation to other adhesives they are used only when their special properties are in demand. Epoxy resins are of limited use when maximum mechanical properties are needed – for example they tend to have a brittle glue line, and are affected by temperatures over 90 °C. The epoxy resins have been modified to improve their performance for specific industrial applications. The list below gives some examples.

Epoxy-nitrile: the addition of nitrile rubber reduces maximum service temperature, but increases peel strength and reduces brittleness.

Epoxy-phenolic; this is more brittle than straight epoxy resin, but the maximum service temperature is increased to 260 °C by the addition of phenolic resins.

Epoxy-polyamide: the addition of a nylon-like resin gives flexibility and high bond strength.

Epoxy-polysulphide: has improved peel strength and elasticity – developed for bonding steel to concrete.

Epoxy-silicone; bond strength is about half that of the straight resin, but the maximum continuous service temperature is around 350 °C although this can be exceeded for short periods.

Note Epoxy resins do not shrink on hardening, so they may be used on delicate structures to avoid warping.

Phenolic resins and their modifications formed by additives similar to those used for epoxy resins are used less often, but they may be seleced for some applications because of their greater resistance to vibratory shock.

Polyurethane adhesives combine very good bond and peel strengths, and they are resistant to vibration and most chemicals, but they are unsuitable for use where the environmental conditions are damp, as the bond is attacked by water.

Cyanoacrylate adhesives are just coming into use; they will form a strong bond with most materials which is resistant to water, oil, and most organic solvents. The bond can however be broken down by steam. Two useful properties of cyanoacrylates are the ability of the joint to set in a matter of seconds and the transparancy of the adhesive.

Poly-ester resins are the resins used to make a strong bond when constructing or repairing glass-fibre laminated structures. When bonding GRP, it is advisable to roughen the surfaces to be joined with a rasp or sandpaper as the surface is often coated with a wax used as a release agent in the original moulding process. Polyesters are used as industrial adhesives for other plastics and as 'optical cement' in the assembly of prisms and lenses. Their property of bonding glass is very useful in laboratory applications.

Rubber-based adhesives

A vast range of adhesives are now based on rubbers, the main groups being latex adhesives and 'straight' rubber solutions.

Latex adhesives consist of natural or synthetic rubber latex in a water based emulsion, which may vary from a milky consistency for spray application to a thick creamy consistency for application with a spreader. The adhesives used on self-adhesive envelopes, and 'Copydex' used for bonding textiles, are of this type. Latex adhesives are modified by the addition of cement or bitumen for use in the building trades. Latex-bitumen adhesives are used for bonding thermoplastic floor tiles and wood-blocks to concrete floors.

'Straight' rubber solutions are produced by dissolving raw rubber in an organic solvent. Sometimes vulcanising agents are added which 'cure' the rubber on drying out or heating (according to type), giving stronger bonds which are not affected so much by heat as 'straight' solutions. *Rubber cements* are made by adding resins to straight rubber solutions to increase the bond strength and to give gap-filling properties.

Rubber cements adhere well to most clean, dry, grease-free surfaces, a property fully exploited in the modern range of *Contact* or *Impact* adhesives. Contact adhesives are rubber cements which are used by coating both surfaces of a joint, and allowing the solvent to almost completely evaporate before bringing the surfaces into contact. A strong bond is formed on contact and no sliding of the joint can be made. If extremely precise joints are to be made with contact adhesives, a jig may be necessary to ensure accurate placement.

Although most rubber-based adhesives are good electrical insulators, specially formulated adhesives are produced which are electrically conductive for electrostatic applications.

For normal work, three types of rubber adhesive only will be used in the laboratory workshop and a choice will be fairly simple. An impact adhesive such as 'Evostik' will provide a strong bond on most surfaces; a latex adhesive such as 'Copydex' will bond fabrics to each other or to porous or nonporous surfaces with a moderate peel strength; 'Cow gum' should be used where low peel strength is essential, for example when mounting photographs for a temporary exhibition.

Table 10.13 Selection of adhesives

Use the following table of materials, and the adhesives which will bond to them, by looking up the two materials to be joined and selecting an adhesive which will bond to both.

Material	Adhesives which bond
Plastics:	
GRP	Epoxy
	Polyester
	Melamine formaldehyde
	Rubber cement (if low peel strength is acceptable)
Acrylic (perspex)	Acrylic cement (perspex in solvent)
	Epoxy
	Rubber cement
Cellulose acetate	Cellulose acetate in solvent
	Balsa cement
	Epoxy
Cellulose nitrate	Cellulose nitrate in solvent
	Balsa cement
	Epoxy
Thermosetting plastic	Epoxy
	Rubber cements
	UF resin
	Polyester resin
Polystyrene	'Plastic cement' (polystyrene in solvent)
	Epoxy
Expanded polystyrene	PVA
	Expanded polystyrene adhesive
Nylon	Epoxy
	Phenolic rubber cement
	Resorcinol
PTFE ('teflon')	Silicone
	Phenolic rubber cement
PVC	Phenolic rubber cement
	Nitrile rubber solution
Polythene	Silicone
	Phenolic rubber cement
Wood (including balsa and hardboard)	PVA
	UF resin
	PF resin
	Epoxy
	Resorcinol
	Balsa cement
	Rubber cements
	Polyurethane
	Polyester
	Latex
	Casein
	Dextrin, starch and gums, can be used to attach paper, but not for structural joints
Metals	Epoxy
	Cyanoacrylate
	Polyester
	Polyurethane
	Phenolic rubber cement
	PVA, if a weak bond is acceptable
Paper	Starch
	Dextrins
	Gums
	PVA
	Balsa cement
	Cow gum
	Cellulose adhesive (polycel)
	Latex
	Rubber solutions
Glass	Polyester
	Epoxy
	Rubber cements
Concrete	Latex-bitumen

Note That oily woods such as teak may have to be scrubbed with weak caustic soda solution and glued with UF resin (epoxy or polyester resins can be used if the wood is first thoroughly dried).

FINISHING MATERIALS

When an article has been constructed in the laboratory workshop, it will be necessary to give the components an appropriate finish. This is not just for decorative reasons; correct finishing includes the smoothing of all surfaces and edges so that handling the equipment is not in any way hazardous. Porous material such as wood will need some impervious finish to act as a barrier to dirt and moisture, and metallic components will probably require some protective coating to prevent rusting or tarnishing.

Glass

Glass will require little finishing other than a good cleaning with window cleaner or industrial methylated spirit unless cut edges will be left exposed in the finished article. Cut edges should have any jagged pieces removed with an old file before they are rubbed down with an oilstone. Edges of glass may be rendered safe by covering with adhesive tape, provided that there are no large projections. Plate glass may be edged with plastic chanelling intended for edging hardboard or forming sliding door grooves (available from D.I.Y. shops).

Metals

All surface blemishes should be removed by filing, using a series of files – second cut, smooth cut, and finally draw-filing with a dead smooth file (see page 189). A graded series of emery cloths, finishing with an oiled FF, will give a good surface. If a high polish is required, the components may be 'buffed'.

Buffing

This is carried out by bringing the article into contact with a cloth mop which is driven at high speed on a polishing spindle, which may be mounted directly on a motor shaft or placed in the chuck of the lathe. To obtain a high polish it is usual to use a series of mops charged with abrasive materials embedded in wax. For most laboratory purposes, a calico mop charged with tripoli wax will suffice. The full range of mops and their appropriate abrasive compounds is given in Table 10.14.

Table 10.14 Mops and Abrasives

Polishing compound		Mop
Course	Carbrax	Stapol
	Tripoli	Stitched calico
	Crocus	Unstitched calico or cotton
Fine	Rouge	Swansdown

If a lathe or polishing motor is not available, polishing mops are available for mounting on an arbor for use in electric drills.

Warning: when using a machine for polishing,

make sure that ties and loose clothing, particularly sleeves, are tucked in or secured. **Always** use light pressure, and work with the metal below the mop with the mop tending to pull the metal away from you (see Figure 10.20). Make sure that no one is standing behind the machine.

Never hold the material to be buffed with a rag. If it gets too hot to hold – *wait!* If for any reason chain or wire must be buffed, hold it firmly against a flat piece of wood. Thin sheet metal may also require the support of a piece of wood.

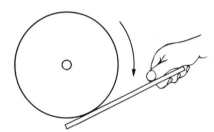

Figure 10.20 Using a polishing machine

Satin finish

A matt or satin finish may be made by using a fine wire brush mounted in the lathe or on a polishing motor spindle. A satin finish can also be produced by hand with fine steel wool; it is usually best to make all the strokes in the same direction. A satinising wheel is available for use on buffing machines to produce a professional finish.

Lacquering

Metal lacquers are available from all good ironmongers. After the metal has been thoroughly cleaned, the laquer is applied with a soft brush. The lacquer dries in a few minutes, and a second coat is advisable as this will cover any 'pinpricks' not covered by the first coat. In an emergency clear nail varnish may be used. Lacquering is usually used to preserve satin and high polished finishes on brass and copper from tarnishing.

Aluminium finishes

If aluminium is polished, the surface will slowly oxidize to give a matt finish. This oxide should not be removed as it forms a barrier against further oxidation. Attractive coloured finishes can be obtained by *anodizing*. (Free leaflets giving full details of the process are available to schools from the British Aluminium Company, which also sends out kits of materials.)

Plastic coating

The metal to be treated is heated and dipped into a low density polythene powder. A kit containing six 5 lb bags of different colours is available from Telcon Plastics Ltd., Farnborough Works, Green Street Green, Orpington, Kent. This is a very simple and effective treatment. It may be possible to share the cost with, or even borrow the materials from, the handicraft department.

Bright finish (steel)

The surface is brought to a good finish by draw-filing, followed by fine emery cloth, terminating with FF grade lubricated with a light oil. This is a time-consuming finish and is liable to corrosion unless a fine oil film is maintained on the surface. Under laboratory conditions, this finish is really only suitable for components made from stainless steel.

Blueing (steels)

The component is heated in a flame whilst a careful watch is kept on the surface colour of the metal, this will change from yellow to brown, then to purple and blue. When the steel has reached the tempering blue colour, it is removed from the flame, and plunged rapidly into oil. This is a cheap and easy method of protecting steel from tarnishing, providing the metal is not to be used in a hostile environment.

Bronzing

Copper and brass may be finished by thorough cleaning and then immersion in a hot solution of sodium sulphide. When satisfactory bronze tints have developed, the metal is washed thoroughly in hot water and then dried. The bronzed finish must be protected with lacquer.

Painting

Steel may be protected by painting with a coat of primer followed by an undercoat. The finishing coat may be gloss, eggshell, or matt oil paint. If galvanised steel is to be painted, it is essential to use a suitable calcium-based primer coat to avoid peeling.

Knurling

This is usually carried out with a knurling tool in the lathe, and consists of impressing a milled or diamond shaped pattern into the surface of the metal. It is a useful finish for metal knobs, heads of adjusting screws, and tool handles as it increases the coefficient of friction.

Wood finishes

The purpose of wood finishes is to seal the grain against the entry of dirt and water. Modern polyurethane varnishes have so many advantages over other finishes that it is hard to see any justification for using other wood finishes in the laboratory.

Polyurethane varnish

'One pot' varnishes, such as 'Ronseal', are available in matt and gloss finishes. The first coat is best applied very thinly with a lint-free cloth. Do not leave the surface 'wet'. When the first coat has hardened in about 2 hours, a second, thicker coat may be applied with a brush. (If the first coat is hard it seems to catalyse the setting of the second coat; if it is not hard the second coat may take a long time to set.) If a high finish is required, the second coat should be lightly rubbed down with flour paper and a third coat applied. If a very water resistant finish is needed, use one of the proprietary polyurethane yacht varnishes. If a coloured, black, or white finish is required, use a polyurethane yacht enamel.

Polyurethane varnish is very resistant to most chemicals and is therefore much more satisfactory in the laboratory than shellac and cellulose finishes which are very badly affected by a wide range of organic solvents, including methylated spirits. A properly applied polyurethane varnish will not be damaged by boiling water.

Emulsion paint

This can occasionally be useful in the laboratory workshop as it produces a semi-matt finish which dries quickly. This can be of value if something has to be finished in a hurry, or if a smooth, non reflecting surface is needed.

Cellulose finishes

These are not recommended for laboratory use as they are damaged by hot water, many laboratory reagents and solvents, and tend to scratch easily.

Button polish (French polish)

This is fairly difficult to apply well, is soluble in alcohols, and is damaged by hot water and many chemicals. If it is necessary to use this type of polish, for example to restore baseboards of apparatus already treated with shellac polishes, then try the proprietory kit marketed for amateurs under the name of 'Furniglas', with which it is relatively easy to get a good finish.

Wax

Traditionally used for bench tops, wax is affected by heat and alkalis as well as by organic solvents. If not damaged by spillage, it gradually darkens as it collects dirt on the surface. New benches should be polyurethene varnished, or topped with 'Formica' or an equivalent plastic laminate.

Plastics

If care is taken when working with plastic sheet, the main surfaces should be undamaged and only the edges will need polishing. Smooth cut edges with fine sand paper supported on a cork block, and polish with a buffing mop and fine abrasive such as crocus compound or rouge compound. If a machine is not available, brisk hand polishing with a metal polish such as 'Brasso' or 'Bluebell' on a soft rag will give a good finish. Turned plastic components and specimens embedded in plastic blocks are, of course, most easily finished and polished in the lathe.

SHEETMETAL WORKING

The laboratory technician is unlikely to be called upon for any but the most basic operations. These will include cutting, bending, and jointing.

Cutting

Cutting will usually be carried out in the laboratory workshop using tinshears, or the more modern type of sheet metal cutter of the Monodex type, whose advantage is that it cuts circles and curves as well as straight edges without distortion. It can also cut and trim inside edges if a pilot hole is drilled first. The only slight disadvantage is that it removes a narrow strip of metal, and this must be allowed for when marking out. It is unsuitable for use with stainless steel.

If tinshears are used, it will usually be necessary to flatten the cut edges with a rawhide mallet and to remove rough edges with a file.

If heavy gauge, or large quantities, of sheet have to be cut, a bench shear may be available in the school or college metalwork room.

Drilling and punching

The position of the hole is marked with a centre punch (always used on a flat metal surface to prevent distortion of the sheet). Twist drills are used for small holes, but if accuracy and an undistorted surface are important, larger holes are best cut with a chassis punch; these are available in at least 25 sizes ranging from $\frac{3}{8}$ in. to 3 in. (9.5 mm–76 mm). A pilot hole is drilled first and the punch assembled through the hole; the screw is then tightened, causing the punch to cut cleanly through the sheet. For laboratories where electronic instruments are made, the $\frac{3}{8}$ in. (9.5 mm), $\frac{7}{16}$ in. (11.1 mm) and $\frac{1}{2}$ in. (12.7 mm) will be found sufficient for mounting potentiometer switch and jack socket bushes. A large size can be very useful for cutting the hole for the barrel of meters and must be selected to suit the make of meter most usually employed. (*Radiospares* sell an excellent range of meters with a small barrel requiring a $1\frac{1}{16}$ in. (27 mm) punch. In addition to their wide range of circular punches, *Q*-Max also make $\frac{11}{16}$ in. (17.5 mm) and 1 in. (25.4 mm) square hole punches, and a $\frac{21}{32}$ in. $\times \frac{15}{16}$ in. (16.7 $\times$ 23.8 mm) rectangular punch. Larger holes can be produced by punching two or more holes in line.

Bending

For the type of constructional work usual in laboratories, simple right angle bending is all that is required. This may be carried out in the vice, or preferably using a pair of bending bars which may be purchased, or home made. An example of what can be achieved is shown in Figure 10.21.

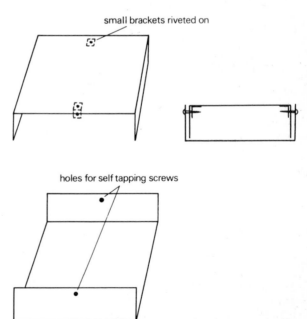

small brackets riveted on

holes for self tapping screws

Figure 10.21 A simple instrument case made from bent sheet metal

11. Hand Tools

Purchasing tools is a problem in any laboratory workshop – the scope of the tools required will be related to the range of work to be undertaken and the money available. Under-equipping is a false economy, as the total cost of hand tools in a well equipped workshop will represent only a fraction of one technician's salary. The time wasted by a technician using an inefficient method on a job is often more expensive than the cost of the special tool which was required.

If initial economies have to be made, make sure, on grounds of efficiency and safety, that necessary tools such as vices and cramps are purchased so that work may be held securely. High on the list of priorities should be a set of good marking-out and measuring tools; these will save the wastage of both time and materials. All cutting tools should be of the finest quality available, and provision must be made for keeping them sharp.

The following basic list of tools for the science workshop will prove adequate for many jobs. If in addition a small metal turning lathe, with screw cutting facilities and a range of accessories, is provided then almost any likely job can be carried out.

Woodworking vice (this should be as large as possible; a rapid action is not essential)
An Engineer's or Fitter's vice (which can be fixed to a 4 in. × 3 in. block of wood if space is short, so that it can be placed in the jaws of the woodwork vice when required)
Four 6 in. G-cramps
Setsquare (metal)
Combination set (need not be high precision – Picador type is suitable)
Scriber
Dot punch (Layout punch)
Centre punch
Dividers
Internal calipers
External calipers
Jenny odd-legs calipers
Vernier calipers
Micrometer (digital read-out types now available)
Steel rule
Combination marking and mortise gauge
Small wheelbrace
Set of drills 1.60 mm–6.40 mm ($\frac{1}{16}$–$\frac{1}{4}$ in.) high speed steel (an extension of the set to 13 mm ($\frac{1}{2}$ in.) may

be considered desirable)
Countersink bit, high speed steel
Screwdrivers, small insulated electrical
  medium insulated electrical
  large woodworking
  Posidrive headed
Wrench, adjustable, preferably Mole type
Pliers, combination
  electrical
  gasfitters
Pincers
Files, round
  triangular
  flat, second cut
    smooth cut
    dead smooth
    rough or bastard cut for soft metals
Hacksaw and blades, preferably with vertical handle and high speed steel blades
Junior hacksaw and blades
Tin snips, straight (or Goscut tool)
Pop rivet kit
Set of Allen keys $\frac{1}{16}$–$\frac{3}{8}$ in.
Oil can
Tennon saw
Coping saw and blades
Hand saw (optional – most timber can be obtained cut to size)
Chisels $\frac{1}{4}$, $\frac{1}{2}$, $\frac{3}{4}$, and 1 in. (metric sizes not essential)
Plane
Surform blockplane
Mallet, woodworkers
Mallet, no. 1 rawhide
Hammers, claw
  ball pein
  cross pein
  tack
Combination oilstone (coarse and fine grit)
Safety spectacles
Side cutting pliers
Carpenters ratchet brace, 10 in.
Six wheel glass cutter
Glass cutting knife
Cork borers and sharpener
Set of spanners, preferably combination type BSF, BSW, metric, and B.A.
Small spirit level
Bradawl
Electrical instrument soldering iron

As many technicians have received little or no training in the use of hand tools, this chapter contains all the basic information for selection and use of the correct handtool for all the operations likely to be found in school and college laboratory workshops. Of necessity it will contain a number of tools not included in the basic tool list above.

THE VICE

Woodworker's vice

These are made in various sizes, with jaws $6-10\frac{1}{2}$ in. wide and opening from $4\frac{1}{2}-15$ in. (Figure 11.1). The jaws must be lined with soft wood which should be renewed from time to time or the edge becomes damaged and the vice unable to grip small pieces of work accurately. This lining reduces the nominal opening by approximately $1\frac{1}{2}$ in.

Some vices have a quick-release lever which enables the vice to be opened or closed without using the screw action. This saves time, but adds considerably to the price.

Other tools, such as a metalwork vice and an electric drill bench stand, can be kept mounted on 3×4 in. blocks of wood and gripped in the woodwork vice when they are needed.

Engineer's parallel vices

These powerful vices (Figure 11.2) are used for gripping metal. Jaw width ranges from $2\frac{1}{4}$ to 6 in., and jaw openings from $2\frac{1}{4}$ to 8 in. Jaw plates are made from cast iron and are replaceable.

The lighter models, are often called mechanic's vices, and the heavier models fitter's vices, which are often supplied with a quick-release mechanism.

When permanently fixed to the bench, engineers vices should be situated over the bench leg for maximum rigidity.

Slip-on fibre grips are available to grip delicate work without marking the surface. Soft grips or vice clamps ('clams') can also be made from sheet copper, aluminium, or lead to protect work from the rough jaw surface (see Figure 11.3). When cutting clamps from sheet metal always leave a piece of metal which can be folded around the jaws to hold the clamps in place.

The vice anvil may be used for *light* work.

The tommy-bar should be tightened only by hand. If the vice is overtightened, a short length of metal tube may be slipped over the tommy-bar to give greater leaverage. Rubber rings placed over the tommy-bar ends will make the vice quieter in use.

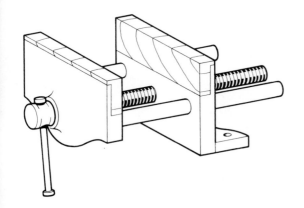

Figure 11.1   A woodworker's vice

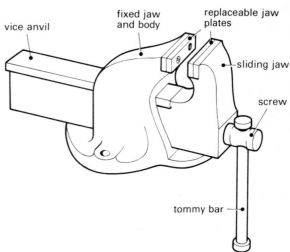

Figure 11.2   An engineer's vice

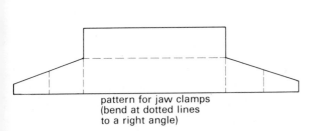

pattern for jaw clamps
(bend at dotted lines
to a right angle)

Figure 11.3   Vice jaw clamp

Machine vice (drill vice)

These are used to hold material steadily and safely in a drilling machine (Figure 11.4). Their specially designed jaws, which are parallel with deep V grooves both vertically and horizontally, will grip tubes, rods, hexagonal rods, and flat bar.

The base is accurately machined for a perfect fit on the machine bed. Slots in the base enable the machine vice to be bolted down with special T bolts to machine beds or to the cross slide of a lathe (used when milling).

Hand vice

A small tool (Figure 11.5) used to hold thin metal safely while drilling, the jaws are parallel only when fully closed, and hence the grip on thick material is poor.

Toolmaker's clamps

These may be used to grip small work for drilling and machining as either a hand vice or bolted to the machine bed or lathe faceplate.

If the tools are to be home-made, use oblong section stock rather than square as all the strain is in one direction.

The jaws should be parallel in use.

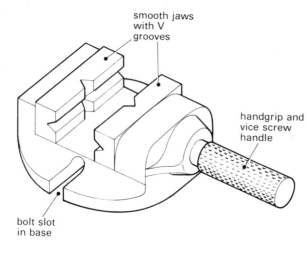

smooth jaws with V grooves

handgrip and vice screw handle

bolt slot in base

Figure 11.4   A drill vice

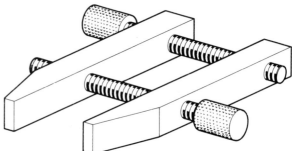

Figure 11.6   A toolmaker's clamp

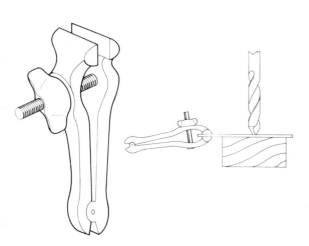

Figure 11.5   A hand vice

SETTING-OUT AND MEASURING INSTRUMENTS

Setsquare (Trysquare)

Used by metalworkers and woodworkers to test internal and external right angles, and in marking-out work (the length of the blade may be extended by laying a steel rule along it); the metalwork type may also be used for wood. Well-made metal trysquares are finished with a groove in the handle at the point of union with the blade. This is to accommodate the burr on roughly cut metal when marking out.

To maintain the accuracy of this tool, it should always be held by the stock (handle) and placed safely in a rack when not in use.

The sizes quoted in catalogues refer to the length of the blade.

Combination set

This is a multi-purpose setting out and measuring tool (see Figure 11.8).

The stock and blade have permanent angles of 90° and 45°. The protractor head can be used in conjunction with the blade to mark out or check any angle, which can be read from the scale. The centre head is used to find the centre of circular material.

The stock usually contains a spirit level.

Dividers

These are used for numerous marking out operations such as circles, parallels, and equidistant points along a line; one leg is held firm and the other is used to scribe a line.

Sharpen as for scribers.

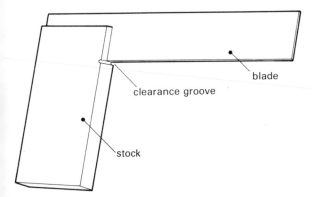

Figure 11.7   A trysquare

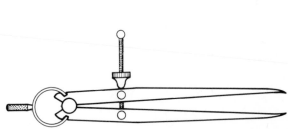

Figure 11.9   Dividers

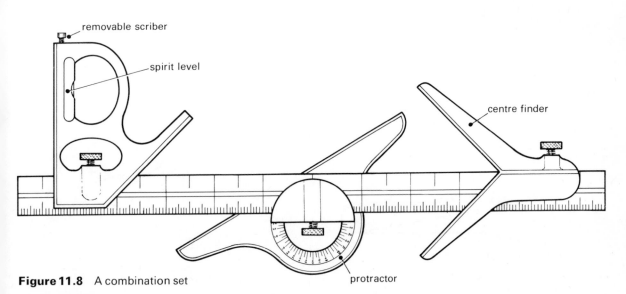

Figure 11.8   A combination set

Calipers

Many types of caliper are used in marking-out, measuring, and checking work (see Figure 11.10).

Adjustment is made either by moving the nut on the bow spring types or by gently tapping firm joint types on the bench. *Internal and external calipers* are used for measurement, one leg being held firm and the other rocked gently for adjustment. Measurements are read off from external calipers with one leg on the end of the rule, the other on the scale. Internal measurements are read off most accurately if the rule is stood on end on the bench, one leg of the calipers placed on the bench, and the other against the scale (see Figure 11.11).

Hermaphrodite calipers, also known as Jenny odd-legs or jennys, are used in marking out lines which are roughly parallel to the edge of the work, and for finding the centre of circular stock.

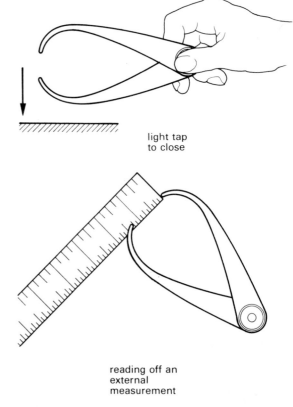

internal calipers   jenny oddlegs   external calipers

Figure 11.10   Calipers

Figure 11.11   Operations with calipers

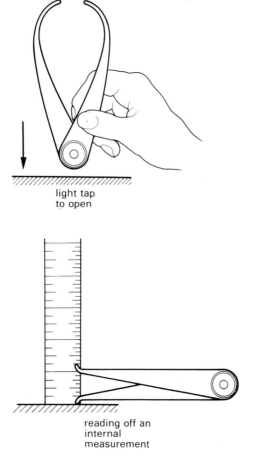

light tap
to open

light tap
to close

reading off an
internal
measurement

reading off an
external
measurement

Vernier caliper gauge

These are precision instruments for taking internal and external measurements (Figure 11.12). Cheaper instruments usually read to $\frac{1}{100}$ in., better instruments to 0.001 in., or 0.02 mm in the case of metric instruments.

Set the sliding jaw to the approximate measurement, and make the final adjustment with the fine adjustment screw. Read the main scale, then add the reading on the sliding vernier scale which coincides with a line on the main scale to the main scale reading. If used for an internal reading, the width of the jaws must be added.

Micrometer

This is a very precise measuring instrument, available in Imperial or metric sizes, which must be used with care to preserve its accuracy.

There is usually a main scale on the sleeve and a vernier scale on the thimble which is easier to read than that of the vernier caliper gauge. New types of micrometers, metric and Imperial, which are digital readout have now appeared with obvious advantages for technicians who do not use a micrometer regularly. The ratchet device allows light- or heavy-handed users to take measurements with the same pressure.

MARKING INSTRUMENTS
Scriber

These are used for marking out on metal, and are made from cast steel, which is then hardened and tempered. In use, the point must always rest against the rule. When sharpening scribers, always grind in line with the axis or the point will break off easily.

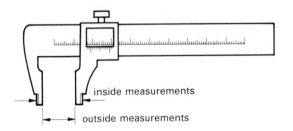

Figure 11.12 A vernier caliper guage

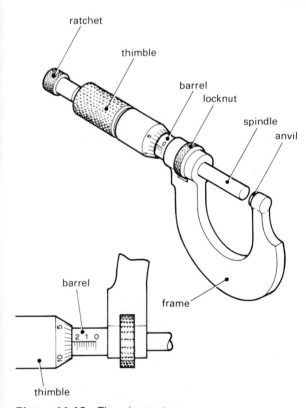

Figure 11.13 The micrometer

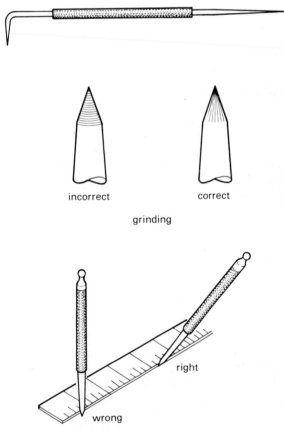

Figure 11.14 Scribers

Dot punch

After a line has been marked with a scriber, some workers like to punch the lines at intervals with a sharp, light punch called a dot punch. The punched line is less likely to be 'lost' during subsequent work.

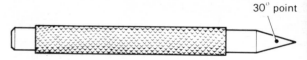

30° point

Figure 11.15   A dot punch

Centre punch

These are used to make an accurate starting point for twist drills. A spring-loaded type is available; if pressed firmly on the work, when the spring is fully compressed, a trip mechanism delivers a 'kick' to the point of the punch.

Bell punches (Figure 11.17) are made with a conical skirt, which enables the centre of metal rod to be punched with reasonable accuracy.

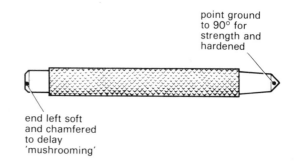

point ground to 90° for strength and hardened

end left soft and chamfered to delay 'mushrooming'

Figure 11.16   A centre punch

Marking gauges

Four basic types are made:

Marking gauges – single pin (Figure 11.18).

Mortice – two pins, one of which is on a brass slide and is adjustable.

Combination gauge – marking and mortice gauges on opposite sides of the same tool.

Cutting gauge – has a fine knife blade in place of the pin. Used to cut veneers; can be useful to cut two parallel lines close together in the veneer of plywood prior to sawing. The saw will not then chip the veneer.

Letter punches and number punches

These can be used to stamp numbers or letters on metal (Figure 11.19). The large sizes can be used on hardwoods. The height of the characters ranges from $\frac{1}{32}$ to $\frac{3}{8}$ in. and the price of a set of numbers varies according to size.

A set of letters and numbers is often available in the school metalwork shop.

Figure 11.19
A number punch

SCREWDRIVERS

Many types are available, in a wide range of sizes. It is important to use the correct size if damage to the head of the screw is to be avoided. The principal types are:

   Plain
   Ratchet
   Spiral (rapid action)
   Short handled (for work in confined spaces, e.g. boxes)
   Insulated (for electrical work)
   Indicator (clear plastic handle contains a neon bulb, which lights up when the blade touches the live (line) wire of the mains supply)

All types are made with standard or Phillips head ends (see Figure 11.20).

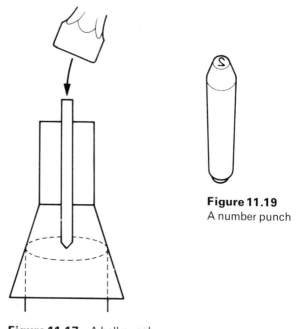

Figure 11.17   A bell punch

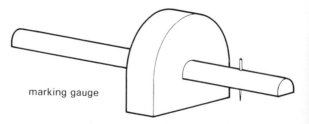

marking gauge

Figure 11.18   A marking gauge

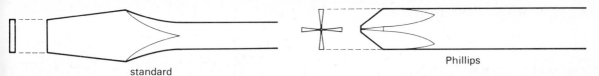

Figure 11.20   Screwdrivers

standard

Phillips

PLIERS AND WIRE-STRIPPERS

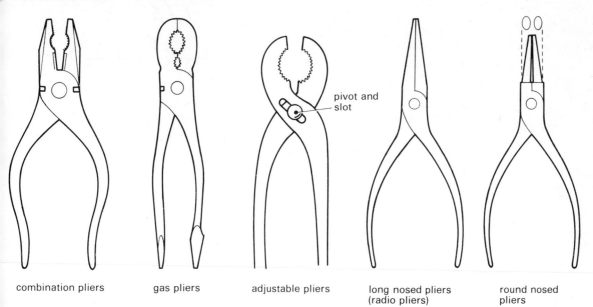

pivot and slot

combination pliers

gas pliers

adjustable pliers

long nosed pliers (radio pliers)

round nosed pliers

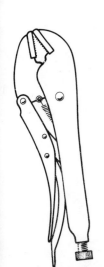
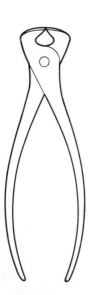

universal grip pliers 'mole' wrench pattern

topcutting pliers

side or diagonal cutting pliers

adjustable wire strippers

precision bib adjustable wire stripper and cutter

SPANNERS AND WRENCHES

Figure 11.22   Spanner types

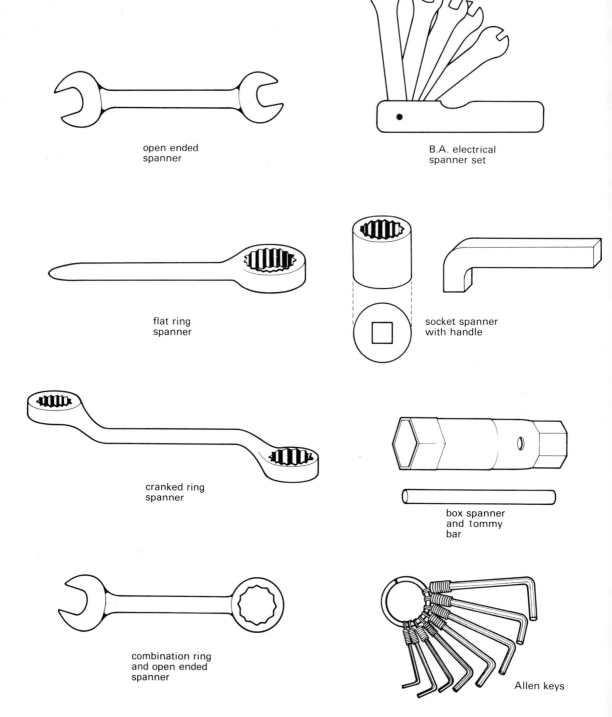

open ended
spanner

B.A. electrical
spanner set

flat ring
spanner

socket spanner
with handle

cranked ring
spanner

box spanner
and tommy
bar

combination ring
and open ended
spanner

Allen keys

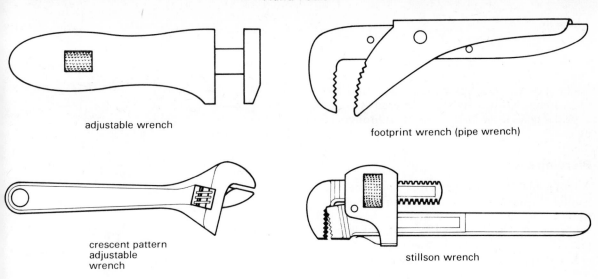

adjustable wrench

footprint wrench (pipe wrench)

crescent pattern
adjustable
wrench

stillson wrench

Figure 11.23 Wrench types

SAWS

Hacksaw

The best pattern has an oval tubular frame and a vertical handle. In some types, spare blades can be stored in this handle. Cheaper types with horizontal handles are available, but they are more difficult to control, and are more tiring to use (see Figure 11.24). Frames are adjustable to take 10 or 12 in. blades.

By turning the tensioning pins through steps of 90°, the blade may be turned through 90°. This enables long cuts parallel to the edge of the metal to be made (otherwise the depth of cut is restricted to approximately 3 in.).

To locate the blade in the frame, place with the teeth pointing *away* from the handle, and then tension with the wing nut. Various types of blade are available. High speed steel blades are probably the most useful; other types are low tungsten steel, with the teeth hardened and the back of the blade soft (sometimes called 'flexible' blades), and 'all hard' blades in which the back is also tempered (often preferred by skilled craftsmen because of their rigidity). Blades are available with 18, 24, or 32 teeth to the inch. Always select a blade which will have at least three teeth in contact with the work, otherwise 'chatter' and broken teeth will result.

Grip material in the vice with the cutting line vertical. Work as close to the vice jaws as possible to minimise 'whip' and 'chatter'. Use the full length of the blade with slow strokes. Rapid strokes draw the temper of the blade by overheating it and increase the risk of breaking the blade by twisting.

Junior hacksaw

This is very useful for small work; the 6 in. blade is tensioned by the spring of the frame.

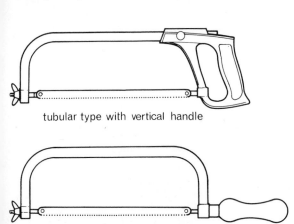

tubular type with vertical handle

horizontal handle type

Figure 11.24 Hacksaws

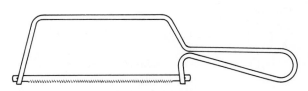

Figure 11.25 Junior hacksaw

Abrafile

This is a tension file which is used as a saw (see Figure 11.26). The blade, which is a thin round file, can be used in an ordinary hacksaw frame if a pair of special links are bought. It is very useful for awkward curves, for cutting plastics, and can be used for cutting irregular holes if the blade is inserted through a drill hole.

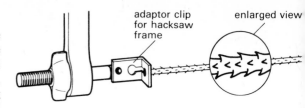

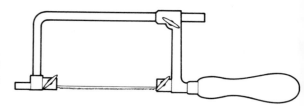

Figure 11.26 Abrafile tension file

Piercing saw

A tool used to cut intricate shapes from sheet metal, with blades like fretsaw blades. File-like blades with teeth all around are also available; these are the easiest to use. The adjustable frame means that large pieces of broken blade can be used.

In use, always keep the blade moving up and down, particularly whilst turning the work. *Never* force the blade forward.

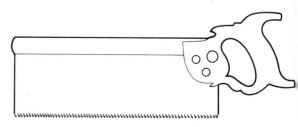

Figure 11.27 A piercing saw

Tenon saw

Made in 8, 10, and 12 in. sizes, the 10 in. saw being generally the most useful.

The blade is thin, but rigidly supported by a brass spine. Tenon saws are used for accurate work; *never* split the last bit of wood by twisting the saw.

Six or eight inch tenon saws with very fine teeth are known as 'dovetail' saws as they are used for cutting very precise joints.

Figure 11.28 A tenon saw

Handsaws

Sizes run from 20 to 26 in. in two inch steps; a 22 in. blade suits most people.

Cross-cut saws have fine teeth and are best at cutting across the grain; rip saws have coarse teeth and are best at cutting along the grain. In the laboratory workshop, compromise with a general purpose handsaw with medium-sized teeth.

Figure 11.29 A hand saw

Coping saw

Very useful for cutting shapes in wood, the blade can be rotated at any angle to the frame to facilitate following curves. Use a bowsaw for larger work.

Hole saws

These are made in a range of sizes from $\frac{1}{2}$ to 4 in., with shanks to fit $\frac{1}{4}$ in. electric drills or carpenters' braces. Good quality blades are intended for cutting holes in galvanised water tanks. Cheap sets with blades from $\frac{3}{4}$ to $2\frac{1}{2}$ in. are available, and are excellent for cutting holes in hardboard and plywood up to 12 mm thick.

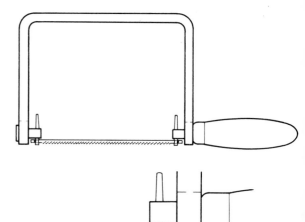

Figure 11.30 A coping saw

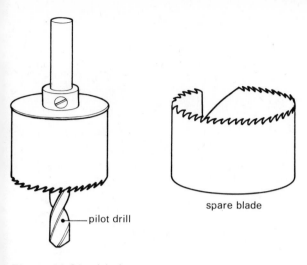

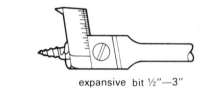

Figure 11.31 A hole saw

spare blade

pilot drill

BRACES

Wheelbraces

These are used with twist bits and countersinks. Two main patterns are available: Stanley type, and the more expensive Leytool type which has the gears encased.

Capacity is up to $\frac{1}{4}$ in. shank, but larger-headed drills with $\frac{1}{4}$ in. shanks are available.

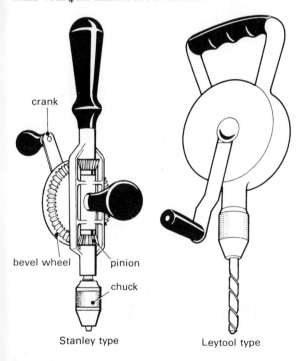

crank

bevel wheel   pinion

chuck

Stanley type   Leytool type

Figure 11.32 Wheelbraces

Carpenters' hand brace

These are used for drilling $\frac{1}{4}$ to 3 in. holes in wood. Plain and ratchet action types are available.

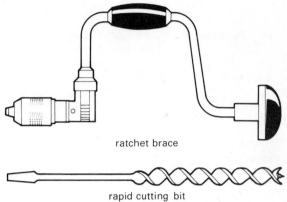

ratchet brace

rapid cutting bit

expansive bit ½"—3"

Figure 11.33 A carpenter's brace

PLANES

Smoothing and jack planes

Used to give a good finish on sawn timber, little or no sandpapering is necessary on most woods if the plane is *really* sharp and used well.

The larger size of the jack plane (Figure 11.34) makes this a better tool for planing long straight edges and big surfaces. The smoothing plane will be large enough for laboratory workshop use. For ease of work and good finish, keep the plane sharp and finely set. Paraffin wax rubbed on the sole of the plane reduces friction and makes the work much easier.

Some timbers, particularly the mahogany group, have grain which tends to tear out whichever way one planes. On such 'wild grained' timber, use a sharp plane set very finely and plane diagonally across the grain.

Plough or grooving plane

Grooving is a very valuable constrictional technique in the laboratory workshop. It may be done with a grooving plane (Figure 11.35), or with a circular saw – both tools are likely to be available in the school woodwork room.

Grooving planes are made with adjustable guides so that grooves can be cut parallel to the edge of the wood, and with depth stops to ensure that the groove is cut evenly to the chosen depth. They will cut only along the grain. If a groove across the grain is needed, clamp some wood across the workpiece to act as a guide, and make two parallel sawcuts with a tenon saw. The grooving plane can then be used to remove the waste wood.

A set of blades ranges from $\frac{1}{16}$ to $\frac{9}{16}$ in. in $\frac{1}{16}$ in. steps. As grooves are difficult to clean up after they have been cut, make sure that the blade is very sharp and the cut is not too deep.

Surform planes and rasps

These do not give as good a finish on wood as a smoothing plane, but they are cheaper and more versatile. They will shape and plane wood, plastics, metals, and even plaster and mortar.

Blades are replaceable. Use only blades in good condition for wood, and keep old blades for rough work.

Use diagonally across the grain to remove as much wood as possible, and then use along the grain with light pressure for final smoothing.

The Surform Planerfile has a reversible handle and can be used as a plane or a file. The smaller blockplane with a $5\frac{1}{2}$ in. blade is a very handy tool. Circular rasps and convex planes are also available, and are useful for rapidly cutting awkward shapes.

Figure 11.35   A grooving plane

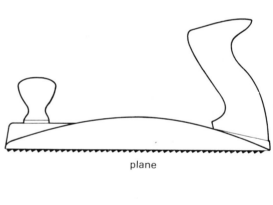

plane

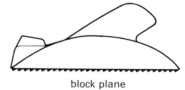

block plane

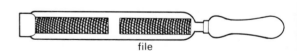

file

Figure 11.36   Surform tools

smoothing plane

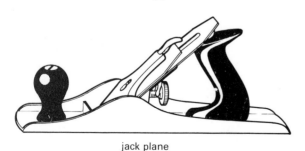

jack plane

Figure 11.34   Planes

FILES

A file is one of the most important hand tools, capable of shaping metals and plastics as no other tool can. To be so versatile, files have to be made in a wide variety of sizes, shapes, cuts, and tooth sizes.

Size

Sizes range from 4 to 16 in. – the length refers to the hardened cutting blade and does not include the tang (see Figure 11.37). (**For safety** the tang must always have a handle; files are sold without handles as they are regarded as disposable when blunt, but the handles can be re-used.)

Shapes and uses

Square   Used for filing drilled holes square, for cutting slots, and for cleaning up internal corners.

Knife   Used for filing in sharp corners.

Three square (triangular)   For corners of less

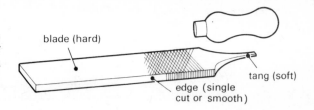

Figure 11.37   Parts of a file

than 90°, for sharpening saws, and, in the laboratory, to notch glass rod and tube when cutting.

Round   Used to enlarge holes and file radii; the smaller sizes are called rat-tail files.

Half round   Used for filing concave surfaces and the inside of large holes.

Hand   The sides are parallel along the entire length, but may taper in thickness towards the end; the most generally used file.

Warding   4 to 8 in. in length, tapering in width but not in thickness. Used for filing narrow slots.

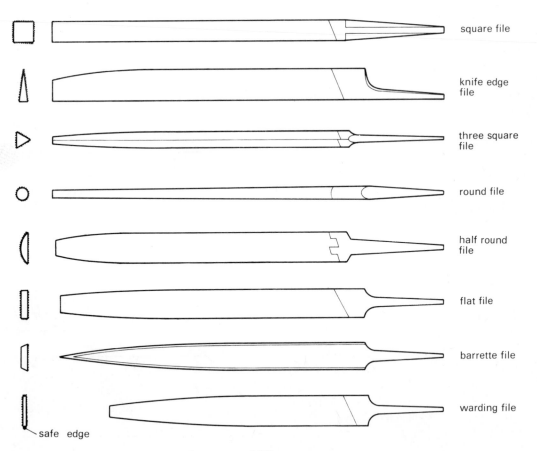

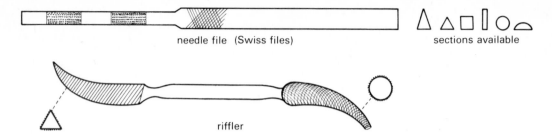

needle file (Swiss files)   sections available

riffler

Figure 11.38   Types of file

Needle files or Swiss files   Available in all the above shapes, these are very fine files for delicate work; their shafts are knurled to form a handle.

Rifflers   Delicate files used for precision work in awkward places; usually double ended and available in a vast range of shapes.

Most files are made with their edges parallel for two-thirds of their length, with the last third tapering – usually in thickness as well as width.

Types of cut

Single cut   Used on file edges and sometimes on the face.

Double cut   The usual type of cut giving the best finish for most purposes.

Dreadnought   Does not clog as easily as other cuts on soft metal such as lead.

Safe edge   One edge of a file is often left uncut so that corners may be filed in such a way that only one face is cut.

Tooth size

Files are made in a variety of grades. The number of teeth per inch for each grade varies with the size of the file (a four inch second cut file would have more teeth per inch than a ten inch smooth file). The grades normally used, and the number of teeth per inch, are (numbers relate to a ten inch file):

Rough	14–22
Bastard	22–32
Second cut	28–42
Smooth	50–65
Dead smooth	70–110

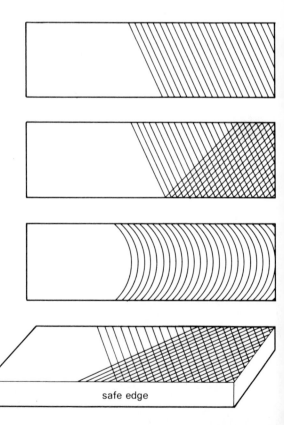

safe edge

188

USING FILES

Always secure work firmly in the vice with the surface to be filed horizontal (it is easier to control the file in this position). The stance should be like that of a boxer (left foot forward for a right-handed person). Hold the file with both hands; use slightly different grip for light and heavy filing (see Figure 11.40).

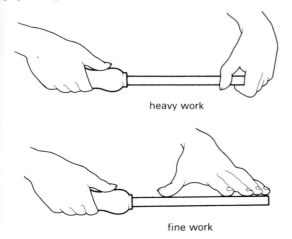

heavy work

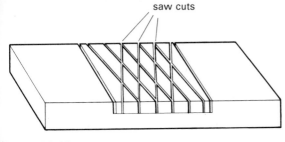

fine work

Figure 11.40   Holding a file

The forearm should be parallel to the ground and the wrist kept rigid. As the file moves across the work it must not rock or the flat surface will be lost. Cut on the forward stroke only. Filing on the return stroke drastically reduces file life. For heavy work, cut diagonally across the work.

If a great deal of metal has to be removed, make criss-cross saw cuts before filing as in Figure 11.41.

saw cuts

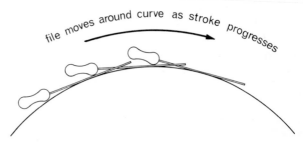

Figure 11.41

Always use new files for soft metals; older ones will still cut steel. Always use old files to remove the sandy skin of a casting, and never attempt to file hardened steel.

Store files in racks. They damage each others teeth very quickly in drawers.

Old files can be tempered and ground to make scrapers for woodturning.

Drawfiling

To obtain a fine grained finish with a file, draw a dead smooth file backwards and forwards across the work (Figure 11.42).

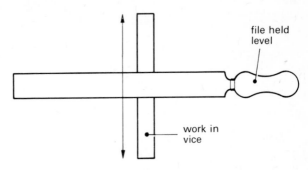

file held level

work in vice

Figure 11.42   Drawfiling

Filing external curves

Most workers find this operation difficult at first and tend to produce 'flats'. The knack is to follow around the curve as the file moves forward (see Figure 11.43), so that successive rows of teeth do not cut the same spot.

file moves around curve as stroke progresses

Figure 11.43

Pinning

'Pinning' or clogging of files occurs with soft metals and is worst with fine grades of file, so use as coarse a file as possible on soft metals. Pinning is reduced if chalk is rubbed into the teeth of the file before use.

Do not use a badly pinned file – stop and clean it with a fine wire brush, known as a 'file card', or with a piece of brass rod which has been filed to the shape of a fine screwdriver.

Never use oil on a file as this tends to make the file skid and cut less efficiently. With needle files even perspiration from the fingers can reduce their efficiency.

HAMMERS AND MALLETS

Mallets are used to strike wooden and plastic chisel handles and to tap wooden joints together without marking the wood. They are generally made of beech and the head is secured to the handle by a wooden wedge.

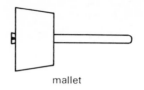

mallet

Hammers are made in many varieties, most of which are for sheet metal work and are not often used in the laboratory. The types of hammers most commonly met with are:

Ball pein
Straight pein
Cross pein

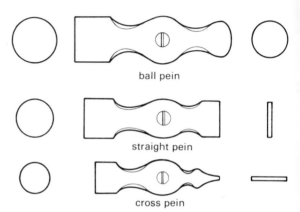

ball pein

straight pein

cross pein

Tack – a very light straight pein hammer.
Jeweller's hammer – a small ball pein hammer with a one ounce head sometimes used in instrument work.
Soft faced hammers – the face may be hard rubber, plastic, lead, brass, or rawhide (uncured pigskin). All are intended for use when the surface being hammered must not be marked. Rawhide hammers are the most common and are made in a range of sizes.

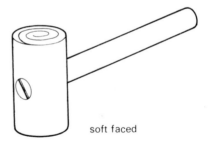

soft faced

Carpenter's claw hammer – intended for driving large nails; the claw is used to remove nails which bend.

Handles are usually made of ash or hickory. The hole in the hammer head tapers both ways and a metal wedge holds the head on the handle. New handles and wedges can be bought from ironmongers. Some manufacturers are now fitting stainless tubular steel handles, with a rubber shock absorbing grip, to their hammers. These are excellent as the heads do not work loose and the handles do not break. Beware of chromium plated handles as these are usually a sign of poor quality tools.

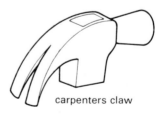

carpenters claw

Figure 11.44   Hammers and mallets

CHISELS

The main types are:

Bevel edged   Used for light paring.

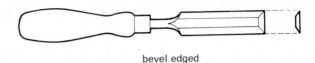

bevel edged

Firmer chisels   Much stronger and used with a mallet for heavier work.

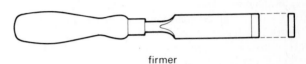

firmer

Mortice chisels   Very thick blade which will not be free to move around in the hole when a mortice is being chopped. Their great strength enables them to be used for levering chips out of the hole without bending.

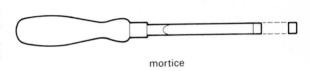

mortice

Cold chisels   Used by the metalworker to cut sheet metal, clean up corners of castings, and occasionally to cut grooves (little application in the laboratory).

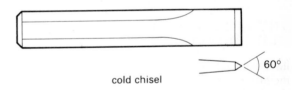

cold chisel

Figure 11.45   Chisels

SCREW EXTRACTORS

These are sold in sets and are used when a screw shears, or its slot becomes burred, and it has to be removed.

Centre punch the screw or stud, drill a hole into the screw with a twist drill, and insert the appropriate extractor until its anti-clockwise thread binds in the hole; use a wrench to remove the screw. Finally, place the removed screw firmly in a vice and turn the extractor clockwise with a wrench.

Figure 11.46   A screw extractor

12. Drills and Grinders

The accuracy and speed with which work may be carried out can be greatly improved and the range of work which can be successfully undertaken is greatly increased if the science workshop is provided with a few machines. Even in schools where the science workshop is not equipped with machinery, the use of machines is often still relevant as there will almost certainly be a well equipped machine shop in boys or mixed schools.

Most school science workshops, do not need the full range of machine tools; a lathe, a drill press, and a polisher/grinder, with suitable accessories, will carry out almost every operation which is needed. Machinery is expensive, and will not be used every day in science workshops; the laboratory technician may well have to carry out all the necessary operations on a medium-priced lathe, fitted with suitable accessories. For a description of the use of the lathe, see Chapter 13.

THE DRILL PRESS (OR POWER HAND-DRILL MOUNTED IN A STAND)

This machine will allow accurate drilling to be carried out with the maximum ease. The drill press (or sensitive drill as it is sometimes called – the name derives from the fact that all strains and break-through are transmitted to the operator through the operating handle) may be a floor standing type with an adjustable table or a bench drill which may have either an adjustable table or a head which can be moved up or down on its column (see Figure 12.1).

For laboratory workshop purposes, a good electric hand drill mounted on a vertical stand can also be used – one of the more robust types such as the Arcoy Bucaneer, which has a $\frac{1}{2}$ h.p. motor and a rigid stand with attachments for cutting mortice, dovetail, and comb joints. Their chief disadvantage is that they have only one speed, which tends to be rather fast for metal drilling, although many models can now be supplied with a reduction gear; some measure of speed control (with some loss of torque) can be obtained by the use of a rheostat, or by using a thyristor speed controller.

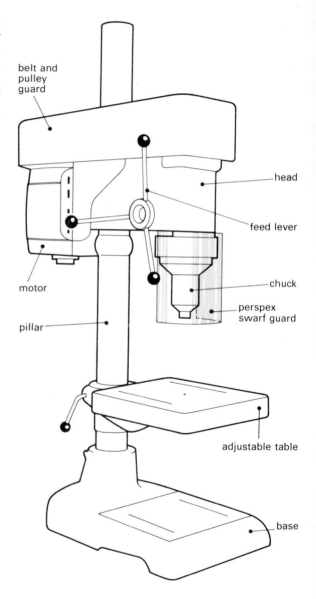

Figure 12.1   Parts of a drill press

Drilling hints

1 Always make a centre punch dot to start small drills.

2 If a large hole is required, drill a small pilot hole first; this increases accuracy and prevents excessive wear of the drill bit.

3 Always use a sharp drill – if it is blunt, it will burn. When grinding twist drills, make sure that the lips are of equal length, or the tip of the drill will tend to 'wander' in the hole and will produce an oversize hole (this can be a deliberate technique when drilling thermoplastics, such as perspex, to avoid excessive friction and binding).

4 Whenever possible hold work in a machine vice or bolt it to the slotted baseplate – a piece of metal leaving a $\frac{1}{2}$ h.p. drill at 1–2000 r.p.m. is a formidable missile. Figures 12.2–12.6 show some methods for holding work while drilling.

5 Choose a suitable speed – high speed for small diameters, slow for large diameters. Use higher speeds for wood, soft metals, and plastics than for steels.

6 Use a suitable cutting fluid on all but the lightest work on metals and plastics (see workshop tables).

7 Check that the chuck key has been removed before switching on, and that the drill is running true before it enters the workpiece.

8 The speed at which the drill is fed through the work is very important – too much pressure may burn the point or bend a small drill, insufficient feed will mean that the drill stops cutting and begins to rub which is one of the most rapid ways of blunting. At the correct feed rate the operator feels the drill continuously biting into the work with minimum pressure applied. If a drill will not cut without 'forcing', it should be reground or the work examined to ensure that un-annealed tool steel is not being drilled.

9 If very accurate holes are necessary, they should be drilled 0.010–0.025 mm (0.005–0.010 in.) undersize for holes up to 12 mm ($\frac{1}{2}$ in.), 0.40–0.75 mm ($\frac{1}{64}-\frac{1}{32}$ in.) oversize for larger holes, and then finished with a handreamer using a suitable cutting fluid.

10 **When drilling metal, always use the plastic swarf-guard on the drill-press, or wear safety spectacles.**

Figure 12.2 Using the machine vice to hold work safely and efficiently. 'V' grooves in the vice jaws allow most work to be held horizontally or vertically with precision

Figure 12.3 A hand vice in use to grip sheet material safely for drilling

Figure 12.4 Using the 'V' block and clamp for accurate drilling of round bar

Table 12.1 Correct drilling speeds in different materials for the commonly used drill sizes.

(The speeds given are for High Speed Steel (HSS) drills; for carbon steel drills, halve the speeds given. Where speeds are out of the available range, use the highest or lowest speed possible.)

	3.0 mm ($\frac{1}{8}$ in.)	6.2 mm ($\frac{1}{4}$ in.)	12.8 mm ($\frac{1}{2}$ in.)	Coolant
Mild steel	2500–3300	1250–1650	625–1225	soluble oil
Tool steel	1000–1250	500–625	250–300	soluble oil
Stainless steel	1250–2500	625–1250	300–625	soluble oil
Cast iron	1250–2000	625–1000	300–500	dry
Brass	3500–7000	1750–3500	875–1750	dry, paraffin, or soluble oil
Copper	5000–6500	2500–3250	1250–1625	dry, paraffin, or soluble oil
Aluminium	5000–8000	2500–4000	1250–2000	dry, paraffin, or soluble oil
Bronze	1250–2500	625–1250	300–625	dry, paraffin, or soluble oil
Zinc alloys	2500–4000	1250–2000	625–1000	dry, paraffin, or soluble oil
Plastics	3000–6000	1500–3000	750–1500	soluble oil or dry

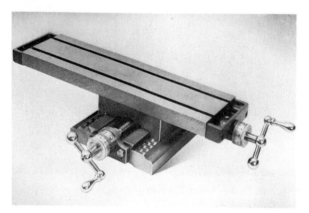

Figure 12.5 A milling table for use on a pillar drill (available as castings from E. W. Cowell Ltd., Watford)

Figure 12.6 End milling in the drilling machine using the Cowell milling table. The drill chuck has been replaced by a collet chuck although the drill chuck can be used. (The chuck guard has been removed for the photograph)

Types of drill

High speed steel (HSS) twist bits are suited to all tasks which are likely to be undertaken in the science workshop. Special drills which may be encountered (see Figure 12.7) are *slow helix drills* for plastics, brass, bronze, and cast iron; *quick helix drills* for copper, aluminium, and zinc; *3 or 4 flute core drills* used to enlarge existing holes, particularly in castings; *HSS countersink bit*; *D bit*, made from hardened and tempered silver steel for bottoming, counterboring, or spot facing in holes drilled with a twist bit; *hand reamer* (may have spiral or straight flutes) used to make an accurate hole from one drilled slightly undersize with a twist drill.

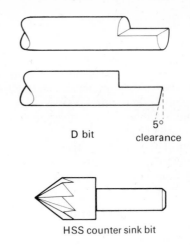

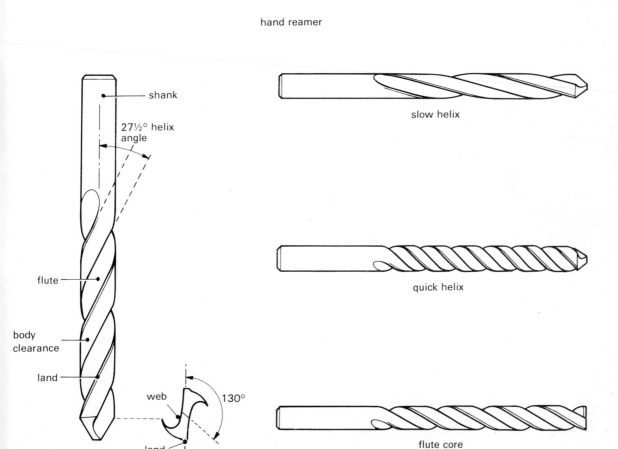

Figure 12.7 Types of drill

Sharpening drills

This is carried out on a grindstone and is most easily done in a special drill sharpening jig (a jig, or the services of a skilled technical assistant, may be available in the school craft department). Observe the following points (see Figure 12.8):

1. Do not burn the drill on the grindstone; dip the tip in water occasionally.
2. Ensure that the cutting lips are of equal length and the tip angle is approximately 118°.
3. There must be a clearance angle of 10°.

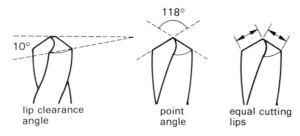

Figure 12.8   Sharpening drills

Twist drill sizes

The home handyman will probably be familiar with only a small range of drill sizes which will probably run from $\frac{1}{16}$ in. to $\frac{1}{4}$ in. or $\frac{1}{2}$ in. in steps of $\frac{1}{16}$ in. The engineer uses a much wider range of drills ranging from $\frac{1}{64}$ in. to 3 in. in steps of $\frac{1}{64}$ in. Various systems of nomenclature have been applied to Imperial size drills – fractional nomenclature, drill gauge numbers, and letter sizes. The British Standards Institute has published a standard system (B.S. 328 A 1963) which replaces Imperial sizes with metric dimensions; a few essential fractional sizes were retained.

Table 12.2 gives a direct comparison of the various systems of nomenclature, but metric sizes should be used whenever possible, even though an interim set can be made up from imperial sizes which for all practical purposes correspond with metric sizes.

The drills listed in Table 12.3 are very close equivalents and should be suitable for all laboratory work. The least accurate is the $\frac{1}{4}$ in. drill which is 0.05 mm undersize when used in place of the 6.40 mm drill, an error of 0.78 per cent which is well within the tolerance of fit in all normal applications.

Table 12.2 British standard twist drill sizes

British Standard (International series)		Old drill gauge, letter, and fractional sizes		
New size (mm)	equivalent (in.)	old size (in.)	decimal equivalent (in.)	fractional
0.35	0.0138	80	0.0135	
0.38	0.0150	79	0.0145	
			0.0156	$\frac{1}{64}$
0.40	0.0157	78	0.0160	
0.45	0.0177	77	0.0180	
0.50	0.0197	76	0.0200	
0.52	0.0205	75	0.0210	
0.58	0.0228	74	0.0225	
0.60	0.0236	73	0.0240	
0.65	0.0256	72	0.0250	
0.65	0.0256	71	0.0260	
0.70	0.0276	70	0.0280	
0.75	0.0295	69	0.0292	
$\frac{1}{32}$	0.0312	68	0.0310	$\frac{1}{32}$
0.82	0.0323	67	0.0320	
0.85	0.0335	66	0.0330	
0.90	0.0354	65	0.0350	
0.92	0.0362	64	0.0360	
0.95	0.0374	63	0.0370	
0.98	0.0386	62	0.0380	
1.00	0.0394	61	0.0390	
1.00	0.0394	60	0.0400	
1.05	0.0413	59	0.0410	
1.05	0.0413	58	0.0420	
1.10	0.0433	57	0.0430	
$\frac{3}{64}$	0.0469	56	0.0465	$\frac{3}{64}$
1.30	0.0512	55	0.0520	
1.40	0.0551	54	0.0550	
1.50	0.0591	53	0.0595	
			0.0625	$\frac{1}{16}$
1.60	0.0630	52	0.0635	
1.70	0.0669	51	0.0670	
1.80	0.0709	50	0.0700	
1.85	0.0728	49	0.0730	
1.95	0.0768	48	0.0760	
			0.0791	$\frac{5}{64}$
2.00	0.0787	47	0.0785	
2.05	0.0807	46	0.0810	
2.10	0.0827	45	0.0820	
2.20	0.0866	44	0.0860	
2.25	0.0886	43	0.0890	
$\frac{3}{32}$	0.0938	42	0.0935	
2.45	0.0965	41	0.0960	
2.50	0.0984	40	0.0980	
2.55	0.1004	39	0.0995	
2.60	0.1024	38	0.1015	
2.65	0.1043	37	0.1040	
2.70	0.1063	36	0.1065	
			0.1094	$\frac{7}{64}$

Metric	Decimal	No.	Decimal	Fraction
2.80	0.1102	35	0.1100	
2.80	0.1102	34	0.1110	
2.85	0.1122	33	0.1130	
2.95	0.1161	32	0.1160	
3.00	0.1181	31	0.1200	
			0.1250	$\frac{1}{8}$
3.30	0.1290	30	0.1285	
3.50	0.1378	29	0.1390	
$\frac{5}{64}$	0.1406	28	0.1405	
3.70	0.1457	27	0.1440	
3.70	0.1457	26	0.1470	
3.80	0.1496	25	0.1495	
3.90	0.1535	24	0.1520	
3.90	0.1535	23	0.1540	
			0.1562	$\frac{5}{32}$
4.00	0.1575	22	0.1570	
4.00	0.1575	21	0.1590	
4.10	0.1614	20	0.1610	
4.20	0.1654	19	0.1660	
4.30	0.1693	18	0.1695	
			0.1719	$\frac{11}{64}$
4.40	0.1732	17	0.1730	
4.50	0.1772	16	0.1770	
4.60	0.1811	15	0.1800	
4.60	0.1811	14	0.1820	
4.70	0.1850	13	0.1850	
			0.1875	$\frac{3}{16}$
4.80	0.1890	12	0.1890	
4.90	0.1929	11	0.1910	
4.90	0.1929	10	0.1935	
5.00	0.1968	9	0.1960	
5.10	0.2008	8	0.1990	
5.10	0.2008	7	0.2010	
			0.2031	$\frac{13}{64}$
5.20	0.2047	6	0.2040	
5.20	0.2047	5	0.2055	
5.30	0.2087	4	0.2090	
5.40	0.2126	3	0.2130	
			0.2187	$\frac{7}{32}$
5.60	0.2205	2	0.2210	
5.80	0.2283	1	0.2280	
$\frac{15}{64}$	0.2344	A	0.2340	$\frac{15}{64}$
6.00	0.2362	B	0.2380	
6.10	0.2420	C	0.2420	
6.20	0.2441	D	0.2460	
$\frac{1}{4}$	0.2500	E	0.2500	$\frac{1}{4}$
6.50	0.2559	F	0.2570	
6.60	0.2598	G	0.2610	
$\frac{17}{64}$	0.2656	H	0.2660	$\frac{17}{64}$
6.90	0.2717	I	0.2720	
7.00	0.2756	J	0.2770	
$\frac{9}{32}$	0.2812	K	0.2810	$\frac{9}{32}$
7.40	0.2913	L	0.2900	
7.50	0.2953	M	0.2950	
			0.2969	$\frac{19}{64}$
7.70	0.3031	N	0.3020	
			0.3125	$\frac{5}{16}$
8.00	0.3150	O	0.3160	
8.20	0.3228	P	0.3230	$\frac{21}{64}$

Metric	Decimal	Letter	Decimal	Fraction
8.40	0.3307	Q	0.3320	
8.60	0.3386	R	0.3390	
			0.3437	$\frac{11}{32}$
8.80	0.3465	S	0.3480	
9.10	0.3583	T	0.3580	
9.30	0.3660	U	0.3680	
$\frac{3}{8}$	0.3750	V	0.3770	$\frac{3}{8}$
9.80	0.3858	W	0.3860	
10.10	0.3976	X	0.3970	
10.30	0.4055	Y	0.4040	
10.50	0.4134	Z	0.4130	
10.80	0.4219		0.4219	$\frac{27}{64}$
11.00	0.4375		0.375	$\frac{7}{16}$
11.50	0.4531		0.4531	$\frac{25}{64}$
11.80	0.4687		0.4687	$\frac{15}{32}$
12.20	0.4844		0.4844	$\frac{31}{64}$
12.80	0.5000		0.5000	$\frac{1}{2}$
13.00	0.5156		0.5156	$\frac{33}{64}$

Table 12.3 An interim set of metric drills made up from imperial drills

Metric	Imperial	Metric	Imperial
1.00	60	7.00	J
1.20	$\frac{3}{64}$	7.20	$\frac{9}{32}$
1.40	54	7.40	L
1.60	$\frac{1}{16}$	7.60	N
1.80	50	7.80	$\frac{5}{16}$
2.00	47	8.00	O
2.20	44	8.20	P
2.40	$\frac{3}{32}$	8.40	Q or $\frac{2}{1}$
2.60	38	8.60	R
2.80	34 or 35	8.80	S
3.00	31	9.00	T
3.20	$\frac{1}{8}$	9.20	$\frac{23}{64}$
3.40	29	9.40	U
3.60	$\frac{5}{64}$	9.60	V
3.80	25	9.80	W
4.00	22	10.00	X or $\frac{23}{64}$
4.20	19	10.20	Y
4.40	17	10.40	Z
4.60	14	10.60	$\frac{27}{64}$
4.80	12 or $\frac{3}{16}$	10.80	
5.00	9	11.00	$\frac{7}{16}$
5.20	6 or $\frac{13}{64}$	11.20	use metric
5.40	3	11.50	$\frac{29}{64}$
5.60	2 or $\frac{7}{32}$	11.80	$\frac{15}{32}$
5.80	1	12.00	use metric
6.00	B or $\frac{15}{64}$	12.20	$\frac{31}{64}$
6.20	D	12.50	use metric
6.40	$\frac{1}{4}$ or E	12.80	$\frac{1}{2}$
6.60	G	13.00	$\frac{33}{64}$
6.80	H		

GRINDING WHEELS AND POLISHERS

Bench grinding machines consist of an electric motor with two wheels, coarse and fine, mounted on the ends of the motor shaft. Sometimes a water reservoir is fitted for dipping the tip of the tool being ground to reduce the chance of 'burning'. The grinder **must** be equipped with guards and tool rests; if the machine is without a spark guard, it is advisable to wear safety spectacles, particularly when trueing the stone with a tool known as a 'dresser' which is used when the stone has had much use, becomes glazed, and fails to grind efficiently, or when the stone has become slightly ovoid and the work bumps.

Polishing machines have tapering left and right hand threads on the motor spindle which allow a range of wire and bristle brushes and calico mops to be fitted. The correct use of the different grades of mop and abrasives is given in Chapter 10.

Combined polishing and grinding machines are available (see Figure 12.9) which are fitted with one stone and a polishing spindle.* Short adaptor spindles are also made by Picador which screw onto the outside of a grinder shaft, enabling it to be used for polishing. The same adaptor can be used in the three-jaw chuck of the lathe. (The lathe can also be used for grinding, but this may well prove to be false economy in the long run as carborundum grit mixed with oil can create havoc with accurate mechanisms and bearings.)

An electric hand drill can be used for grinding and polishing with a horizontal bench clamp as an accessory; this may be screwed to a block of wood so that it can be inserted in the woodwork vice when required.

Long hair, ties, loose clothing, and sleeves are a great hazard when using grinders and polishers.

* Some authorities have outlawed the combined polisher as too dangerous. If one *is* used in the laboratory workshop it should be wired to a safety stop switch, only one person should use it at a time, and the mop spindle must be adequately guarded.

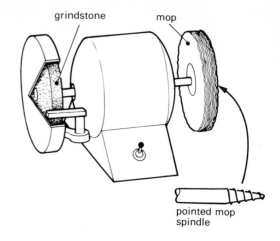

grindstone   mop

pointed mop spindle

Figure 12.9   A combined grinder and polisher

Figure 12.10   A hacksawing machine with capacity for metal up to 50 mm diameter which can be built from castings available from E. W. Cowell Ltd., Watford

13. The Lathe

This is the most versatile machine tool available. The range of work which can be carried out is limited only by the ingenuity of the operator. In addition to its basic turning function, the lathe has many uses – the ingenious craftsman looks upon it as a very rigidly mounted source of power in which tools and work-piece can be moved very accurately relative to each other. The following are some of the important operations which can be carried out on one machine, provided the appropriate attachments are available:

Wood turning
Metal turning:
    facing
    parallel turning
    taper turning
    screw cutting
    knurling
    boring
    free turning
Spring winding
Drilling
Milling
Polishing
Grinding
Coil winding
Thread cutting
Gear cutting
Spinning
Sanding

Attachments may also be available for sawing wood and metal and for planing wood.

Not all of these are of equal importance in the science workshop; some processes, such as gear cutting, would perhaps not be used sufficiently often to justify the cost of the necessary attachments. In this chapter we will be concerned with the basic operation of the lathe. For advanced techniques the reader should consult one of the many specialized texts available.

TYPES OF LATHE

Plain lathe; no screw cutting or automatic movement of the tool is available – virtually obsolete for metalwork.

Woodturning lathe; rather similar to the plain lathe, but runs at the higher speeds desirable for wood turning. It can also be used for metal spinning and, with attachments, for sawing and planing timber.

Back Geared Screw Cutting (B.G.S.C.) lathe. This is the type of machine required for the science workshop as it will perform all the operations listed above, with the probable exception of planing. It is designed primarily for metal turning – its maximum speed of around 2200 r.p.m. will be a little slow for turning wood of small diameter unless the operator is prepared to obtain his final finish from sandpaper rather than the gouge. The metal turning lathe is sometimes called an 'engine lathe' – this is a relic from the nineteenth century when the lathe was used chiefly to turn parts and finish castings for steam engines.

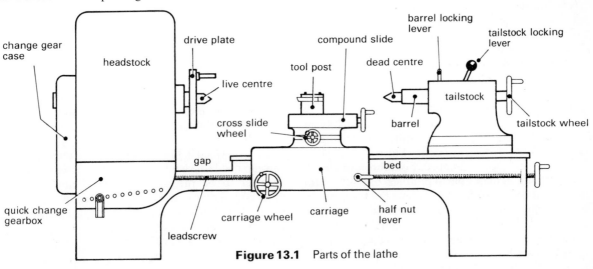

Figure 13.1   Parts of the lathe

199

The headstock

This contains a shaft called the *mandrel* which is connected to the drive mechanism by a set of stepped pulleys, or gears, or by a combination of pulleys and gears to give a range of direct drive speeds. Behind the mandrel lies the lay shaft on an eccentric bearing. By moving the back gear lever with the lathe switched off, the back gears on the lay shaft mesh with the gear wheels on the mandrel. By releasing a lever on the bullgear of the mandrel one of the mandrel gears, linked to the stepped pulley, becomes free to rotate without moving the mandrel. When the lathe is switched on, this gear transmits movement to a first gearwheel on the lay shaft; a second gearwheel on the layshaft then transfers the power to the fixed gearwheel on the mandrel (see Figure 13.2). This simple mechanism, known as the *backgear*, enables a speed reduction to be made. On the 'Myford Super 7' lathe, which is one of the most popular models for science workshops, the following speeds are available from a 1420 r.p.m. motor:

ungeared 2150, 1480, 1020, 700, 515, 425, 290, 200, 130, 90

back geared 80, 55, 40, 25.

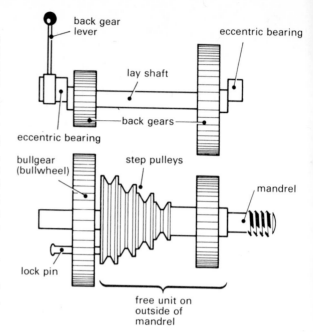

Figure 13.2   Back gear mechanism

Leadscrew, changewheels, and gearbox

Screw cutting lathes have a long screw running the length of the lathe bed; this is the *leadscrew* which moves the cutting tool along the workpiece automatically at a predetermined rate. For simple screw cutting lathes, the rate of feed is altered by changing the arrangement of gears within the changegear case. As this can require reference to charts or calculations to determine the correct combination of change gears, most modern screw cutting lathes have a quick-change gearbox fitted as standard, or at least available as an extra. The 'Super 7' is supplied with a quick change gearbox which offers a lever selection of 48 feed or screw cutting speeds, covering all the common British Standard, B.A., and metric threads; the addition of one or two changewheels to the gearbox further extends this range.

The main use of the power feed is probably in automatic traverse of the tool during parallel and taper turning, enabling a very fine finish virtually free from toolmarks to be obtained. For laboratory work it is often useful to be able to produce 'one-off' worm drives and vernier adjustment screws but the availability of a commercial product should always be considered; B.A. studding and terminals are readily available and are cheaper than the cost of turning them in school.

The bed

This is the backbone of the lathe and should be a very heavy casting. The surface is very accurately ground, and every effort must be made to protect this from damage – when removing chucks and centres from the mandrel it is good practice to place a piece of hardboard over the bed; when mounting irregular work, always turn the lathe by hand to check that no part of the work touches the bed before switching on the power. Lathes are made with either *plain* or *gap beds*, and the size of a British lathe is expressed as the maximum *radius* of work which can be swung over the bed. (Continental and American lathes express their sizes as the maximum *diameter* of work which can be swung over the bed. A British $3\frac{1}{2}$ inch lathe would be termed a 7 inch lathe in the United States.) The gap bed lathe enables work of somewhat greater diameter to be turned on the faceplate – a very important feature in a small workshop lathe; for example, the 'Myford Super 7' lathe admits work of $3\frac{1}{2}$ in. radius between centres, but work up to 10 in. diameter provided that it is not more than $1\frac{1}{2}$ in. deep can be machined in the gap.

The tailstock

This casting has a baseplate machined and scraped to a perfect fit with the bed on which it slides. The tailstock can be locked on the bed by the movement of a lever. On most lathes the tailstock is made of two castings; the upper casting which carries the centre can be moved across the other by means of set screws, thus bringing the dead centre (see page 202) out of line with the live centre. This important facility makes the turning of long tapers a practical proposition without using an expensive attachment.

The *tailstock barrel* is a tight sliding fit in the tailstock casting and can be moved forward and back by the tailstock wheel which operates a square-threaded screw in the rear of the barrel. At the other end of the barrel is a tapered hole used to mount various centres, chucks, die-holders, and taper shanked drills. The barrel may be engraved with a scale which enables holes to be bored to an accurate depth. A barrel locking lever prevents any movement when turning between centres.

The carriage and slides

The carriage enables either the workpiece or the tool to be moved along the bed manually, by turning the carriage wheel, or automatically, by engaging the leadscrew with the half nut lever. The *cross slide* moves the tool across the axis of the work and the *compound slide* (sometimes called the *topslide*) allows accurate movements across or along the work at any angle. The movement of the compound slide is limited to about 3 in. on most small lathes.

The carriage has a number of 'T' slots parallel to the lathe bed which can be used for bolting down work or accessories; work is mounted in this way to have holes accurately machined by a cutter mounted on a boring bar between the centres, and for this reason the carriage is sometimes called the *boring table*.

The toolpost

Tools for turning metal and plastic are usually mounted rigidly on a toolpost bolted to the compound slide. A number of patterns are available.

The American type grips the tool in a slotted post with the tool resting on a packing piece with a curved lower surface which in turn lies in a 'boat' with a hollow upper surface. This arrangement allows rapid setting of the tool height, but has some serious disadvantages the most important being that the post has to be mounted some distance from the edge of the topslide which makes work close to the chuck difficult

if the chuck jaws are not to strike the topslide; the tool can be given more overhang, but there is the consequent risk of 'chatter'. A further disadvantage is that adjustment of tool height by means of a boat alters the cutting angle of the toolface.

The simple Myford toolrest is effective but suffers from the disadvantage that tools are mounted directly on the topslide, which is liable to damage after prolonged use, but this is not a serious problem on the laboratory lathe. The box toolpost is a better arrangement; it has none of the foregoing disadvantages and can be loosened from the topslide to adjust the horizontal angle of the tool to the workface.

The most useful holder for the school workshop is the fourway toolpost which enables four tools of different shapes to be mounted at once, the tool or angle of presentation to the work being changed by movement of a locking lever.

HOLDING WORK IN THE LATHE

Methods employed for holding work in the lathe depend on the material, the type of work, and the available facilities. There are two main methods of mounting work – *between centres*, or by attachment to the mandrel by the use of a *faceplate* or some kind of *chuck*.

Turning between centres

A small hole should be drilled at each end of the work as centrally as possible, using a special centre drill (Slocombe bit) (Figure 13.3). With these holes as points of contact, the workpiece is mounted between the lathe centres with the tailpiece adjusted so that the work turns freely without any 'chatter'. The drive is by means of a catchplate (screwed onto the mandrel) and a carrier (see Figure 13.4).

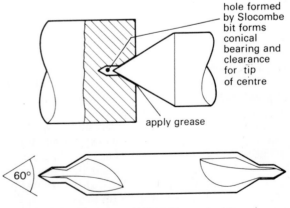

hole formed by Slocombe bit forms conical bearing and clearance for tip of centre

apply grease

60°

Figure 13.3 The centre drill or Slocombe bit

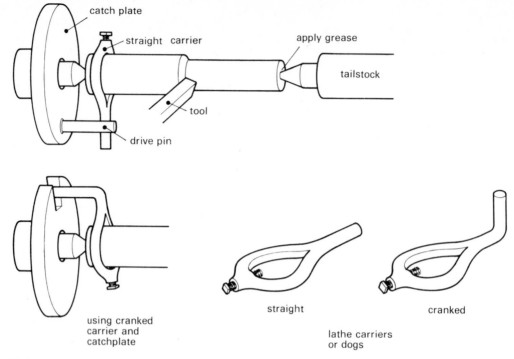

Figure 13.4 Drive and support mechanisms for turning between centres

There are several types of centre. The *driving* or *soft centre* fits onto the mandrel; it is made of unhardened steel (as the centre revolves with the work it receives little wear and being unhardened it can be turned in the lathe to make it absolutely accurate). Because the driving centre revolves with the work it is often referred to as the *live centre*.

The work is supported at the tailstock end by a hardened centre known as the *dead centre*, since it does not revolve. On occasions special centres are used in the tailstock, the *half-centre* (see Figure 13.5) permits the facing of a shaft end and the *revolving centre* (which spins on a ballrace – see Figure 13.6) is used when spinning sheet metal and for between-centre turning when friction must be eliminated, such as with plastics. Care is required when using the revolving centre with long work as this is likely to expand more than short work. If a dead centre is used it must be greased to prevent 'burning'.

When turning long and relatively slender work, the cutting tool tends to 'bow' the workpiece and it may be necessary to use a *steady*. The travelling steady, which is mounted on the saddle, is the most useful type (see Figure 13.7), particularly when turning long threads. The work must be set up to run true before adjusting the steady to it (Figure 13.8). Any attempt

to true the work by pressure from the steady will result in damage to the steady and/or cause the work to turn out of the round.

For *wood turning*, the prong centre is generally used to drive the work, with a ring centre at the tailstock end (Figure 13.9).

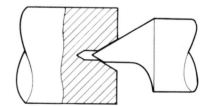

Figure 13.5 Half centre

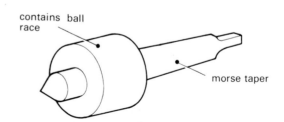

Figure 13.6 Revolving centre

Figure 13.7   A travelling steady being used to support a long shaft during turning

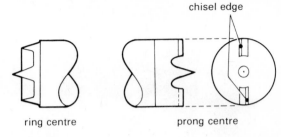

ring centre   prong centre

Figure 13.9

Faceplates

Faceplates are used when the workpiece would be difficult or impossible to hold by other means. Castings with one machined face are often bolted to the slots of the faceplate for subsequent operations. If the workpiece lacks suitable holes for bolting down, it may be held with *faceplate dogs* (Figure 13.10).

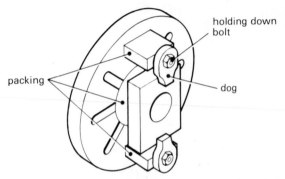

Figure 13.10   Workpiece mounted on a face plate

It is often easier to set up work on the faceplate roughly on the bench and then adjust the work in the lathe with the aid of a Dial Test Indicator (D.T.I.) (Figure 13.8) or a steel pointer set in the tool holder. Paper placed between the faceplate and the work often improves the grip. When work has to be mounted eccentrically, appropriate balance weights should be bolted on to prevent vibration. Occasionally work may have to be attached by means of an angle plate, 'V' blocks, or toolmakers clamps (see Figures 13.11, 13.12, and 13.13).

Chucks

The most commonly used type of chuck is the *self centring three jaw chuck* (Figure 13.14); these are provided with two sets of jaws, *inside* and *outside* for larger diameter work. The jaws are numbered so that they are always inserted into the chuck in the same position.

Figure 13.8   Testing with a Dial Test Indicator (DTI) or 'clock' to ensure that the work is running concentrically

203

Figure 13.11 Casting mounted on a face plate using a bolt and two dogs. Note the block of iron acting as a counterbalance

Figure 13.12 Using an angle plate to mount an awkward casting on the face plate

Figure 13.13 Small component mounted on the face plate using a toolmaker's clamp and two dogs

The *four jaw independent chuck* (Figure 13.15) is used for gripping work of irregular shape, each jaw being tightened independently, and for turning work 'off centre'. Work can be very securely gripped in the four jaw chuck and it should always be used for large work, where the cutting strains tend to be great, in preference to the three jaw chuck.

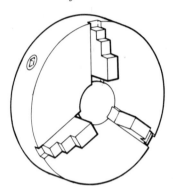

Figure 13.14 A three-jaw chuck

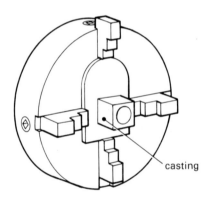

casting

Figure 13.15 A four-jaw chuck

Collets

Collets, or *split chucks* as they are sometimes called, are invaluable because of their ability to centre small, symmetrical work – such as mild and silver steel rods of round, hexagonal, or square cross-section – accurately and instantly. When a number of components are to be machined from a rod, the material

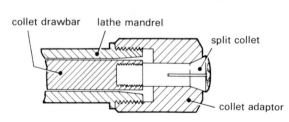

collet drawbar   lathe mandrel   split collet   collet adaptor

Figure 13.16 Collet for a small lathe

can be fed in a continuous length through the mandrel to the collet (Figure 13.16). Sets of collets to fit material of 2–13 mm diameter in 0.5 mm steps are available for most small lathes.

Drill chucks

Drill chucks for use on the lathe have a morse taper to fit the tailstock; the work is held on the mandrel by means of a chuck or faceplate. Large or awkward workpieces are sometimes bolted to the cross slide and the drill chuck fitted in the mandrel.

Whatever type of chuck is used, it is essential that the chuck key is never left in the chuck; some well-designed keys are fitted with a spring loaded collar and are ejected from the chuck as soon as the hand is removed from them.

Mandrels

Mandrels (Figure 13.17) (not to be confused with the lathe mandrel) and *spiggots* (Figure 13.18) are used to hold work which would otherwise be impossible to mount, or where other devices would mark a finished surface. Mandrels and spiggots are often made specially for a job; they have a very slight taper to allow a friction fit.

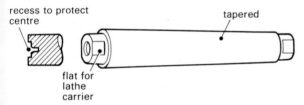

Figure 13.17 A mandrel

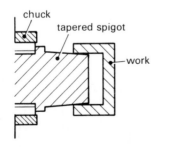

Figure 13.18 A spiggot or stub mandrel

LATHE TOOLS

Cutting tools used in the lathe may be in the form of a large solid tool, or tool bits which are held in a holder. In the laboratory workshop, High Speed Steel (H.S.S.) tools will normally be used. Tungsten tipped tools, which can be operated at higher speeds, are also available, but these need to be sharpened on diamond wheels. Although special tools may be ground for specific jobs, the great majority of work may be undertaken with the standard range of tools illustrated in Figure 13.19.

The *roughing tool* is for coarse cut and is followed with a *roundnosed tool* or with the *finishing tool* for a good finishing cut; the *roundnosed tool* is also used for brass turning.

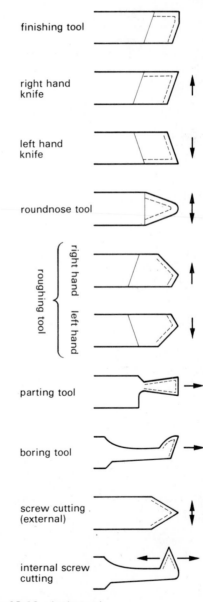

Figure 13.19 Lathe tools

Left and right hand knife tools are for cutting shoulders.

The *parting tool* is for cutting off finished work.

The *boring tool* is used when turning holes to an exact size.

Screwcutting tools are made for internal and external threadcutting and have their points ground to the appropriate form (55 degrees for Whitworth; 60 degrees for American and metric threads).

When grinding tools, it is important to make cutting and clearance angles to suit the material to be cut (see Figure 13.20 and Table 13.1).

Table 13.1 Grinding angles for various materials

Material	Top rake	Side and front clearance
Aluminium	25 °–35 °	
Mild steel	23 °–30 °	
Carbon steel	15 °–25 °	5 °–10 °
Cast iron	7 °–12 °	
Bronze	5 °–10 °	
Brass	5 °–10 °	

(negative rakes for bronze and brass)

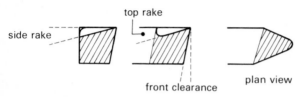

Figure 13.20 Grinding angles

Figure 13.21 Wood turning tools

For wood turning a *deep throated gouge* is the most useful tool (see Figure 13.21), but, this requires some skill to handle safely. The *turning chisel* is used to 'true up' straight tapers and cylinders and the *parting knife* to cut off finished work. When choosing wood turning tools, avoid the small tools produced as 'presents' for amateurs as these are too light and too short for safe working. A well-designed tool will measure at least 18 in. (46 cm) overall and allow the handle and right hand to rest on the hip when working thus providing the operator full control of the tool should it 'dig in'; short tools are likely to be

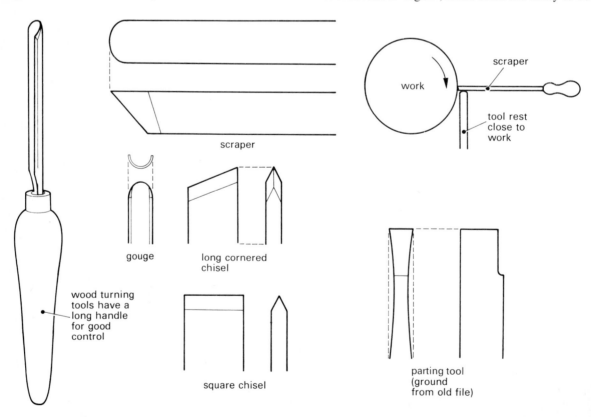

scraper

work

scraper

tool rest close to work

gouge

long cornered chisel

square chisel

wood turning tools have a long handle for good control

parting tool (ground from old file)

thrown across the room. Until some skill is acquired, wood is perhaps most safely turned with a *scraper* which can be ground from an old file. Scrapers are used just below the centre point of the work and rub off rather than cut the surplus wood. Old files used as scrapers must always be fitted with a handle.

LATHE OPERATIONS

Turning between centres

This is the method used for turning long shafts. One advantage is that work can be removed and turned end for end if necessary, the component remaining true-running all the time.

To set the work up, it is usual to set in a three jaw chuck and to drill accurate centres at each end using a centre drill held in the tailstock drill chuck. Remove the chucks, place the soft centre (driving centre) in the lathe mandrel, and turn true if necessary. Fit a hardened dead centre to the tailstock. Fit the appropriate size of carrier to one end of the bar, with a piece of soft packing such as copper or brass placed between the set screw and the bar. Cranked carriers are fitted with their end in a slot in the catchplate which screws on to the mandrel – straight carriers are driven by the pin of the catchplate (see Figure 13.22). The pin and carrier should be bound together with wire or string, taking care that the pin *pushes* the carrier arm around when the lathe is running.

Figure 13.22 When turning between centres, the work is driven by the catch plate which engages with the lathe carrier clamped on the workpiece

When the work has been mounted between the centres, bring the dead centre up to the centre hole and tighten so that the work turns freely without chatter. Always apply grease to the dead centre hole.

For straight turning (cylindrical work) move the tool along the work manually, using the apron handwheel, or automatically, using the leadscrew.

Taper turning

Short tapers are turned with the work mounted between centres and the topslide set to the required angle. The length of taper is determined by the amount of movement available on the topslide.

Long tapers are machined between centres by 'setting over' the tailstock centre by a calculated amount. The tool is set to cut parallel (see Figure 13.23).

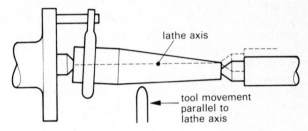

Figure 13.23   Turning long tapers

The above methods will suffice for all taper cutting likely to be needed in the laboratory. Special *taper turning attachments* are made for most lathes for use when production runs are to be undertaken. With these attachments, tapers of greater angles can be produced and taper boring is possible.

Turning on a mandrel

Mandrels may be made for special jobs as required. They are also available as accessories for most lathes in diameters ranging from $\frac{3}{16}$ in. to $1\frac{1}{2}$ in. Professionally-made mandrels are made from hardened, tempered, and ground steel and have a slight taper of about 0.003 in. over a six inch length. Each end is centre drilled and recessed to protect the centre edges from damage. A flat is provided at each end to accommodate the carrier. Work which has been bored or reamed is held on the mandrel by friction. The method is often used for turning accurate collars and gear wheel blanks.

Facing

This term applies to any operation where a 'face' has to be cut at right angles to the lathe axis. A *right or left hand knife* or *facing tool* is used, in conjunction with the cross slide or topslide movement. Work which is too large to mount on the faceplate may be faced by mounting it on a vertical slide on the boring table (cross slide) and machining with a milling cutter in the lathe. Large shafts may be centred using a half centre in the tailstock and small shafts in a three jaw chuck or preferably in a collet.

Boring and drilling

Holes up to 1 in. diameter are usually drilled with a twist drill mounted in the tailstock drill chuck. Drill bits over $\frac{1}{2}$ in. may have a stepped shank so they can be used in $\frac{1}{2}$ in. chucks. Large twist drills can often be obtained cheaply as government surplus or bankrupt stock; it is not generally appreciated that the shanks, even of H.S.S. drills, are not hardened and can be turned down to fit a $\frac{1}{2}$ in. chuck. If deep holes have to be drilled, the turned-down bit can be inserted in a hole drilled in a mild steel rod and the joint brazed or hard soldered; the brazing heat will not damage the hardened drill tip if this is inserted into a large potato.

For occasional jobs a home-made drill can be used. These are of two types (Figure 13.24). *'Flat' drills* can be made from flat cast steel, or silver steel stock, or even an old file; the end of the material is brought to a red heat and splayed out by hammering. The drill is then ground and given the same backing-off clearance as twist drills. Finally the tip is hardened. Flat drills are mounted in the tool post.

D bits are made from a length of high speed or silver steel of the same diameter as the hole to be drilled. The cutting edge is formed by filing down or grinding the steel until there is a flat surface approximately 0.005 in. above the centre line. Clearance is given to the bit by grinding the end of the tool at 10°. The D bit is particularly useful for drilling deep straight holes by following a pilot hole drilled $\frac{1}{32}$ in. undersize. A start for the D bit is given by drilling to a depth equal to four times its diameter with a twist drill.

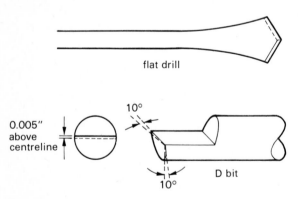

Figure 13.24

Large holes which cannot for some reason be drilled may be bored out on the lathe. If the work can be mounted on a faceplate or chuck, a boring tool held in the toolpost may be used. If the work cannot be held on the faceplate or chuck, it may be bolted to the cross slide or to an angle plate on the slide and bored by means of a fly cutter mounted in the mandrel chuck. If the hole is a large one which goes right through the workpiece, it can be machined with a H.S.S. cutter mounted in a boring bar held between centres. A boring bar can easily be made as shown in Figure 13.25.

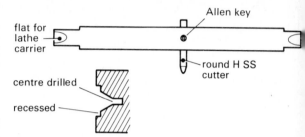

Figure 13.25 A cutter bar

If an angle plate or a vertical slide is available in addition to a small machine vice, the lathe can be used as a jig for accurate drilling of holes into round material if the drill is mounted in the self centring chuck. For drilling holes at right angles through the centre of round stock, the V drill pad is a useful accessory available for Myford lathes (see Figure 13.26).

Figure 13.26 Accurate cross-drilling by means of a 'V' grooved tailstock pad. The work is held in place with a toolmaker's clamp

Screw cutting and screw threading

For *screw threading*, the lathe is used as a jig to present the workpiece accurately to ordinary taps or dies. Male threads can be started satisfactorily by holding an ordinary die holder against the barrel of the tailstock and feeding it towards the work with the tailstock wheel. If much work of this type is anticipated it would be as well to purchase a *tailstock die holder* (Figure 13.27). For die cutting threads the chuck should be turned by hand. Female threads are formed with the tap held in the tailstock drill chuck, or in a tap wrench, the cut being started by pressure from the back centre (Figure 13.28). A cutting oil must always be used and taps withdrawn frequently to clear chips – failure to clear chips invariably leads to broken taps.

Figure 13.27 Cutting a screw thread using a tailstock dieholder. The lathe is best rotated by hand

Figure 13.28 Tapping in the lathe. The work must be rotated by hand

For *screw cutting* the thread is cut by a single pointed tool ground to the correct angle. The pitch of the thread is determined by the leadscrew moving the cutting tool in a certain relationship to the speed of the headstock, this relationship or ratio being determined by the gears used. On older lathes the gear wheels themselves have to be changed to the correct ratio for a given thread; the lathe handbook will give details, and in most cases a chart inside the change wheel cover gives gear trains to cut a wide variety of threads. Most modern Imperial lathes are fitted with a Norton gear box and the selection of the correct ratio for a given thread is only a matter of moving a lever. Imperial lathes are usually supplied with gears ranging from 20 to 120 teeth in steps of five, the 40 and 60 tooth wheels being duplicated; a 127 tooth wheel is supplied to cut most metric threads.

Prepare work for screw cutting by turning to the correct diameter, and, if possible, turning the end to the core diameter. At the end of the portion to be threaded, cut a groove to the core depth for a distance equal to twice the pitch of the thread (see Figure 13.29). Material of any length should be supported with a centre.

The tool is now set up accurately using the thread gauge (Figure 13.30). If a fine thread is to be cut, the tool may be fed in directly with the topslide at right angles to the axis. For a coarse thread, the tool is fed so that it cuts on one edge only, by setting-over the topslide at an angle to the centre line; the amount of set-over is equal to half the angle of the thread form – $27\frac{1}{2}°$ for Whitworth threads and 30° for metric threads.

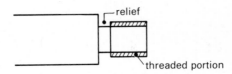

Figure 13.29

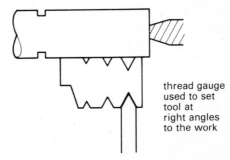

Figure 13.30 Thread gauge used to set tool at right angles to the work

Set the topslide micrometer collar to zero and bring the tool up to the work with the cross slide. Move the tool a little to the right of the work and adjust the top slide ready to give a light cut. Start the machine on a slow speed and engage the half nut with the leadscrew. At the end of the cut withdraw the tool and reverse the lathe without disengaging the halfnut; use the topslide to increase the cut on the next forward traverse. The thread is complete when it reaches the core diameter.

If the lathe is fitted with a *thread indicator* it is possible to disengage the half nut from the leadscrew and pick up the thread accurately. A thread indicator is essential for cutting multi-start threads.

With experience, threads may be cut at higher speeds.

Knurling

This is the process of imparting a roughened surface to the work to provide a finger or hand grip; the pattern imparted may be straight or diamond. The workpiece is rotated against an engraved wheel of hardened steel, which is forced against the workpiece radially with the cross slide. This imposes a considerable strain on the lathe; with the light lathes found in laboratory workshops it is advisable to use a straddle knurling tool (Figure 13.31) which will put far less strain on the lathe.

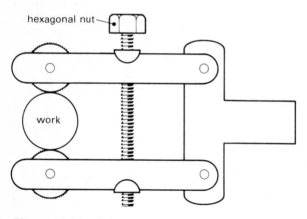

Figure 13.31   A straddle knurling tool

Mount the knurling tool in the tool post and bring it into contact with the work (Figure 13.32). At the start, the wheel should only be in contact with about $\frac{1}{16}$ in. of the work and sufficient pressure applied to establish the pattern. The knurling tool can then be fed along the work using the apron hand wheel. After the first cut, return the tool to the starting point and feed in 0.004 in. with the cross slide. Knurling is complete when all the diamonds are properly formed. Over-knurling will cause parts of the pattern to break away and must be avoided. A good finish to the component can be provided with a chamfer or turned step.

Copious lubricant is essential for knurling to wash away the fine metal particles formed by the operation: if lubrication is insufficient these particles will be rolled into the work and will spoil it. For work up to 1 in. diameter, a lathe speed of 75 r.p.m. is suitable.

Figure 13.32   Knurling in the lathe. This type of knurling tool is most suitable for light lathes as it places no strain on the mandrel

Milling

Light milling can be carried out successfully on a small lathe. It involves the use of a rotating cutter to machine work which is held stationary in a machine vice. Maximum scope is provided by the addition of a compound milling slide (vertical slide) which permits the work to be presented to the cutter at any angle in any plane. If a milling slide is not available, simple operations may be carried out using T bolts and packing to secure the work in the correct position.

The main types of milling operation are *end milling*, *sawing*, and *flycutting*.

End mills (Figure 13.33) are held in a chuck, preferably a collet chuck: they are used to machine edges and surfaces flat as well as for cutting keyways and slots. The tool usually has six or more cutting edges, except for slot drills (Figure 13.34) which have only two cutting edges. With a light feed and cut it is possible to use a standard end mill cutter for slotting.

Sawing is carried out with a slit saw mounted on an arbor (Figure 13.35), supported between centres in the lathe. The author uses a home made arbor which fits the Morse taper of the headstock and is supported by the tailstock centre. The dimensions of this arbor, which takes blades with a $\frac{5}{8}$ in. hole, are given in Figure 13.35.

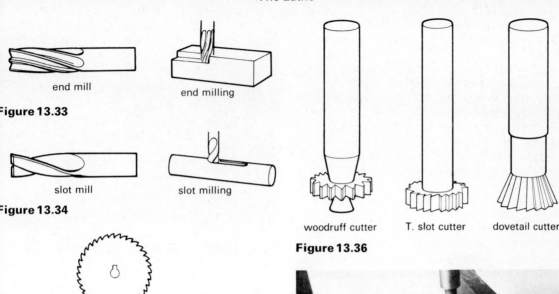

Figure 13.33

Figure 13.34

slitting saw

Figure 13.36

home-made arbor

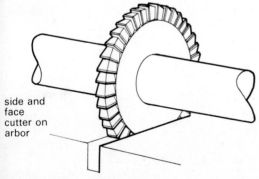

side and
face
cutter on
arbor

Figure 13.35 Sawing

The *Woodruff cutter* (Figure 13.36) is a special type
of saw which can be used for two types of operation –
cutting the seatings for the semicircular woodruff
keys, and milling undercuts such as the T slots used
on machine tables.

Figure 13.37 Milling a slot using a machine vice
mounted on the lathe boring table with an angle plate

Flycutting provides an inexpensive method of
milling as the tool holders and cutters can be
homemade (Figure 13.38). The work is bolted on the
cross slide, with the cut controlled by the leadscrew
and the feed by the cross slide. The process is
particularly useful for machining large surfaces.

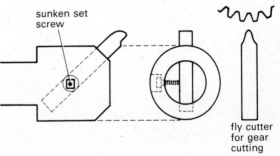

Figure 13.38 Flycutter holder for lathe chuck

If a flycutter is mounted on a boring bar, it can be used to mill concave surfaces, or if the cutter is profiled, grooves of special shapes. Flycutters are ground from H.S.S. toolbits or, if accurate profiles are needed, filed from silver steel and hardened.

Extension of the milling operations which can be undertaken on the lathe are possible if a milling head driven by a belt and independent motor is fitted to the cross slide, and the work mounted in the headstock (Figures 13.39 and 13.40). With such an arrangement it is possible to use the headstock for indexing (that is, rotating the work by an accurate degree), thereby making possible the machining of polygons, hexagons, and squares, as well as some simple gear cutting and the milling of circular grooves. If much gear cutting is contemplated, the purchase of the dividing head designed for the lathe should be considered. A home-made dividing head is shown in Figures 13.42 and 13.43.

Figure 13.41 A simple filing jig. The hardened rollers enable accurate flats to be filed on a round bar. The work can be indexed on the headstock to give different sections (square, triangular etc.)

Figure 13.39 Home-made milling attachment for use in the Myford lathe. The spindle will accept all lathe chucks and collets

Figure 13.42 Home-made dividing head to show sector plates

Figure 13.40 Milling a keyway in the lathe using the home-made attachment shown in Figure 13.39

Figure 13.43 Home-made dividing head mounted on the lathe

Spinning

Spinning is the process of shaping a disc of metal into a hollow vessel by levering it towards a shaped former or spinning 'chuck' (see Figure 13.44). The disc is held against the former with a wooden pressure pad in contact with a revolving centre (the life of the pad will be increased if it is fitted with a brass backing plate). For one-off and small production runs the formers are turned from hardwoods with a close grain such as beech. If deep shapes are to be spun, it is advisable to use two formers. Leverage is applied with a rounded spinning tool (Figure 13.45) pressed against a loose fulcrum mounted in a toolrest held in the tool post. Aways spin at the highest speed available and coat the disc with grease, tallow, or soap to reduce friction and prevent scoring. Start spinning by pressing the polished end of the tool against the disc close to the pad, moving it slowly outwards. Cover only a small area at first and as the metal spins down to the former move the fulcrum to the left to provide a better leverage. If the metal shows a tendency to wrinkle, use a backing stick to support the edge. Too much pressure will cause the metal to become grooved. If the shape is a deep one, the metal will 'work harden' and will need annealing from time to time. Aluminium in 16, 18, or 20 S.W.G. or copper in 20 S.W.G. are the most suitable materials for spinning. Since the metal stretches during spinning, a disc of suitable size must be chosen to give the correct size of finished article; as a rule of thumb, the diameter of the disc should be equal to the average diameter of the vessel added to its height.

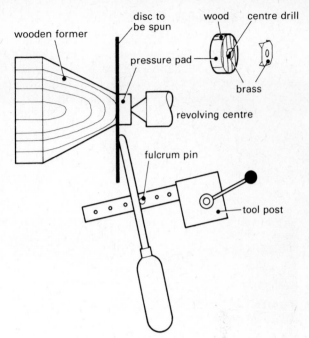

Figure 13.44 Spinning

Wood turning

The capacity of the metal turning lathe for wood turning is limited by its relatively low speed, which makes the turning of softwood difficult. For the limited range of wood turning (usually knobs and handles) required in the laboratory workshop, good results can be obtained using hard woods with the lathe on maximum speed. For a good finish, the gouges and chisels must be really sharp, and scrapers must be given a good cutting 'burr'.

A wood turning tool is always hand-held and supported on a rest. The cutting edge of a gouge should always be just above the centre line of the work and the bevel should rub on the centre line (see Figure 13.45) – this will prevent accidents caused by digging in. A scraping tool should always be held below the centre line.

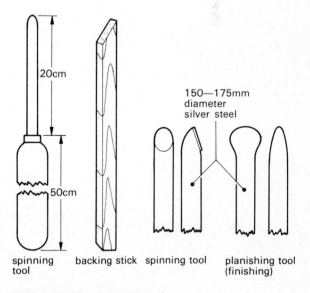

Figure 13.45 Spinning tools

Grinding and polishing

These operations can be carried out on the lathe using spindles or arbors, but great care must be taken to prevent excessive quantities of abrasive materials causing undue wear to the moving parts of the lathe. If the lathe motor has a cooling fan, fluff from polishing mops must be removed frequently to prevent overheating.

Figure 13.46 Winding a spring on the lathe

Shaping in the lathe

Shaping is a method of removing metal by a planing action; neither the tool nor the work revolve. The main use of shaping in the lathe is in the cutting of long keyways in shafts.

Chuck the shaft firmly and support with the tailstock centre; lock the mandrel by engaging the backgear but leave the bull wheel locked to the pulleys (some lathes are provided with a spindle lock). A travelling steady should be used to support long shafts. Mount a parting tool or shaping cutter in the toolpost with its cutting edge facing the headstock. Run the cutter along the work using the lathe carriage manually – take only a shallow cut as this operation places great strain on the lathe. The cut should begin and finish off the work – if this is not possible, a hole should be drilled at the end of the keyway into which the tool may run.

BIBLIOGRAPHY

(All titles are published by Model and Allied Publications Ltd., P.O. Box 35, Hemel Hempstead, Hertfordshire HP1 1EE.)
MARSHALL A. W., *Gear Wheels and Gear Cutting*
SPAREY L. H. (1972), *The Amateur's Lathe*
WESTBURY E. T. (1964), *Lathe Accessories*
WESTBURY E. T. (1967), *Milling in the Lathe*

14. Workshop Tables

CONVERSION FACTORS

* Denotes exact conversion factor.

Length

1 inch	$= 25.4$ mm*
1 foot	$= 304.8$ mm or 0.3048 m*
1 yard	$= 0.9114$ m*
1 fathom	$= 1.8288$ m*
1 chain	$= 20.1168$ m*
1 nautical mile (UK)	$= 1.853$ km
1 nautical mile (international)	$= 1.852$ km*
1 mile	$= 1.60934$ km

Area

1 square inch	$= 6.4516$ cm^2*
1 square foot	$= 0.092\,903$ m^2
1 square yard	$= 0.836\,127$ m^2
1 square mile	$= 2.58999$ km$^2 = 258.999$ ha (hectares)
1 acre	$= 4046.86$ m$^2 = 0.404\,686$ ha

Volume

1 minim (UK)	$= 0.059\,193\,8$ cm^3
1 fluid drachm (UK)	$= 3.55163$ cm^3
1 gallon (UK)	$= 4.54609$ dm^3 (4.54596 l)
1 gallon (US)	$= 3.78541$ dm^3 (3.78531 l)
1 pint (UK)	$= 568.261$ cm^3 (0.5682 l)
1 bushel (UK)	$= 0.036\,368\,7$ m^3 (36.3687 dm^3)
1 bushel (US)	$= 0.035\,239\,1$ m^3 (35.2391 dm^3)

Mass

1 pound (avoir.)	$= 0.453\,592$ kg
1 cwt (UK)	$= 50.8023$ kg
1 cwt (US)	$= 45.3592$ kg
1 ton (UK)	$= 1016.05$ kg
1 ton (US)	$= 907.185$ kg
1 grain	$= 0.064\,798\,9$ g $= 64.7989$ mg
1 dram (avoir.)	$= 1.77185$ g
1 drachm (apoth.)	$= 3.88793$ g
1 ounce (troy or apoth.)	$= 31.1035$ g
1 ounce (avoir.)	$= 28.3495$ g

Force

1 lbf	$= 4.44822$ N
1 tonf	$= 9.96402$ kN
(1 kgf	$= 9.80665$ N)

Stress (pressure)

1 lbf in.$^{-2}$	$= 6.89476$ kN m^{-2} (divide by 100 to obtain pressure in bars)

Moment of inertia

1 lbf ft^{-2}	$= 0.042\,140\,1$ kgf m^{-2}
1 tonf in.$^{-2}$	$= 15.4443$ MN m^{-2}

Heat and specific heat

1 cal	$= 4.1868$ J
1 Btu	$= 1.05506$ kJ
1 Btu lb^{-1} deg F^{-1} 1 Chu lb^{-1} deg C^{-1} 1 cal g^{-1} deg C^{-1}	$\left. \right\} = 4.1868$ kJ kg^{-1} deg C^{-1}

Electrical energy

1 kWh	$= 3.6$ MJ*

Mechanical power

1 horsepower	$= 0.7457$ kW
1 metric horsepower	$= 0.735\,499$ kW

Velocity

1 in. min^{-1}	$= 0.042\,333$ cm s^{-1}
1 ft min^{-1}	$= 0.00508$ m s^{-1} $= 0.3048$ m min^{-1}*

Surface tension

1 dyne	$= 10^{-5}$ N*

Thermal conductivity

1 cal cm cm^{-2} s^{-1} deg C^{-1}	$= 4.1868$ W cm cm^{-2} deg C^{-1}
1 Btu ft ft^{-2} s^{-1} deg F^{-1}	$= 1.73073$ W m^{-1} deg C^{-1}

Energy

1 erg	$= 10^{-7}$ J*
1 hp h	$= 2.68452$ MJ
1 Therm	$= 105.506$ MJ

Angles

1 grade	$= 1/100$ th right angle
1 degree	$= 1.11111$ grade

Plane angle

1 rad (radian)	$= 57.2958°$
1 degree	$= 0.017\,453\,3$ rad
1 minute	$= 2.90888 \times 10^{-4}$ rad
1 second	$= 4.84814 \times 10^{-6}$ rad

Velocity of rotation

1 rev min^{-1}	$= 0.104\,720$ rad s^{-1}

Table 14.1 Celsius (centigrade) to fahrenheit

°C	°F	°C	°F	°C	°F	°C	°F
0	32	35	95.0	70	158.0	290	554
1	33.8	36	96.8	75	167	300	572
2	35.6	37	98.6	80	176	320	608
3	37.4	38	100.4	85	185	340	644
4	39.2	39	102.2	90	194	360	680
5	41	40	104.0	95	203	380	716
6	42.4	41	105.8	100	212	400	752
7	44.6	42	107.6	105	221	420	788
8	46.4	43	109.4	110	230	440	824
9	48.2	44	111.2	115	239	460	860
10	50	45	113.0	120	248	480	896
11	51.8	46	114.8	125	257	500	932
12	53.6	47	116.6	130	266	550	1022
13	55.4	48	118.4	135	275	600	1112
14	57.2	49	120.2	140	284	650	1202
15	59.0	50	122.0	145	293	700	1292
16	60.8	51	123.8	150	302	750	1392
17	62.6	52	125.6	155	311	800	1472
18	64.4	53	127.4	160	320	850	1562
19	66.2	54	129.2	165	329	900	1652
20	68.0	55	131.0	170	338	1000	1832
21	69.8	56	132.8	175	347	1100	2012
22	71.6	57	134.6	180	356	1200	2192
23	73.4	58	136.4	185	365	1300	2372
24	75.2	59	138.2	190	374	1400	2552
25	77.0	60	140.0	195	383	1500	2732
26	78.8	61	141.8	200	392	1600	2912
27	80.6	62	143.6	210	410	1700	3092
28	82.4	63	145.4	220	428	1800	3272
29	84.2	64	147.2	230	446	1900	3452
30	86.0	65	149.0	240	464	2000	3632
31	87.8	66	150.8	250	482	2500	4532
32	89.6	67	152.6	260	500		
33	91.4	68	154.4	270	518		
34	93.2	69	156.2	280	536		

Celcius to Fahrenheit: $(°C \times \frac{9}{5}) + 32 = °F$

Fahrenheit to Celcius: $(°F - 32) \times \frac{5}{9} = °C$

Table 14.2 Fractions of an inch and inches to millimetres

inches	mm	inches	mm
$\frac{1}{16}$	1.6	1	25.4
$\frac{1}{8}$	3.2	2	50.8
$\frac{3}{16}$	4.8	3	76.2
$\frac{1}{4}$	6.4	4	101.6
$\frac{5}{16}$	7.9	5	127.0
$\frac{3}{8}$	9.5	6	152.4
$\frac{7}{16}$	11.1	7	177.8
$\frac{1}{2}$	12.7	8	203.2
$\frac{9}{16}$	14.3	9	228.6
$\frac{5}{8}$	15.9	10	254.0
$\frac{11}{16}$	17.5	11	279.4
$\frac{3}{4}$	19.1	12	304.8
$\frac{13}{16}$	20.6		
$\frac{7}{8}$	22.2		
$\frac{15}{16}$	23.8		

Table 14.3 Millimetres to inches (decimal equivalent)

mm	inches	mm	inches
.1	.00394	22.0	.86614
.2	.00787	23.0	.90551
.3	.01181	24.0	.94488
.4	.01575	25.0	.98425
.5	.01968	26.0	1.02362
.6	.02362	27.0	1.06299
.7	.02756	28.0	1.10236
.8	.03149	29.0	1.14173
.9	.03543	30.0	1.18110
1.0	.03937	31.0	1.22047
2.0	.07874	32.0	1.25984
3.0	.11811	33.0	1.29921
4.0	.15748	34.0	1.33858
5.0	.19685	35.0	1.37795
6.0	.23622	36.0	1.41732
7.0	.27559	37.0	1.45669
8.0	.31496	38.0	1.49606
9.0	.35433	39.0	1.53543
10.0	.39370	40.0	1.57480
11.0	.43307	41.0	1.61417
12.0	.47244	42.0	1.65354
13.0	.51181	43.0	1.69291
14.0	.55118	44.0	1.73228
15.0	.59055	45.0	1.77165
16.0	.62992	46.0	1.81102
17.0	.66929	47.0	1.85039
18.0	.70866	48.0	1.88976
19.0	.74803	49.0	1.92913
20.0	.78740	50.0	1.96850
21.0	.82677		

Metric timber sizes

The following tables relate to the standard metric sizes of timber sold in Britain. Table 14.4 shows the availability of cross section sizes in sawn timber (planed all round, PAR). Table 14.5 compares old Imperial sizes with metric sizes; it can be seen that most of the metric sizes are only slightly smaller. Table 14.6 shows the stock lengths available; softwood will be available in lengths beginning at 1.8 metres, increasing by increments of 300 millimetres. The metric lengths will thus be shorter than the existing Imperial lengths in feet.

Table 14.4 Availability of metric timber sizes

Thickness in mm	75	100	125	150	175	200	225	250	300
16	●	●	●	●					
19	●	●	●	●					
22	●	●	●	●					
25	●	●	●	●	●	●	●	●	●
32	●	●	●	●	●	●	●	●	●
38	●	●	●	●	●	●	●	●	
44	●	●	●	●	●	●	●	●	●
50	●	●	●	●	●	●	●	●	●
63		●	●	●	●	●	●		●
75		●	●	●	●	●		●	●
100		●		●		●			●
150				●		●			●
200						●			
250								●	
300									●

Table 14.6 Stock timber lengths (metric)

Metric lengths	Equivalent in feet and inches	
1.8 m	5 ft	10$\frac{7}{8}$ in.
2.1 m	6	10$\frac{5}{8}$
2.4 m	7	10$\frac{1}{2}$
2.7 m	8	10$\frac{1}{4}$
3.0 m	9	10$\frac{1}{8}$
3.3 m	10	9$\frac{7}{8}$
3.6 m	11	9$\frac{3}{4}$
3.9 m	12	9$\frac{1}{2}$
4.2 m	13	9$\frac{3}{8}$
4.5 m	14	9$\frac{1}{8}$
4.8 m	15	9
5.1 m	16	8$\frac{3}{4}$
5.4 m	17	8$\frac{5}{8}$
5.7 m	18	8$\frac{3}{8}$
6.0 m	19	8$\frac{1}{8}$
6.3 m	20	8

Table 14.5 Comparison of metric and Imperial timber sizes

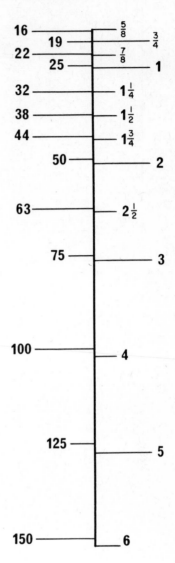

Metric paper sizes

The ISO A series (BS 4000; 1968) is based on the A0 size which has an area of 1 m^2, the ratio of the sides being based on the proportion 2:1, so that the A0 size is 841 × 1189 mm (see Figure 14.1). Smaller sizes are obtained by dividing the longer sides by 2. The A1 size is therefore 594 × 841 mm

A1 replaces the present full imperial
A2 replaces the present half imperial
A4 replaces the present foolscap

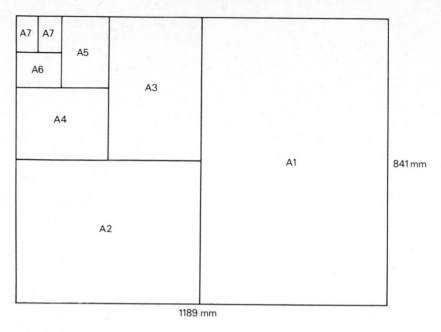

Figure 14.1   Metric paper sizes

Table 14.7 Metric paper sizes

Designation	mm	inches
A0	841 × 1189	33 × 47
A1	594 × 841	$23\frac{1}{2}$ × 33
A2	420 × 594	$16\frac{1}{2}$ × $23\frac{1}{2}$
A3	297 × 420	$11\frac{3}{4}$ × $16\frac{1}{2}$
A4	210 × 297	$8\frac{1}{4}$ × $11\frac{3}{4}$
A5	148 × 210	$5\frac{3}{4}$ × $8\frac{1}{4}$
A6	105 × 148	$4\frac{1}{8}$ × $5\frac{3}{4}$
A7	74 × 105	$2\frac{7}{8}$ × $4\frac{1}{8}$

Heat treatment of carbon steels

Normalising
Process used to refine the grain structure and relieve internal stress.

Heat above the upper critical temperature and allow to cool in air.

Annealing
Process used to soften steel for machining.

Heat above the upper critical temperature and allow to cool very slowly in the furnace.

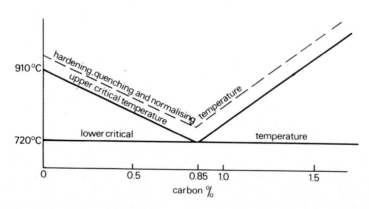

Figure 14.2   Critical temperatures for various steels

Hardening

Makes the steel very hard and brittle.

Heat above the upper critical temperature and quench in water or oil.

Case Hardening

A method of hardening the outer skin of the steel by increasing carbon content.

Heat to red heat and plunge in to charcoal dust; repeat the process several times before bringing to hardening temperature and quenching in water.

Tempering

Hardened tool steels are inclined to be brittle. Tempering is a process which reduces the brittleness without loosing too much of the hardness. The compromise between hardness and toughness varies for different tools.

Table 14.8 Tempering colours on steel

Type of tool	Colour	Temperature (°C)
Scrapers; tools for use on brass	pale straw	220
Drills; small lathe tools	straw	230
Drills; hammers; reamers	dark straw	240
Shears; scissors; dies and taps; punches	brown	255
Axes; woodworking tools	brown/purple	265
Knives; sets; cold chisels	purple	280
Smiths tools; circular saws; screwdrivers	blue	295
Springs; hand saws	dark blue	305
Rules	pale blue	340

Table 14.9 Flame temperatures

Combustible	Diluent	% combustible	Flame temperature (°C)
Methane	air	10.0	1875
propane	air	4.15	1925
butane	air	3.2	1895
iso-butane	air	3.2	1900
acetylene	air	9.0	2325
	oxygen	18.0	2927
		33.0	3007
		44.0	3137
		50.0	2927

To temper tools after hardening, first polish one surface so that the oxide layer can be carefully observed. Then apply heat well back from the cutting edge, and observe the oxide colour until the blue band reaches the cutting edge. Remove the tool from the heat, and notice the succession of colours that creep back from the tip; these colours are used as temperature indicators. Plunge the tool into water the moment the tip attains the appropriate colour for the degree of tempering required (see Table 14.8).

Table 14.10 Wire gauge sizes

Gauge number	Standard Wire Gauge (S.W.G.)	American Brown and Sharp Gauge	Birmingham Wire Gauge
1	0.300	0.2893	0.300
2	0.276	0.2576	0.284
3	0.252	0.2294	0.259
4	0.232	0.2043	0.238
5	0.212	0.1819	0.220
6	0.192	0.1620	0.203
7	0.176	0.1443	0.180
8	0.160	0.1285	0.165
9	0.144	0.1144	0.148
10	0.128	0.1019	0.134
11	0.116	0.0907	0.120
12	0.104	0.0808	0.109
13	0.092	0.0720	0.095
14	0.080	0.0641	0.083
15	0.072	0.0571	0.072
16	0.064	0.0508	0.065
17	0.056	0.0453	0.058
18	0.048	0.0403	0.049
19	0.040	0.0359	0.042
20	0.036	0.0320	0.035
21	0.032	0.0285	0.032
22	0.028	0.0253	0.028
23	0.024	0.0226	0.025
24	0.022	0.0201	0.022
25	0.020	0.0179	0.020
26	0.018	0.0159	0.018
27	0.016	0.0142	0.016
28	0.0148	0.0126	0.014
29	0.0136	0.0113	0.013
30	0.0124	0.0100	0.012
31	0.0116	0.0089	0.010
32	0.0108	0.0080	0.009
33	0.0100	0.0071	0.008
34	0.0092	0.0063	0.007
35	0.0084	0.0056	0.005
36	0.0076	0.0050	0.004

Table 14.11 Tinplate gauge

Gauge name	Mark	Approx. S.W.G.	Overall size
No. 1 common	IC	29	
No. 1 cross	IX	28	
No. 1 two-cross	IXX	27	28 in. ×
No. 1 three-cross	IXXX	26	20 in.
No. 1 four-cross	IXXXX	25	

Table 14.12 Cutting speeds for machine tools

Material worked	Metres per minute	Feet per minute
Tool steel	13–20	45–65
Mild steel	25–35	90
Cast iron	13–20	45–65
Brass	50–80	150–240
Copper	50–80	150–240
Aluminium alloy	65–140	180–400
Bronze	13–23	40–70
Zinc alloys	30–50	90–150
Stainless steel	13–30	40–90
Plastic	35–60	100–180

The figures in Table 14.12 are a guide for High Speed Steel tools for turning, milling, and drilling. With carbon steel tools the speed should be halved. Twice or three times these speeds may be used with carbide tipped tools.

For machine reaming use $\frac{1}{5}$ of the speeds in Table 14.12. For knurling use $\frac{1}{3}$ of the speeds in Table 14.12.

For shaping use $\frac{1}{2}$–$\frac{2}{3}$ of the speeds in Table 14.12.

To calculate the revolutions per minute required, use the following formula:

$$\text{r.p.m.} = \frac{\text{cutting speed in metres per minute} \times 1000}{\pi \times \text{diameter in millimetres}}$$

Cutting lubricants

Tool steel	soluble oil
Mild steel	soluble oil
Stainless steel	soluble oil
Cast iron	dry
Brass	dry, soluble oil or paraffin
Copper	dry, soluble oil or paraffin
Aluminium alloy	dry, soluble oil or paraffin
Bronze	dry, soluble oil or paraffin
Zinc alloy	dry, soluble oil or paraffin
Plastics	dry

Table 14.13 Tapers

Designation of taper	Taper per foot on diameter (in.)	Taper per inch on diameter (in.)	Included angle of taper	Diameter at gauge plane (in.)	Plug depth (in.)	Diameter at plug line (in.)
Morse taper (the most common taper on school machines; will continue to be used even after metrication is complete)						
No. 1 morse	0.5986	0.0499	2°51.45′	0.475	$2\frac{1}{8}$	0.369
No. 2 morse	0.5994	0.0499	2°51.68′	0.700	$2\frac{5}{16}$	0.572
No. 3 morse	0.6023	0.0502	2°52.52′	0.938	$3\frac{3}{16}$	0.778
No. 4 morse	0.6233	0.0519	2°58.51′	1.231	$4\frac{1}{16}$	1.020
No. 5 morse	0.6315	0.0526	3°00.87′	1.748	$5\frac{3}{16}$	1.475
No. 6 morse	0.6256	0.0521	2°59.19′	2.494	$7\frac{1}{4}$	2.116
Brown and Sharp taper						
No. 1 B&S	0.5020	0.0418	2°23.79′	0.2392	$\frac{15}{16}$	0.2000
No. 2 B&S	0.5020	0.0418	2°23.79′	0.2997	$1\frac{3}{16}$	0.2500
No. 3 B&S	0.5020	0.0418	2°23.79′	0.3752	$1\frac{1}{2}$	0.3125
Jarno taper						
Nos. 2 to 20	0.6000	0.0500				
Quick-release tapers (milling machines)						
30	3.500	0.2917	16°35.56′	1.250	1.875	0.7031
40	3.500	0.2917	16°35.56′	1.750	2.678	0.9661
50	3.500	0.2917	16°35.56′	2.750	4.000	1.5833
60	3.500	0.2917	16°35.56′	4.250	6.375	2.3906

(for further details see BS 1660; 1963 [parts 1 and 3])

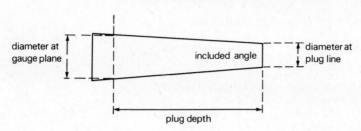

Figure 14.3 Tapers

Table 14.14 Woodworking machinery peripheral or surface speeds

Timber type	Cutting speed in metres per minute						
	Drilling	Turning	Bandsaw	Jigsaw	Circular saw	Sanding	Planing
Hardwood	80	150	650	30	2500	850	1250
Softwood	120	300	1000	100	3000	1000	1750
Very soft wood	160	450	1250	200	4000	1250	2000

Table 14.15 Grinding wheel speeds (peripheral)

Metal tool grinder	850–1250 metres per minute
Woodwork tool grinder	125–200 metres per minute
Horizontal grindstone	125–175 metres per minute
Sandstone	125–150 metres per minute

Table 14.17 Melting points of metals and alloys (°C)

Cast iron	1200–1400
Wrought iron	1600–1700
Tool steel	1200–1400
Mild steel	1200–1350
Aluminium	700
Copper	1083
Brass (including gilding metal)	930–1010
Tin	231
Zinc	419
Lead	327
Silver	800–910

Table 14.16 Solders and fluxes

Material	Flux	Solder alloy (% composition)					
		Tin	Lead	Zinc	Aluminium	Bismuth	Antimony
Aluminium	Stearin	85		10	5		
Brass Copper	Zinc chloride	65	35				
Electrical work	Resin in solder core	60	39.5				0.5
Galvanised sheet Zinc	Dilute hydrochloric acid	50	50				
Lead	Tallow	30	70				
Britannia metal Pewter	Tallow or Gallipoli oil	25	25			50	
Tinplate	Zinc chloride	50	50				

Table 14.18 Alloys – composition and uses

Alloy	Composition		Properties/uses
Steels			
Manganese steel	3–14% manganese		Highly resistant to wear; non magnetic
Molybdenum steel	0.7% molybdenum		Tough but malleable/spades
Nickel steel	6–30% nickel		Strong; corrosion resistant/car engine valves; turbine blades
Invar	36% nickel		Expansion very slight even when heated/pendulums; survey tapes
Nickel-chrome steels	1–5% nickel 2% chromium		Corrosion resistant; self hardening in air/heavily loaded machines
Stainless steels	4–15% chromium		Corrosion resistant/cutlery; tools
Tungsten steels	10–18% tungsten		Resistant to frictional heat/cutting tools; wire-drawing dies
High speed steels	0.75% carbon 0.3% silicon 0.3% manganese 15% tungsten 3% vanadium 4% chromium		Chief property is the ability to hold a cutting edge with minimal wear/modern cutting tools
Brasses – copper and zinc based			
Gilding metal	Copper	90	Jewellery and art metalwork
	Zinc	10	
Standard brass	Copper	70	Presswork
	Zinc	30	
Free cutting brass	Copper	60	Easy to machine at high speeds; does not bend
	Zinc	37	
	Lead	3	
Naval brass	Copper	62	Resistant to salt water corrosion
	Zinc	37	
	Tin	1	
Muntz metal	Copper	60	Hot water fittings
	Zinc	40	
Bronzes – copper and tin based			
Bell metal	Copper	78	Sonorous and resonant/bell casting
	Tin	22	
Gun metal	Copper	90	Load carrying bearings
	Tin	10	
Other copper based alloys			
Nickel silver	Copper	63	Tableware; cheap jewellery
	Zinc	27	
	Tin	10	
Phosphor bronze	Copper	92	Bearings
	Zinc	7.5	
	Phosphorus	0.5	
Coinage bronze	Copper	97	Coinage
	Zinc	2.5	
	Tin	0.5	
Tin, lead, and white metal alloys			
Fine solder	Tin	60	Quick-setting soft solder/electrical work
	Lead	40	
Tinman's solder	Tin	50	Soft solder/general purposes
	Lead	49	
	Antimony	1	
Plumbers'	Tin	30	Has a long pasty phase/wiped joints in plumbing
	Lead	69	
	Antimony	1	

Pewter	Tin	80	Art metalwork
	Lead	20	
Brittania metal	Tin	90	Art metalwork
	Antimony	10	
Type metal	Tin	3	Expands slightly just before solidifying to give a sharp impression of the mould/printers' type
	Lead	82	
	Antimony	15	
Wood's metal	Tin	14	Melting point 65 °C/low temperature castings and fusible plugs
	Lead	24	
	Cadmium	12	
	Bismuth	50	
Babbitt metal	Tin	90	White metal bearings for high speed machines
	Copper	3	
	Antimony	7	
Aluminium alloys			
LM4	Aluminium	92	Most common casting aluminium alloy
	Silicon	5	
	Copper	3	
LM6	Aluminium	88	Very free flowing/intricate or thin castings
	Silicon	12	
Duralumin	Aluminium	94.5	Age hardens in five days/sheet metal work; pressing; turning (but not for casting)
	Copper	4	
	Silicon	0.5	
	Manganese	0.5	
	Magnesium	0.5	
Zinc alloys			
Zinc plus	0.1–3% copper		Low temperature castings which machine to a good finish; strong enough to make press-tools for mild steel sheet
	4% alluminium		
	0.05–0.1%		
	magnesium		

Shrinkage

When making patterns for metal casting allow for the
following shrinkage rates:

	Shrinkage per foot of pattern (in.)	*Shrinkage per 300 mm of pattern (mm)*
Zinc alloys	$\frac{1}{8}$	3
Brass	$\frac{3}{16}$	5
Aluminium	$\frac{5}{32}$	4
Cast iron	$\frac{1}{10}$	2.5

15. Laboratory Tables

Table 15.1 Relative humidities of aqueous solutions of H_2SO_4, NaOH, and $CaCl_2$ at 25 °C
Concentrations are expressed as percentage of anhydrous solute in solution, by mass.

Humidity (%)	H_2SO_4	NaOH	$CaCl_2$	Humidity (%)	H_2SO_4	NaOH	$CaCl_2$
100	0.00	0.00	0.00	50	43.10	28.15	35.64
95	11.02	5.54	9.33	45	45.41	29.86	37.61
90	17.91	9.83	14.95	40	47.71	31.58	39.62
85	22.88	13.32	19.03	35	50.04	33.38	41.83
80	26.79	16.10	22.25	30	52.45	35.29	44.36
75	30.14	18.60	24.95	25	55.01	37.45	—
70	33.09	20.80	27.40	20	57.76	40.00	—
65	35.80	22.80	29.64	15	60.80	43.32	—
60	38.35	24.66	31.73	10	64.45	47.97	—
55	40.75	26.42	33.71	5	69.44	—	—

Table 15.2 Relative humidities from a wet and dry bulb hygrometer (ventilated type)

Depression of wet bulb (°C)	Dry bulb temperature (°C)								
	0	5	10	15	20	25	30	35	40
1	81	87	88	89	90	92	93	93	94
2	64	72	76	80	82	85	86	87	88
3	46	59	66	71	74	77	79	81	82
4	29	45	55	62	66	70	73	75	76
5	13	33	44	53	59	63	67	70	72
6	—	21	34	44	52	57	61	64	66
7	—	9	25	36	45	50	55	59	61
8	—	—	15	28	38	44	50	54	56
9	—	—	6	20	30	38	44	50	52
10	—	—	—	13	24	33	39	44	48

More complete data can be found in 'Hygrometric Tables' of the Meteorological Office (Air Ministry), published by HMSO.

Table 15.3 Efficiency of drying agents

Drying agent	Milligrams of residual water per litre of gas dried at 25 °C
P_2O_5	2×10^{-5}
BaO	1×10^{-4}
$Mg(ClO_4)_2$	5×10^{-4}
KOH (fused)	2×10^{-3}
H_2SO_4	3×10^{-3}
$CaSO_4$	4×10^{-3}
$CaCl_2$ (granular)	0.14–0.25

Table 15.4 Relative Atomic Masses

(Scaled to the relative atomic mass $^{12}C = 12$ exactly)

Values quoted in the table, unless marked *, are reliable to at least ± 1 in the fourth significant figure. A number in parentheses denotes the atomic mass number of the isotope of longest known half-life.

Atomic number	Name	Symbol	Relative atomic mass	Atomic number	Name	Symbol	Relative atomic mass
1	Hydrogen	H	1.008	53	Iodine	I	126.9
2	Helium	He	4.003	54	Xenon	Xe	131.3
3	Lithium	Li	6.941*	55	Caesium	Cs	132.9
4	Beryllium	Be	9.012	56	Barium	Ba	137.3
5	Boron	B	10.81*	57	Lanthanum	La	138.9
6	Carbon	C	12.01	58	Cerium	Ce	140.1
7	Nitrogen	N	14.01	59	Praseodymium	Pr	140.9
8	Oxygen	O	16.00	60	Neodymium	Nd	144.2
9	Fluorine	F	19.00	61	Promethium	Pm	(145)
10	Neon	Ne	20.18	62	Samarium	Sm	150.4
11	Sodium	Na	22.99	63	Europium	Eu	152.0
12	Magnesium	Mg	24.31	64	Gadolinium	Gd	157.3
13	Aluminium	Al	26.98	65	Terbium	Tb	158.9
14	Silicon	Si	28.09	66	Dysprosium	Dy	162.5
15	Phosphorus	P	30.97	67	Holmium	Ho	164.9
16	Sulphur	S	32.06*	68	Erbium	Er	167.3
17	Chlorine	Cl	35.45	69	Thulium	Tm	168.9
18	Argon	Ar	39.95	70	Ytterbium	Yb	173.0
19	Potassium	K	39.10	71	Lutetium	Lu	175.0
20	Calcium	Ca	40.08*	72	Hafnium	Hf	178.5
21	Scandium	Sc	44.96	73	Tantalum	Ta	180.9
22	Titanium	Ti	47.90*	74	Tungsten	W	183.9
23	Vanadium	V	50.94		(Wolfram)		
24	Chromium	Cr	52.00	75	Rhenium	Re	186.2
25	Manganese	Mn	54.94	76	Osmium	Os	190.2
26	Iron	Fe	55.85	77	Iridium	Ir	192.2
27	Cobalt	Co	58.93	78	Platinum	Pt	195.1
28	Nickel	Ni	58.70	79	Gold	Au	197.0
29	Copper	Cu	63.55	80	Mercury	Hg	200.6
30	Zinc	Zn	65.38	81	Thallium	Tl	204.4
31	Gallium	Ga	69.72	82	Lead	Pb	207.2*
32	Germanium	Ge	72.59*	83	Bismuth	Bi	209.0
33	Arsenic	As	74.92	84	Polonium	Po	(209)
34	Selenium	Se	78.96*	85	Astatine	At	(210)
35	Bromine	Br	79.90	86	Radon	Rn	(222)
36	Krypton	Kr	83.80	87	Francium	Fr	(223)
37	Rubidium	Rb	85.47	88	Radium	Ra	(226)
38	Strontium	Sr	87.62*	89	Actinium	Ac	(227)
39	Yttrium	Y	88.91	90	Thorium	Th	232.0
40	Zirconium	Zr	91.22	91	Protactinium	Pa	(231)
41	Niobium	Nb	92.91	92	Uranium	U	238.0*
42	Molybdenum	Mo	95.94*	93	Neptunium	Np	(237)
43	Technetium	Tc	(97)	94	Plutonium	Pu	(244)
44	Ruthenium	Ru	101.1	95	Americium	Am	(243)
45	Rhodium	Rh	102.9	96	Curium	Cm	(247)
46	Palladium	Pd	106.4	97	Berkelium	Bk	(247)
47	Silver	Ag	107.9	98	Californium	Cf	(251)
48	Cadmium	Cd	112.4	99	Einsteinium	Es	(254)
49	Indium	In	114.8	100	Fermium	Fm	(257)
50	Tin	Sn	118.7	101	Mendelevium	Md	(258)
51	Antimony	Sb	121.8	102	Nobelium	No	(259)
52	Tellurium	Te	127.6	103	Lawrencium	Lr	(260)

IUPAC 1975

Table 15.5 Solubility of Nitrogen, Oxygen, Carbon Dioxide, and Air in Water

The table gives the volume of gas, in cm^3 dissolved in 1000 cm^3 of water at a total pressure (i.e. partial pressure of gas plus aqueous vapour pressure at the temperature stated) of 760 mm of mercury. (It is assumed that one mole of the gas occupies 22 411.5 cm^3 at s.t.p.)

Temperature (%C)	Nitrogen	Oxygen	Carbon dioxide	Air	Temperature (%C)	Nitrogen	Oxygen	Carbon dioxide	Air
0	23.53	48.63	1704	29.18	16	17.30	34.83	1025	20.14
1	23.03	47.48	1642	28.42	17	17.04	34.23	997	19.75
2	22.54	46.38	1586	27.69	18	16.77	33.65	970	19.38
3	22.08	45.32	1533	26.99	19	16.53	33.10	945	19.02
4	21.61	44.28	1483	26.32	20	16.29	32.56	921	18.68
5	21.18	43.30	1438	25.68	21	16.07	32.01	898	18.34
6	20.74	42.63	1395	25.06	22	15.85	31.48	873	18.01
7	20.32	41.46	1352	24.47	23	15.65	30.96	848	17.69
8	19.91	40.60	1306	23.90	24	15.45	30.45	825	17.38
9	19.53	39.77	1264	23.36	25	15.24	29.96	803	17.08
10	19.17	38.96	1223	22.84	26	15.05	29.49	782	16.79
11	18.82	38.21	1186	22.34	27	14.87	29.04	762	16.50
12	18.49	37.47	1150	21.87	28	14.69	28.60	742	16.21
13	18.17	36.77	1119	21.41	29	14.51	28.17	725	15.92
14	17.86	36.10	1087	20.97	30	14.36	27.77	707	15.64
15	17.58	35.45	1057	20.55					

Table 15.6 Ignition temperatures of some organic substances, and limits of flammability of gases

Substance	Ignition temperature in air at atmospheric pressure (°C)	Limits for flammability as percentage by volume of gas in air at atmospheric pressure and 25 °C	
		Lower	Upper
Acetaldehyde $\}$ see below			
Acetone			
Acetylene			
Aniline			
Benzaldehyde, $C_6H_5.CHO$	180		
Benzene, C_6H_6	740	1.4	5.5
Butane, C_4H_{10}		1.5	6
n-butyl alcohol (butan-1-ol), C_4H_5OH	450		
Butylene, C_4H_8		1.7	9
Carbon monoxide, CO	644–658	6.3	71.2
Ethanol (acetaldehyde), CH_3CHO	185	4	57
Ethanol, C_2H_5OH	558	3.7	13.7
Ethane, C_2H_6	520–630	3.3	10.6
Ethene (ethylene), C_2H_4	542–547	3.4	14.1
Ethoxyethane (diethyl ether), $(C_2H_2)_2O$	343	1.6	7.7
Ethyne (acetylene), C_2H_2	406–440	2.5	82
Glycerol (propane-1,2,3-triol), $C_3H_5(OH)_3$	500		
Hexane, C_6H_{14}	487	1.3	
Hydrogen, H_2	580–590	6.2	71.4
Hydrogen sulphide, H_2S		4.3	45

Isopropyl alcohol (propane-2-ol), C_3H_7OH	590	2.7	
Kerosene	295		
Methane, CH_4	650–750	5.8	13.3
Methanol, CH_3OH		6	36
Methylbenzene (toluene), $C_6H_5.CH_3$	810	1.3	6.8
Pentane, C_5H_{12}		1.3	4.9
Pentyl alcohol (amyl alcohol), $C_5H_{11}OH$	409		
Phenylamine (aniline), $C_6H_5NH_2$	770		
Propane, C_3H_8		2	7
Propanone (acetone), $(CH_3)_2CO$	700	2.5	13
Propyl alcohol, C_3H_7OH	505	2.6	
Propylene, C_3H_6		2.2	9.7
Town gas		7	21

Table 15.7 Approximate pH of solutions of acids and bases

Acid	Concentration	pH	Base	Concentration	pH
Hydrochloric	M (N)	0.1	Sodium hydroxide	N	14.0
Hydrochloric	0.1 M (0.1 N)	1.1	Sodium hydroxide	0.1 N	13.0
Hydrochloric	0.01 M (0.1 N)	2.0	Sodium hydroxide	0.01 N	12.0
Sulphuric	0.5 M (N)	0.3	Potassium hydroxide	N	14.0
Sulphuric	0.05 M (0.1 N)	1.2	Potassium hydroxide	0.1 N	13.0
Sulphuric	0.005 M (0.01 N)	2.1	Potassium hydroxide	0.01 N	12.0
Sulphurous	0.05 M (0.1 N)	1.5	Lime	saturated	12.4
Phosphoric(V)	0.1 N	1.5	Trisodium phosphate	0.1 N	12.0
Oxalic (ethanedioic)	0.1 N	1.6	Sodium carbonate	0.1 N	11.6
Tartaric	0.1 N	2.2	Ammonia	N	11.6
Malic	0.1 N	2.2	Ammonia	0.1 N	11.1
Citric	0.1 N	2.2	Ammonia	0.01 N	10.6
Methanoic (formic)	0.1 N	2.3	Magnesium oxide	saturated	10.5
Lactic	0.1 N	2.4	Calcium carbonate	saturated	9.4
Ethanoic (acetic)	N	2.4	Borax	0.1 N	9.2
Ethanoic (acetic)	0.1 N	2.9	Sodium hydrogencarbonate	0.1 N	8.4
Ethanoic (acetic)	0.01 N	3.4			
Carbonic	saturated	3.8			
Boric	0.1 N	5.2			

Osmotic pressure

In a dilute solution, let:

the volume of the solvent be v cm^3

the mass of substance
dissolved be g grams

the molecular mass of the
substance be m relative molecular
mass units

the temperature be t °C.

Then the osmotic pressure of the solution, p,
(in mmHg) is given by:

$$p = \frac{g \times (273 + t) \times 760 \times 22\,400}{273\,mv}.$$

Table 15.8 Identification colours for gas cylinders

(British Standards 349 : 1932. Amendments 1938, 1942, 1947, 1952, 1957. For non-medical purposes.)
The American code is entirely different.

Gas	Ground colour of cylinder	Colour of bands
Acetylene (ethyne)	Maroon	None
Air	Grey	None
Ammonia	Black	Red and yellow*
Argon	Blue	None
Carbon dioxide (for temperate use)	Black	None
Carbon dioxide (for tropical or marine use)	Black	White or aluminium paint
Carbon monoxide	Red	Yellow
Chlorine	Yellow	None
Chlorine (cylinders fitted with dip pipes)	Yellow	Black
Coal gas	Red	None
Dichlorodifluoromethane	Mauve (neck end) Grey (bottom end)	None
Ethyl chloride (chloroethane) (flammable)	Grey	Red
Ethylene (ethene)	Mauve	Red
Ethylene oxide (epoxyethane)	Mauve	Red and yellow
Helium	Medium brown	None
Hydrocyanic acid	Blue	Yellow
Hydrogen	Red	None
Methane	Red	None
Methyl bromide	Blue	Black
Methyl chloride (inflammable)	Green	Red
Neon	Medium brown	Black
Nitrogen	Grey	Black
Oxygen	Black	None
Phosgene	Black	Blue and yellow*
Sulphur dioxide	Green	Yellow

* Note that the red or blue band should be adjacent to the valve fitting and the yellow band between that and the ground colour of the cylinder.

Table 15.9 Oxygen consumption of animals

Animal	Temperature (°C)	Average wet mass	Oxygen consumption (cm³/g/hour)
Amoeba proteus	20		0.20
Paramecium caudatum	21	0.001 mg	0.50
Anemonia sulcata	18	10 g	0.0134
Lumbricus	18	10 g	0.06
Chaetopterus	15		0.031
Arenicola	12		0.008
Mytilus	22		0.055
Asterias rubens	15		0.030
Astacus	20	32 g	0.070
Asellus aquaticus	17	50 mg	0.35
Butterfly, Vanessa sp.			
resting	20	300 mg	0.60
in flight	20		100.0
Calliphora larva	20	80 mg	1.3
Tenebrio larva	25	100 mg	0.4
Anguilla vulgaris	25		0.128
Rana temporaria (spring)	19	30 g	0.26
Rana temporaria (winter)	19	30 g	0.08
Lacerta viridis	25	40 g	0.18
Gallus (hen)			0.63
Canary			2.90
Mouse (resting)		20 g	2.5
Mouse (running)		20 g	200.0
Hamster (awake)		500 g	2.90
Hamster (hibernating)			0.07
Dog		10 kg	0.580
Man (resting)		70 kg	0.20
Man (hard work)			40.0

Microscope magnification and resolution

The magnification of a microscope is calculated by dividing the tube length, l, by the focal length of the objective, f, to obtain the *primary magnification*, which is then multiplied by the magnifying power of the eyepiece:

Total magnification =
$$\frac{\text{tube length, } l}{\text{focal length of objective, } f} \times \frac{\text{magnification of}}{\text{eyepiece}}$$

The tube length is the distance from the top of the eyepiece to the top of the objective.

Thus the magnification of a microscope having a tube length of 160 mm and fitted with a no. 1 eyepiece ($\times 5$) and a 16 mm focal length objective will be $160/16 \times 5 = 50$.

Table 15.10 gives the approximate magnification of the majority of lens systems likely to be encountered. The standard modern tube length of 160 mm is assumed throughout.

Table 15.10 Magnification of some lens systems

Focal length of objective (f)	Primary magnification (1/f)	Eyepiece No. 0 (×4)	No. 1 (×5)	No. 2 (×6)	No. 3 (×8)	No. 4 (×10)	No. 5 (×12)	No. 6 (×15)
50 mm (2 in.)	×3.2	13	16	19	26	32	38	48
25 mm (1 in.)	×6.4	26	32	38	51	64	76	96
16 mm ($\frac{2}{3}$ in.)	×10	40	50	60	80	100	120	150
8 mm ($\frac{1}{3}$ in.)	×20	80	100	120	160	200	240	300
6 mm ($\frac{1}{4}$ in.)	×26.6	106	133	160	213	266	319	399
4 mm ($\frac{1}{6}$ in.)	×40	160	200	240	320	400	480	600
3 mm ($\frac{1}{8}$ in.)	×53.5	212	265	318	424	530	636	795
2 mm ($\frac{1}{12}$ in.)	×80	320	400	480	640	800	960	1200
1.7 mm ($\frac{1}{15}$ in.)	×94.1	376	470	564	752	940	1128	1410

A typical outfit for school use would be No. 2 and No. 4 eyepieces with 16 mm, 4 mm, and possibly 2 mm oil immersion objectives.

Table 15.11 Characteristics of some common objective lenses

Focal length (mm)	NA	Resolution (μm)	Magnification	Useful limit of magnification (1000 × NA)	Working distance (mm)	Depth of focus (μm)
50	0.12	2.2	3.2	120	30	—
25	0.20	1.36	6.4	200	14–20	—
16	0.25	1.05	10	250	4–8	10
4	0.75	0.38	40	750	0.2–0.6	1–2
2 (oil immersion)	1.25	0.23	94.1	1250	0.10–0.16	0.5

Resolution and working distances

The lens barrel of the objective is usually engraved with either the magnification or the focal length, and with another figure – the numerical aperture of the lens (NA). This describes a very important optical property of the objective – the higher the NA for any given focal length, the greater is the ability of the lens to resolve fine detail. As NA increases more light passes through the lens; the resolving power increases in direct proportion, but the depth of field decreases inversely. The useful limit of magnification of a lens is $1000 \times NA$.

Lenses of high NA are expensive to produce and are seldom found on school equipment. A lens of high NA must be used with a condenser of similar NA. Table 15.11 lists the principal characteristics of the more commonly used objectives.

Table 15.12 Refractive indices

Substance	Refractive index
Acrylic resin	1.49–1.51
Air	1.000
Canada balsam (dry)	1.535
Canada balsam in xylol	1.524
Cedarwood oil (unthickened)	1.520
Cell constituents (cleared)	1.540
Cellosolve (Ethex, ethoxyethanol)	1.405
Celluloid	1.49–1.50
Cellulose acetate	1.49–1.50
Clove oil	1.533
'Euparal'	1.484
Glass	
borosilicate crown	1.5160
dense barium crown	1.5881
light flint	1.5787
extra dense flint	1.6469
Glycerol	
(100%)	1.473
(50%)	1.397
Lactic acid	1.414
Lacto-phenol	1.440
'Lenzol' (Gurr's)	1.510
Liquid paraffin	1.471
Mersol (Flatters and Garnett)	1.511
Micrex dry (Gerrards)	1.585
Micrex in xylene	1.525
'Microps 163'	1.633
Neutral mountant (Gurr's)	1.510
Quartz, fused	1.458
Distilled water	1.458
Sea water	1.343
Xylol	1.497
Toluene (methylbenzene)	1.497
Ethanol	1.367
Methanol	1.323
Gelatine	1.530

Table 15.13 Density of various substances

Substance	Density $(g\ cm^{-3})$	Substance	Density $(g\ cm^{-3})$
Elements		*Liquids (20 °C)*	
Aluminium	2.70	Ethanoic (acetic) acid	1.05
Carbon (graphite)	2.25	Propanone (acetone)	0.79
Carbon (diamond)	3.52	Ethanol	0.79
Copper	8.30–8.95	Methanol	0.81
Iron	7.03–7.90	Benzene	0.90
Lead	11.01–11.35	Tetrachloromethane	1.60
Magnesium	1.74	Trichloromethane (chloroform)	1.49
Mercury	13.55	Ether	0.74
Platinum	21.37	Glycerol	1.26
Silver	10.42–10.60	Olive oil	0.92
Sulphur	2.0–2.1	Paraffin (light)	0.83–0.86
Tin	6.97–7.30	Paraffin (heavy)	0.87–0.90
Tungsten	18.6–19.1	Sea water	1.025
Zinc	6.92–7.19	Lactic acid	1.44
		Sulphuric acid,	
Alloys		concentrated	1.83
Brass	8.20–8.70	M/2 (Normal)	1.0304
Bronze	8.74–8.89	charged battery	1.28–1.30
Phosphor-bronze	8.80	discharged battery	
Steel	7.60–7.83	Hydrochloric acid,	
		concentrated	1.20 (40%)
Solids, miscellaneous (bulk)		molar (Normal)	1.0162
Asbestos	2.0–2.8	Nitric acid,	
Bone	1.7–2.0	concentrated	1.52
Cardboard	0.7	molar (Normal)	1.0322
Cork	0.22–0.26	Sodium hydroxide, molar	1.0414
Gelatine	1.3	Potassium hydroxide, molar	1.048
Glass (common)	2.4–2.8	Ammonium hydroxide,	
Glass (flint)	2.9–5.9	concentrated	0.8805
Ice (0 °C)	0.917	molar (Normal)	0.9926
Paper	0.7–1.15		
Paraffin wax	0.87–0.91		
Rubber	0.91–1.19		
Wax (sealing)	1.8		

Density of gases relative to air (= 1.0000)

Oxygen	1.1053
Nitrogen	0.9672
Carbon dioxide	1.5290
Hydrogen	0.06952
Ammonia	0.5963
Methane	0.5544

Table 15.14 Relative levels of typical noises

Noise	Decibels	Relative energy	Sound pressure (pascal)	Examples
	120	1 000 000 000 000	20	Threshold of feeling
Deafening	110	100 000 000 000		Jet aircraft at 150 m (500 ft) Inside boiler making factory Pop music group
	100	10 000 000 000	2	Motor horn at 5 m (16 ft)
Very loud	90	1 000 000 000		Inside tube train Busy street Workshop
	80	100 000 000	0.2	Small car at 7.5 m (24 ft)
Loud	70	10 000 000		Noisy office Inside small car Large shop
	60	1 000 000	0.02	Radio set, full volume
Moderate	50	100 000		Normal conversation at 1 m (3 ft) Urban house Quiet office
	40	10 000	0.002	Rural house
Faint	30	1000		Public library Quite conversation Rustle of paper
	20	100	0.0002	Whisper
Very faint	10	10		Quiet church Still night in the country Sound-proof room
	0	1	0.00002	Threshold of hearing

Table 15.15 Musical notes frequency (vibrations per second)

The frequency of the notes of the pianoforte covers the band from 26 to 4096 vibrations per second.

A	26	G	96	F	341	E	1280
B	30	A	106	G	384	F	1365
C	32	B	120	A	426	G	1536
D	36	C	128	B	480	A	1706
E	40	D	144	C	512	B	1920
F	42	E	160	D	576	C	2048
G	48	F	170	E	640	D	2304
A	53	G	192	F	682	E	2560
B	60	A	213	G	768	F	2730
C	64	B	240	A	853	G	3072
D	72	middle C	256	B	960	A	3413
E	80	D	288	C	1024	B	3840
F	85	E	320	D	1152	C	4096

16. Sources of Equipment, Materials, and Information

The following lists of suppliers should enable the laboratory technician to obtain most of the materials and equipment needed for the teaching laboratory. The list is not exhaustive; the aim has been to produce a comprehensive list without too much duplication.

The catalogues of some suppliers are often informative, and should form part of the reference shelf in every preparation room. A selection of those from the list of general suppliers should form a useful basis to which the catalogues of other more specialised firms can be added.

Many of the names below will be well known to teachers and technicians, and for these the address only is given. Where a firm has specialised products, these are indicated in the right hand column.

A more comprehensive list of addresses and sources is available in *Useful Addresses for Science Teachers* by R. W. Wilson (London: Edward Arnold, 1974).

GENERAL LABORATORY FURNISHERS AND SUPPLIERS

1 Arnold E. J. & Son Ltd., Butterley Street, Leeds 10. *General.*
2 Baird & Tatlock Ltd., P.O. Box 1, Romford, Essex, RM1 1HA. *General equipment and laboratory furniture specialists.*
3 Cussons G., Ltd., 102 Great Clowes Street, Manchester, M7 9RH. *General.*
4 Educational and Municipal Equipment (Scotland) Ltd., Blackaddie Road, Sanquar, Dumfrieshire, D94 6DE. *Laboratory furniture.*
5 Educational Supply Association Ltd., Pinnacles, P.O. Box 22, Harlow, Essex. *Norstedts science kits; general.*
6 Elliott, E. J. Ltd., Treforest Industrial Estate, Pontyprydd, Glamorgan.
7 Fisons Scientific Apparatus Ltd., Bishop Meadow Road, Loughborough, Leicestershire, LE11 0RG. *Adhesive tape with metric scales.*
8 Gallenkamp A & Co. Ltd., P.O. Box 290, Technico House, 6 Christopher Street, London, E.C.2. *General.*
9 Grant, Donald, Ltd., Westfield Road, Edinburgh, EH11 2QF. *Modular furniture.*
10 Griffin Biological Labs Ltd., 113 Lavender Hill, Tonbridge, Kent. *Cell mounts.*

11 Griffin & George Ltd., (Southern Region and Head Office), Ealing Road, Alperton, Wembley, Middlesex. *General.*
12 (Northern Region), Ledson Road, Baguley, Withenshawe, Manchester 23
13 (Scottish Region), Braeview Place, Nerston, East Kilbride, Glasgow.
14 Harris Biological Supplies, Oldmixon, Weston Super Mare, Avon.
15 Harris, Philip, Ltd., Lynn Lane, Shenstone, Staffs. Tel: 480077.
16 Hogg, J. N., 19 The Parade, Birmingham 1. *Mineral wool as substitute for asbestos wool.*
17 Horwell, Arnold R., 2 Grangeway, Kilburn High Road, London, N.W.6. *Disposable labware, surgical and microbiological chuck-type inoculating loop.*
18 Jennings, R. W. & Co. Ltd., Scientech House, Main Street, East Bridgford, Nottingham. *General.*
19 North of England School Furnishing Company Ltd., East Mount Road, Darlington, DL1 1LG. *Laboratory furniture.*
20 Morgan & Grundy Ltd., High Street, Cowley, Uxbridge, Middlesex. *Laboratory furniture; fume cupboards.*
21 Shandon Southern Instruments Ltd., Frimley Road, Camberley, Surrey, GU16 5ET. Tel: 0276 63401. *General, physics; chromatography.*
22 Sintacel Ltd., 15 Charterhouse Street, London, E.C.1. *Laboratory furniture, overhead service booms.*

SUPPLIERS OF SCIENTIFIC INSTRUMENTS AND SPECIALISED EQUIPMENT

Electronics and physics equipment, instruments, and meters

23 Advance Electronics Ltd., Instruments Division, Roebuck Road, Hainault, Ilford, Essex. *Electronics equipment.*
24 Advance Teaching Machines, Stair, Lamberhurst, Tumbridge Wells, Kent. *Linear air tracks.*
25 A.E.I., Cawston, Rugby, Warks. *Transducers.*
26 Airmac Ltd., Cressex, Coronation Road, High Wycombe, Bucks. *Nucleonic and electronic equipment.*
27 Allen, P. W. & Co. Ltd., Lighting Division, 253 Liverpool Road, London, N.1. Tel: 607 4665. *Ultraviolet lighting.*
28 Amalgams Ltd., Tinsley Park Road, Sheffield 9, Yorks. Tel: 42521. *Pyrometers, furnaces.*

29 Anders Electronics Ltd., Anders Meter Service, 43–56 Bayham Place, London, N.W.1. Tel: 0-387 9092. *Meters.*

30 A.P.T. Electronic Industries Ltd., Chertsey Road, Byfleet, Weybridge, Surrey. Tel: 41132. *Power units.*

31 Avo Ltd., Avocet House, Archcliffe Road, Dover, Kent. *Avo meters, ammeters, electrical indicating instruments.*

32 Barr & Stroud Ltd., Caxton Street, Anniesland, Glasgow, W.3. Tel: 041-954-9601. *Also* Kinnaird House, 1 Pall Mall East, London, S.W.1. Tel: 01-930 1541. *Optical filters, infra-red glasses, binoculars*

33 Barratt & Co. Ltd., 1 Mayo Road, Croydon, CRO 2QP. *Regular lists of used scientific equipment.*

34 Bausch & Lomb Ltd., Aldwych House, Aldwych, London, W.C.2. Tel: 01-405 9968. *Microscopes including cheap plastic, lasers.*

35 Beck, R. & J. Ltd., Greycaine Road, Bushey Hill Lane, Watford, Herts. *Optical instruments, interferometers.*

36 Broakdeal Electronics Ltd., Market Street, Bracknell, Berks. *Low noise amplifiers.*

37 Cambridge Scientific Instrument Co., Chesterton Road, Cambridge, CB4 3AN. *Scientific instruments.*

38 Chandos Intercontinental, High Street, New Mills, Stockport, Cheshire. *Instruments.*

39 Chandos Products (Scientific) Ltd., Chandos Works, High Street, New Mills, Stockport, Cheshire, SK12 4AN. Tel: New Mills (Derby) 44344.

40 Cressall Manufacturing Co. Ltd., Cheston Road, Aston, Birmingham, B7 5EF. Tel: 021-327-3571. *Rheostats and resistors.*

41 Croydon Precision Instruments Ltd., Hampton Road, Croydon, Surrey. Tel: 01-684-4025. *Potentiometers, Cropico bridges.*

42 Curtis Manufacturing Co. Ltd., 26–28 Paddenswick Road, Hammersmith, London, W.6. Tel: 01-748 4583. *Low voltage supply units.*

43 Dawe Instruments Ltd., Concord Road, Acton, London, W.3. Tel: 01-992 6751. *Wide range of scientific instruments including sound level meters.*

44 Daystrom Ltd., Bristol Road, Gloucester. *Heathkits.*

45 Derritrom Electronics Ltd., Sedlescombe Road North, Hastings, Sussex. Tel: 51372. *Russian instruments.*

46 D.E.W. Ltd., 254 Ringwood Road, Ferndown, Dorset. *Dewtron wave trap. Dew boxes.*

47 Dick, Son, and Lewis Ltd., Guildford Place, Taunton, Somerset. *Stephenson ray boxes.*

48 Ealing Beck Ltd., Greycaine Road, Bushey Hill Lane, Watford, Herts.

49 Ealing Scientific Ltd., Greycaine Road, Bushey Hill Lane, Watford, Herts. Tel: 22272.

50 Electrano Ltd., Pengarreg, Ystrad Meuring,Cards. *Electricity demonstration laboratory.*

51 Electro Mechanisms Ltd., 220–221 Bedford Avenue, Slough, SL1 4RY, Bucks. Tel: 27242. *Strain gauge tutor, strain gauges.*

52 Electronic Instruments Ltd., Hanworth Lane, Chertsey, Surrey. Tel: Chertsey 62671. *Electronic apparatus and equipment.*

53 Electronics (Croydon) Ltd., 102–103 Tamworth Road, Croydon, Surrey. *Growlux fluorescent tube and control gear.*

54 Electrosil Ltd., P.O. Box 37, Pallion Sunderland, Co. Durham. Tel. Sunderland 7184. *Cheap ampere hour recording meters – very useful in school project work.*

55 Electrothermal Engineering Ltd., 270 Neville Road, London, E.7. Tel: 472 9911.

56 Endecotts Ltd., 9 Lombard Road, Morden Factory Estate, London, S.W.19. Tel: 542 8121. *Test sieves.*

57 Epsylon Industries Ltd., 1 Mount Road, Hanworth, Middlesex. Tel: 898 3541.

58 E.S.I. Nuclear, 2 Church Road, Redhill, Surrey. *Scaler timer and ratemeters for schools.*

59 Shandon.

60 Evans Electroselenium Ltd., St. Andrew's Works, Halstead, Essex. Tel: 2461. *Eel colorimeters, photometers, and galvanometers.*

61 Ferranti Lasers, Gallenkamp, P.O. Box 290, Technico House, Christopher Street, London EC2P 2ER. *Lasers*

62 Flan Microwave Instruments Ltd, Dumere Road, Bodmin, Cornwall. Tel: 3161. *Microwave instruments.*

63 Fractional H.P. Motors Ltd., Millmarsh Lane, Brimsdown, Enfield, Middlesex. *Fracmo fractional H.P. motors.*

64 Frank, Charles Ltd., 145 Queen Street, Glasgow. *Optical equipment.*

65 Greenpar Engineering Co. Ltd., Station Works, Harlow, Essex. Tel: 27192. *Electronic equipment, passive probe systems for oscilloscopes.*

66 Harris Electronics Ltd., 138 Gray's Inn Road, London, W.C.1. Tel: 01-837 7937. *Test meters.*

67 Hotfoil Ltd., Heathmill Road, Wombourne, Wolverhampton, Staffs. Tel: Wombourne 2541. *Electrical heating tapes.*

68 Ideas for Education Ltd., 87a Trowbridge Road, Bradford on Avon, Wilts. *Low price vacuum meter. 'Cheka plug' mains test plug.*

69 Jencons (Scientific) Ltd., Mark Road, Hemel Hempstead, Herts, HP2 7DE. Tel: Hemel Hempstead 64641/2/3. *Various, including, classroom planetarium.*

70 Legg (Industries) Ltd., Meridale Street, Wolverhampton, Staffs. Tel: 24081/3. *Power units.*

71 Linear Products Ltd., Moorfield Road, Leeds. Tel 630126. *Amplifiers.*

72 Linstead Electronics Ltd., Roslyn Works, Roslyn Road, London, N.15. Tel: 802 5144. *Low frequency signal generators. Power units.*

73 Marconi Co. Ltd., Research Laboratories, Great Baddow, Chelmsford, Essex. Tel: 53221.

74 Meccano Ltd., Binns Road, Liverpool 13. Tel: 051-228 2701.

75 Megatron Ltd., 165 Marlborough Road, London, N.19. Tel: 272-3739/5975. *Light meters for field-work.*

76 Microwave Instruments Ltd., West Shriton Industrial Estate, North Shields, Northumberland. *Microwave instruments.*

77 Mullard Ltd., Education Service, Mullard House, Torrington Place, London, W.C.1. Tel: 580 6633.

78 Newmarket Transistors Ltd., Exning Road, Newmarket, Suffolk. Tel: 3381/4. *Transistor manifacturer.*

79 Nuclear Enterprises Ltd., Sighthill, Edinburgh 11. Tel: 031-443 4060. *Scintillators.*

80 Offord, C. E. Ltd., Hurst Green, Etchingham, Surrey. *Open University sound level meter.*

81 Optica Ltd., Higham Lodge, Black Horse Lane, London, E.17. Tel: 01-572 2603. *Spectrophotometers.*

82 Oxford Instrument Co. Ltd., Osney Mead, Oxford, OX2 0NX. Tel: 0865 41456. *Power supplies.*

83 Parkin-Elmer Ltd., Beaconsfield, Bucks, HP9 1QA. Tel: Beaconsfield 6161. *Infra-red spectrometer.*

84 Parkinson Instruments, Crompton Parkinson, P.O. Box 46, Northampton, NN1 1LG. *Instruments and motors.*

85 Paton Hawksley Ltd., Rockhill Laboratories, Wellsway, Keynsham, Bristol. *Diffraction gratings.*

86 Polaroid Ltd., Queensway House, Queensway, Hatfield, Herts. *Polarising materials.*

87 Polaron Equipment and Instrument Co. Ltd., 60/62 Greenhill Crescent, Holywell Industrial Estate, Watford, Herts. Tel: 37144. *Spectrometers.*

88 Pye Unicam Ltd., York Street, Cambridge, CB1 2PX. Tel: 0223 58866. *Instruments.*

89 Racal Instruments Ltd., Duke Street, Windsor, Berks. Tel: 69811.

90 Radford Electronics Ltd., Ashton Vale Trading Estate, Bristol 3. Tel: 662301.

91 Radionic Products Ltd., St Lawrence House, Broad Street, Bristol. *Electronic sets.*

92 Rekab Instruments Ltd., 263 High Holborn, London, W.C.1. *Physics instruments and 'Baker' microscopes.*

93 Research Electronics Ltd., Knowl Road, Mirfield, Yorks. Tel: 494242. *Physics and nucleonic instruments.*

94 Research Instruments Ltd., Ace Works, Cumberland Avenue, London, N.W.10. *Micromanipulators.*

95 Rollo Industries, St. Andrews Works, Bonnybridge, Stirling. Tel: 2469. *Dynamic trolleys.*

96 Ross Ensign Ltd., 4 Clapham Common North Side, London, S.W.4. *Binoculars.*

97 Rustrak Instruments, Division of Gulton Europe Ltd., Brighton, BN2 4JU. *Very low priced chart recorders.*

98 Sangamo Weston Ltd., 22 St. George's Court, New Oxford Street, London, W.C.1. *Various meters.*

99 Scientific and Research Instruments Ltd., (S.R.I.), Faircroft Way, Edenbridge, Kent, TN8 6HE. *Excellent ink recording kymograph & spirometer.*

100 Scientific Teaching Apparatus Ltd., Colouhoun House, 27–37 Broadwick Street, London, W.1. Tel: 437 9641. *Physics equipment including Leybold Teaching Apparatus.*

101 S.D.C. Electronics (Sales) Ltd., 34 Arkwright, Astmoor Trading Estate, Runcorn, Cheshire. *DeC solderless matrix boards.*

102 S.E. Laboratories (Engineering) Ltd., North Feltham Trading Estate, Feltham, Middlesex. Tel: 890 5876.

103 Serinco, 13 Main Street, Leslie, Fife. *Linear air tracks.*

104 Service Trading Co. Ltd., 57 Bridgeman Road, Acton, London, W.4. Tel: 995 1560. *Ultra violet lamps.*

105 Shaw Moisture Meters, Rawson Road, Bradford, Yorks. Tel: 33582. *Hygrometers moisture meters.*

106 Sifam Electrical Instrument Co. Ltd., Woodland Road, Torquay, Devon. Tel: 63822. *Electrical measuring and temperature control instruments.*

107 J. Simble & Sons, 76, Queen's Road, Watford, Herts. Tel: 26052. *Tools and wide range of Piccidor equipment.*

108 Still & Cameron Ltd., 66 Rugby Avenue, Wembley, Middlesex. *Glass helices and glass springs for measuring forces.*

109 Taylor Eletrical Instruments Ltd., Archcliffe Road, Dover, Kent, CT17 9EN. Tel: 202620.

110 Telequipment Ltd., 313 Chase Road, Southgate, London, N.14. Tel: 882 1166. *Oscilloscopes.*

111 Teltron Ltd., 32–36 Telford Way, Acton, London, W.3. Tel: 01-743 0103. *Atomic physics apparatus; demonstration X-ray apparatus; Teltron tubes.*

112 Thomas Industrial Automation, Altrincham, Cheshire. *School electrometer.*

113 Turner, J. W. & Son, Buckland Street, Everton, Liverpool, L17 7DS. *Homogenisers, centrifugers, pumps.*

114 Twentieth Century Electronics Ltd., King Henry's Drive, New Addington, Croydon, Surrey. Tel: Lodge Hill 2121. *Geiger counters.*

115 Unicam Instruments, York Street, Cambridge.

116 Unilab Division of Rainbow Radio, Clarendon Road, Blackburn, Lancs. Tel: 57643. *Instruments and Unilab modular physics equipment.*

117 Walden Precision Apparatus, (W.P.A.) Ltd., Shire Hall, Saffron Walden, Essex. *pH meters, colorimeters. Chart recorder and environmental monitoring equipment at reasonable prices.*

118 Weir Electrical Instrument Co. Ltd., Bradford-on-Avon, Wilts. Tel: 204315. *Physics equipment.*

119 Weir Electronics Ltd., Dunbar Road, South Bersted, Bognor Regis, Sussex. Tel: 5991. *Electronic equipment.*

Biological equipment

120 Ammonite Ltd., Llandow Industrial Estate, Cowbridge, Glamorgan. *Plastic fossil casts.*

121 Biddolf, A. Charles Ltd., Green Belt, London Road, Merstham, Redhill, Surrey, RH1 3AL. *Microscopes and biological equipment.*

122 Bowman, E. K. & Co. Ltd., 32–34 Holmes Road, Kentish Town, London, N.W.5. Tel: 01-485 1653. *Cages for all laboratory animals.*

123 British Trust for Ornithology, Beech Grove, Tring, Herts. *Zoological balances.*

124 Butterfly Farm Ltd., The, Bilsington, Ashford, Kent. *Entomological equipment and set insects.*

125 Cenlab Ltd., Tan House Lane, Widnes, Lancs. *Ovens and incubators.*

126 Christie, L., 137 Gleneldon Road, Streatham, London, S.W.6. *Entomological equipment.*

127 Curfew Incubators Ltd., Chersey, Surrey. *Incubators.*

128 Dyos Plastics Ltd., 242 Tolworth Rise South, Surbiton, Surrey. *Disposable petri dishes and tissue culture dishes.*

129 Entech Services Ltd., 6 Mersey View, Liverpool 27. *Low priced mercury vapour moth traps.*

130 Freshwater Biological Association, Ferry House, Rar Sawrey, Ambleside, Westmorland. *Nets for freshwater ecology.*

131 Gerrard & Haig, Gerrard House, Worthing Road, East Preston, Littlehampton, Sussex. Tel: Rustington 4151. *Biological equipment.*

132 Grant Instruments (Cambridge) Ltd., Barrington, Cambridge, CB2 5QZ. Tel: 0763-60811. *Various, including waterbaths and incubators.*

133 Hockney Engineers Ltd., Derwent Place, Bath Road, Leeds. *Large aquaria frames and stands.*

134 Kernick & Son Ltd., Chemistry House, Monthermer Road, Cardiff, Glamorgan, CF2 4XZ. Tel: 26128. *Sterilizers and incubators.*

135 Lee, Robert Ltd., 8 George Street, Uxbridge, Middlesex. Tel: 33181. *Beekeeping equipment.*

136 Lister, Frank Ltd., Emstead Walks, Ovenden, Halifax, Yorks. *Wire rabbit cages.*

137 Loftus, W. R. Ltd., 12–16 The Terrace, Torquay, Devon. *Vinometer to measure the alcohol produced during fermentation.*

138 Longworth Scientific Instrument Co. Ltd., Radley Road, Abingdon, Berkshire. Tel: Abingdon 982/3. *Longworth small mammal traps.*

139 MacFarlane Robson Ltd., Hedgefield House, Blaydon-on-Tyne, Co. Durham. *Plastic slide holders.*

140 Mayfair Incubators Ltd., Churt, Farnham, Surrey. *Incubators.*

141 Microslides (Oxford) Ltd., 7 Little Clarendon Street, Oxford. *Microscope slides.*

142 North Kent Plastics, Home Gardens, Dartford, Kent. Tel: Dartford 21488. *Laboratory animal cages in polypropylene and polycarbonate.*

143 Plastic Box Co. Ltd., King's Street, Market Rasen Lincs. Tel: 2531. *Plastic boxes for biological collecting equipment.*

144 Sterilin Ltd., 12–14 Hill Rise, Richmond, Surrey *Plastic disposable specimen tubes.*

145 Stevens H. G. & Co. Ltd., 16 Coverdale Road Cricklewood, London, N.W.2. *Soil pH meter.*

146 Sudbury Technical Products Ltd., Sudbury House Tylney Road, Bromley, Kent. *Sudbury soil testing kit and pH equipment.*

147 Taylor E. H. Ltd., Beehive Works, Welwyn Garden City, Herts. Tel: Welwyn 4401. *Beekeeping equipment and bees.*

148 Things of Science, Advisory Centre for Education 32 Trumpington Street, Cambridge, CB2 1QY *Freshwater and air environmental monitoring kits.*

149 Watkins and Doncaster, The Naturalist, Four Throws, Hawkhurst, Kent. *Robinson moth trap and biological collecting equipment.*

150 Worldwide Butterflies, Over Crompton, Sherborne Dorset. *Entomological supplies.*

Chemical equipment

151 Aimer Products Ltd., 56–58 Rochester Place, London, N.W.1. Tel: 485 6466. *Repetitive delivery pipettes and burettes.*

152 Analytical Measurements Ltd., Spring Corner, Feltham, Middlesex. Tel: 01-890 5079. *pH meters.*

153 Auto Ice Ltd., 9 Curzon Street, London, W.1. *Ice making machines.*

154 Beckman Instruments Ltd., Eastfield Trading Estate, Glenrothes, Fife, KY7 4NG. Tel: 0592-771234. *pH meters and various ion sensitive probes.*

155 Boro' Labratory and Appliance Co. Ltd., 1 Station Buildings, Catford, London, S.E.6. Tel: 01-690 2901. *Rectangular polythene aspirators.*

156 Chemlab Instruments Ltd., 1b, Seven Kings Road Ilford, Essex. *Automatic dispensing pipette.*

157 Dent & Hellyer Ltd., 27 The Vale, Acton, London, W.3. *Autoclaves.*

158 Elga Products Ltd., Lane End, High Wycombe, Bucks. Tel: 0494 881393. *Water nutrifiers, Elgastat deioniser.*

159 Esco Rubber Ltd., 14–16 Great Portland Street London, W1N 5AB. *Rubber and plastic tubing and laboratory products.*

160 Haig, W. G. & Co. Ltd., Exelo Works, Margate Road, Broadstairs, Kent, CT10 2PS. Tel: 0843 61365/6. *Semi-micro glassware.*

161 Jobling Laboratory Division, Tilling Drive, Stone, Staffs., ST15 0BG. *General-glassware including Pyrex.*

162 Joyce, Loebl & Co. Ltd., A8 Princes Way, Team Valley Estate, Gateshead NE11 0UJ, Co. Durham. Tel: Low Fell 877891. *Colorimeters.*

163 Labgear Ltd., Cromwell Road, Cambridge. Tel: 47301.

164 Measuring & Scientific Equipment Ltd., Manor Royal, Crawley, Sussex. *Balances, centrifuges.*

165 Med-Lab Ltd., 2–6 Agard Street, Derby. Tel: 49094 & 49095. *Semi-micro kits.*

166 Oertling L., Ltd., Cray Valley Works, St. Mary Cray, Orpington, Kent. Tel: Orpington 25771. *Chemical balances.*

167 Phillips Chromatography Ltd., M.E.L. Equipment Co. Ltd., Manor Royal, Crawley, Sussex. *Chromatography equipment.*

168 Reeve Angel & Co. Ltd., 70 Newington Causeway, London, S.E.1.6BD. Tel: 4076126. *Makers of Whatman filter and chromatography papers.*

169 Schuco Scientific, Halliwick Court Place, Woodhouse Road, London, N12. Tel: 3681642. *Pi-pump rack and wheel operated pipette bulb.*

170 Sierex Ltd., 15–18 Clipstone Street, London, W.1. Tel: 5802464. *Glassware washing machines.*

171 Stanton Instruments, Coppermill Lane, London, S.W.17. Tel: 9467731. *Chemical balances.*

172 Willen Bros., 55/59 Dudden Hill Lane, London, N.W.2. Tel: 4597535/7982. *Plastic syringes.*

173 Xlon Products Ltd., Glynn Street, London, S.E.11. Tel: 01-7358551. *Plastic funnels and labware.*

Balances, microscopes, and other general laboratory equipment

174 A.E.I., 33 Grosvenor Place, London, S.W.1. *X-ray diffraction equipment.*

175 The Balance Consultancy, Avon Court, Castle Street, Trowbridge, Wilts. Tel: Trowbridge 4733 or 5292. *Balance specialists.*

176 Bassett & Goodwin Ltd., 14–16 Combe Rise, Saltdean, Sussex. *Astronomical mirrors and accessories.*

177 Bir-Vac Ltd., Swindon Road, Cheltenham, Glos. *High Vacuum equipment.*

178 Carl Zeiss Jenna Ltd., 93/97 New Cavendish Street, London, W1A2AR. *Microscopes.*

179 Casella & Co., Regent House, Britannia Walk, London, N.W.1. Tel: 01-2538581. *Meteorological instruments. 24 hour clockwork recording drums which can be modified for biological recording.*

180 Deitz (Instruments) Ltd., 48 Park Street, Luton, LU1 3HP. *Microscopes.*

181 Edwards High Vacuum Ltd., Manor Royal, Crawley, Sussex. Tel: 28844. *Vacuum pumps. Speedivac.*

182 Engis Ltd., Parkwood Trading Estate, Parkwood, Maidstone, Kent. *Metallurgy equipment.*

183 Gillet & Sibert Ltd., Lynx House, 50 Vicarage Crescent, London, S.W.11 3LA. Tel: 2280117. *Lynx & Conference microscopes and microprojector.*

184 Leach (Rochester) Ltd., 277 High Street, Rochester, Kent. *Microscopes.*

185 McArthur Microscope, C. E. Offord (Microscopes), Hurst Green, Etchingham, Sussex. *Field microscope which can replace the lens of a 16 mm camera or a Television camera. Plastic version sold as the 'Open University' microscope.*

186 Negretti & Zambra Ltd., Stockdale, Aylesbury, Bucks. Tel: Aylesbury 5931. *Meterological instruments, including cheap hand wind meter.*

187 NGN Ltd., Kirk Road, Church, Accrington, Lancs. *Vacuum pumps.*

188 Norwood Instruments Ltd., New Hill Road, Honley, Huddersfield, Yorks. Tel: Huddersfield 61318. *Engineering laboratory equipment.*

189 Panax Equipment Ltd., Holmethorpe Industrial Estate, Redhill, Surrey. *Radioactivity equipment.*

190 Perry, Charles Ltd., London Road, Bishop Stratford, Herts. *Microscopes.*

191 Prior, W. R. & Co. Ltd., London Road, Bishop Stortford, Herts. *Microscopes, optical equipment.*

192 Pyser-Britex (Swift) Ltd., Fircroft Way, Edenbridge, Kent, TN8 6HA. Tel: 073271 4111. *Micro projector and Swift microscopes.*

193 Research Engineers Ltd., Orsman Road, Shoreditch, London, N.1. *Compressor and vacuum pumps.*

194 Shandon Southern Instruments Ltd., Frimley Road, Camberley, Surrey, GU16 5ET. Tel: 0276 63401. *Ohaus balances.*

195 Swift, James & Son Ltd., Joule Road, Basingstoke, Hants. *Microscopes.*

196 Tensometer Ltd., 81 Morland Road, Croydon, Surrey. *Materials testing equipment.*

197 Torsion Balance Co. Ltd., Vale Road, Windsor, Bucks. *Balances.*

198 Vickers Instruments Ltd., Haxby Road, York, YO3 7SD. *And* Breakfield, Coulsdon, Surrey, CR3 2UE. *Biolux and Vickers Steremag microscopes.*

199 Watson Marlow Ltd., Falmouth, Cornwall. *Pumps.*

Laboratory furniture

200 Marler Haley Ltd., 76 High Street, Barnet, Herts. Tel: 01-449 9611. *Folding display stands.*

201 Leigh Trading Co. Ltd., 2 Mill Mather Lane, Leigh, Lancs. *Steel shelving.*

202 Lawtons Ltd., 60 Vauxhall Road, Liverpool, L69 3AU. Tel: 051-2271212. *Cardboard stock boxes to fit metal shelving supplied flat – leaflet available.*

203 Gratnells Ltd., 256 Church Road, London, E107JQ. Tel: 566 9021. *Storage systems.*

204 Fix Equipment Ltd., Kembel House, Basingstoke, Hants. *Storage cabinets.*

205 Dexion Ltd., Dexion House, 2–4 Empire Way, Wembley, Middlesex. Tel: 902 1281. *Slotted angle.*

206 Brown, N. C. Ltd., Cuba Steelworks, Bury, Lancs. Tel: 2924. *Efficient industrial steel storage shelves, drawers, and bins.*

CHEMICALS, STAINS, AND REAGENTS

207 A.H.S. Ltd., Station Road, Didcot, Berkshire. *Sera.*

208 Anderman & Co. Ltd., Battlebridge House, 87–95 Tooley Street, London, S.E.1. *General, also 'Suprapur' and 'Merk' brands. Non-bleeding test papers. Silicagel for thin layer chromatography.*

209 Belgrave (Mercury) Ltd., 5 Belgrave Gardens, St. John's Wood, London, N.W.8. Tel: 01-6243826.

210 Boots Pure Drug Co. Ltd., Thane Road West, Nottingham. Tel: 56111. *General.*

211 British Drug Houses Ltd., BDH Laboratories, Poole, Dorset. Tel: Parkstone 5502. *General.*

212 British Oxygen Co. Ltd., Hammersmith House, Hammersmith, London, S.W.6. Tel: 01-748 2020. *Gases in cylinders.*

213 Cambrian Chemicals Ltd., 59 Macks Road, Bermondsey, London, S.E.16. *General.*

214 Gerrard & Haig, Gerrard House, Worthing Road, East Preston, Littlehampton, Sussex. Tel: Rustington 4151. *Stains and reagents. Canada balsam in applicator tubes.*

215 Gurr, Edward, Ltd., Michrome Laboratories, 42 Upper Richmond Road, London, S.W.14. *Michrome brand biological stains and reagents.*

216 Hopkin & Williams Ltd., Freshwater Road, Chadwell Heath, Essex. Tel: 01-590 7700. *'AnalaR' reagents, amino acids. Dangerous chemicals supplied in small quantity packs, e.g. bromine in 1 cm^3, 2 cm^3 and 2.5 cm^3.*

217 Ilford Photographic Ltd., Ilford, Essex. Tel: 01-478 3000. *Photographic chemicals.*

218 Imperial Chemical Industries Ltd., ICI House, Millbank, London, S.W.1. Tel: 01-834 4444. *Solid carbon dioxide.*

219 Johnson Matthey & Co. Ltd., 81 Hatton Garden, London, E.C.1. Tel: 01-405 6989. *Silver, gold, and platinum foils and wires.*

220 Johnsons of Hendon Ltd., 335 Hendon Way, London, N.W.4. *Photographic reagents.*

221 Koch-Light Laboratories Ltd., Colnbrook, Bucks, SL3 0BZ. *General, rare-earths and enzymes.*

222 Labgear Ltd., Cromwell Road, Cambridge. Tel: 47301. *Radioactive material*

223 Lamb, Raymond, Ltd., 25 St. Stephen's Road, London, W.13. Tel: 01-997 3827. *Waxes and reagents for histology and cytology.*

224 May & Baker Ltd., Laboratory Chemicals Division, Dagenham, Essex, RM10 7XS. Tel: 01-592 3060. *General.*

225 Medical-Pharmaceutical Developments Ltd., Ellen Street, Portslade, Brighton, BN41 1EQ. *'Decon 75' laboratory glassware washing detergent.*

226 Med-Lab Ltd., 2–6 Agard Street, Derby. Tel: 49094 and 49095. *General.*

227 Methylating Co. Ltd., Dagenham Dock, Dagenham, Essex. *5 gallon drums of industrial meths.*

228 Oxoid Ltd., Southwark Bridge Street, London, SE1F. Tel: 01-928 4515. *'Oxoid' culture media and school microbiological kits.*

229 Panax Equipment Ltd., Homethorpe Industrial Estate, Redhill, Surrey. *Radioactive material.*

230 Proctor & Gamble Ltd., P.O. Box 1EF, City Road, Newcastle-upon-Tyne. Tel: 0632 857141. *Disclosing tablets.*

231 Radiochemical Centre, White Lion Road, Amersham, Bucks. Tel: 02404 4444. *Radioactive material.*

232 Radley, John (Labs) Ltd., 220–222 Elgar Road, Reading, Berks. *Radioactive material in tablet form.*

233 Reeve–Angel Ltd., Laboratory Chemicals Division, Gaunt Street, London, S.E.1. Tel: 01-407 6128. *Successor to 'Harringtons Chemicals'. Good range of biochemicals.*

234 Ridsdale & Co. Ltd., Nerwham Hall, Newby, Middlesbrough, Yorks, TS8 9EA. Tel: 0642 37216. *Analytical reagents in tablet form.*

235 Rubert & Co. Ltd., Acru Works, Demmings Road, Cheadle, Cheshire. Tel: 061-428 5855. *'Acru' resin for plastic embedding of museum specimens.*

236 Speciality Gases Ltd., 15 Barnfield Trading Estate, Tipton, Staffs. Tel: 021-557 2443. *Aerosols of wide range of pure gases for calibrations.*

237 Speedy Carbon Dioxide Service Ltd., 16 Clement Street, Parade, Birmingham. *Carbon dioxide.*

238 Tintometer Sales Ltd., Waterloo Road, Salisbury, Wilts. Tel: 0722. *Water analysis reagents.*

239 Wrigley, Paterson & Co. Ltd., Superlin Works, Isleworth, Middlesex. *Animal room disinfectant.*

LABORATORY GLASSWARE AND CONTAINERS

240 Anderman & Co. Ltd., Lab Supplies Division, Central Avenue, East Moseley, Surrey, K8OZ. *Glassware and porcelain.*

241 Chance Glassworks Ltd., Spon Lane, Smethwick, Birmingham. Tel: 021-553 5552. *Optical glass, slides and cover glasses.*

242 Davey & Moore Ltd., Lockfield Avenue, Brimsdown, Enfield, Middlesex. *'Davisil' glassware.*

243 Elliot, J. J. Ltd., E-Mill Works, Treforest Industrial Estate, Pontypridd, Glamorganshire. *'E-Mil' glassware.*

244 Flaig, W. G. & Sons Ltd., Exelo Works, Margate Road, Broadstairs, Kent. *'Exelo' glassware.*

245 Jobling, James A. & Co. Ltd., Wear Glass Works, Sunderland, Co. Durham. *'Pyrex' glassware.*

246 Process Plant Division, Newstead Industrial Estate, Trentham, Stoke-on-Trent. Tel: 514856.

247 Moncrieff, John Ltd., North British Glassworks, St. Catherine's Road, Perth. *'Monax' glassware.*

248 Pelling and Cross Ltd., 104 Baker Street, London, W.1. *'Squeesytainers' or Air Evac bottles.*

249 Quickfit & Quartz Ltd., Stone, Staffs. *'Quickfit' ground-glass joint glassware.*

250 Quadrant Glass Co. Ltd., The Pinnacles, Harlow, Essex. *Scientific glassware.*

251 Scientific Glass Blowing Co. Ltd., 163 Higginshaw Lane, Royton, Oldham, Lancs. Tel: 061-633 1628.

252 Springham, G. & W. Ltd., 35 South Road, Temple Field, Harlow, Essex. *Stopcocks, separating flasks.*

PLANT, ANIMAL, AND GEOLOGICAL MATERIAL

253 Biddolph, Charles, Green Belt, London Road, Mertstham, Surrey. *Frogs and other amphibians.*

254 Bleak Hall Bird Farm (Luton) Ltd., 45 Waller Street, Luton, Beds. *Frogs and other amphibians.*

255 Samuel Dobie & Son Ltd., Upper Dee Mills, Llangollen, LL20 8SD. *Wide range of seeds, particularly greenhouse plants.*

256 Dunn, T. R. & Son Ltd., Broad Oak Farm, High Legh, Knutsford, Cheshire. *Fertile eggs and day old chicks.*

257 Dunn's Farm Seeds Ltd., Salisbury, Wilts. *Farm seeds.*

258 Field Aquarium, 94 Burdett Road, Bow, London, E.3. *Marine and tropical fish.*

259 Fox Biology (Dunbar) Ltd., Woodrush Laboratories, Woodrush Brae, Dunbar, East Lothian. *Wide range of biological material.*

260 Fox Scientific Animal Service, Home Farm, Aldenham Park, Elstree, Herts. *Mice, rats, and pelleted laboratory animal foods.*

261 Freshwater Biological Association, Ferry House, Far Sawrey, Ambleside, Westmorland. *Freshwater animals.*

262 Gemrocks Ltd., 20–23 Holborn, London, E.C.1. *Minerals.*

263 Keston Foreign Bird Farm, Bramble-tye, Keston, Kent. *Birds.*

264 Laboratory Animal Centre, Medical Research Council Laboratories, Woodmansterne Road, Carshalton. Surrey. *List of accredited breeders from whom it is safe to buy animals for school use.*

265 L. & C. Worm Farm & Biological Supplies, 54 Gladstone Road, Wallasey, Cheshire. *Worms and locusts.*

266 Lee, Robert Ltd., 8 George Street, Uxbridge, Middlesex. *Bees.*

267 Mackie, A. J., Skirmet, Henley, Oxon. *Venus fly traps.*

268 Marine Biological Association, The Laboratory, Citadel Hill, Plymouth, Devon. *Marine material, living and preserved.*

269 Midland Fishery, Nailsworth, Stroud, Glos. *Trout eggs (Jan–March).*

270 Perry's Hardy Plant Farm, Enfield, Middlesex. *Aquatic plants.*

271 Ralph, K., 20 Melampus Street, Barrow-in-Furness, Lancs. *Stick insects and silkworm eggs.*

272 Royal Zoological Society of Scotland, Murrayfield, Edinburgh 12. *Invertebrates.*

273 Sanders, Orchids Ltd., Selsfield, East Grinstead, Surrey. *Orchid plants.*

274 Shirley Aquatics Ltd., Monkspath, Shirley, Solihull, Warwicks. *Fish and aquatic plants.*

275 Spillers Ltd., The Old Change House, Cannon Street, London, E.C.4. Tel: 01-248 5700. *Animal foods, including rabbit pellets.*

276 Taylor, F. E., D.E.S. Laboratories, Ivy Farm, Knockholt, Kent. *Plastercasts of fossil horses and Archaeopteryx; genetics mice.*

277 Thompson & Morgan Ltd., London Road, Ipswich, Suffolk. *Seeds of many plants of botanic interest.*

278 Tuck, A. & Sons Ltd., The Lab Animal Breeding Station, Rayleigh, Essex. *Guinea-pigs and mice.*

279 Wriggley Worm Farm, Blanakeil Village, Dunness, Sutherland. *Worms.*

TOOLS, WORKSHOP MATERIALS, AND COMPONENTS

280 Art Veneers Ltd., Industrial Estate, Mildenhall, Suffolk. Tel: 712550. *Very wide range of special timbers.*

281 Aston & Full Ltd., Hubbard Works, Monk's Risborough, Prince's Risborough, Aylesbury, Bucks. *Paper beakers for mixing resins etc.*

282 Bi-Pre-Pak Ltd., 222–224 West Road, Westcliff-on-Sea, Essex, SS0 9DF. *Transistors, photocell, and photoresistors.*

283 Bloore, G. H. Ltd., 480 Honeypot Lane, Stanmore, Middlesex. *Perspex.*

284 Bonds Ltd., 357 Euston Road, London, N.W.10. Tel: 01-387 5371. *Metal cut to length; all screws and fittings.*

285 British Cellophane Ltd., Bath Road, Bridgewater, Somerset. *Polythene and polypropylene film.*

286 Buck & Ryan Ltd., Tool Merchants, 101 Tottenham Court Road, London, W.1. Tel: 01-636 7475. *Most tools and machines.*

287 BXL Industrial Products Group, Manningtree, Essex. *Acetate sheet, Bexoid.*

288 Coolag Ltd., Charlestown Works, Glossop, Derbyshire. Tel: 3227. *Expanded polystyrene.*

289 Cooper, Charles, Ltd., 12 Hatton Wall, Hatton Garden, London, E.C.1. Tel: 01-405 5897. *Artifex elastic bonded abrasives.*

290 Cowell, E. W. Ltd., Sydney Road, Watford, Herts. Tel: 29664. *Sets of castings for bench drill up to $\frac{1}{2}$ in. Milling table, planer, power hacksaw.*

291 D.E.W. Ltd., Ringwood Road, Ferndown, Dorset. *Dewboxes — alkathene instrument boxes sold in 6 foot lengths, very versatile and cheap.*

292 Dexion Ltd., P.O. Box 7, Empire Way, Wembley Park, Middlesex, HA9 0JW. Tel: 01-903 7281. *Slotted angle iron and Dexion Speediframe.*

293 Elford Plastics Ltd., Quebec Street, Elland, Yorks. Tel: 3053. *Expanded polystyrene.*

294 English, H. W. Ltd., 469 Rayleigh Road, Hutton, Brentwood, Essex. *Lenses.*

295 General Woodwork Supplies Ltd., 78 Stoke Newington High Street, Stoke Newington, London, N.16. Tel: 01-254 6052. *Wide range of hardwoods cut to customer's size and mailed.*

296 Glazer Plastics Ltd., 156 High Road, Willesden Green, London, N.W.10. Tel: 01-459 0295. *Perspex, cut to size.*

297 Griffin & George Ltd., Ealing Road, Alperton, Wembley, HA0 1HJ. *Full range of plastics and small machines for the school workshop.*

298 Haith, John E., Park Street, Cleethorpes, Lincs. *Twilweld for cages.*

299 Hallam, Sleigh, and Chesterton Ltd., Widney Dorlec Division, Oldfield Road, Maidenhead, Bucks. *Miniature framing for instrument cases etc.*

300 H.B. Fillers Ltd., Newcastle-under-Lyme, Staffs. *Casting waxes and investment plaster.*

301 Henry's Radio Ltd., 303 Edgeware Road, London, W.2. Tel: 01-723 1008. *Wide range of electron components including electromagnetic counter ex-equipment.*

302 Home Radio Ltd., 187 London Road, Mitcham, Surrey. *Wide range of components; Japanese meters.*

303 Hughes and Holmes Ltd., 2 Snow Hill, Wolverhampton. *Screws and fittings.*

304 Ians Ltd., Cails Buildings, Quayside, Newcastle-upon-Tyne. *Acetate sheet.*

305 Imhof-Bedio Ltd., Ashley Works, Cowley Mill Road, Uxbridge, Middlesex. *'Imlock' case framing.*

306 ITT Components Group (Europe), (Standard Telephones & Cables Ltd. retail sales), Edinburgh Way, Harlow, Essex. *Thermistors and most electronic components.*

307 W. Kennedy Ltd., Station Works, West Drayton, Middlesex. *Low price power hacksaw.*

308 Kennion, Charles, Ltd., 2 Railway Place, Hertford, Herts. Tel: 2573. *Brass and steel in most sections and brass blanks for gear cut to size. Steel and phosphor bronze balls. Threaded black plastic balls for machine handles. Instrument maker's end mills from 0.035 in.*

309 Lakeland Plastics (Windermere) Ltd., 26 Alexandra Road, Windermere, Cumbria, LA27 2BZ. Tel: 2255. *Plastic bags and sheet.*

310 Maclanes Ltd., 40 North Street, Horsham, Sussex. *Aluminium tubing.*

311 Measurement Research Ltd., New Road, Horwich End, Whaley Bridge, Via Stockport. *Strain gauges.*

312 Meccano-Triang Ltd., Binns Road, Liverpool, L13 1DA.

313 Miller, F. W., 68 Nansen Road, Birmingham 11. *Bi metal strip.*

314 Mole, E. N. & Co. Ltd., 3 Tolpit Lane, Watford, Herts, WD1 8LU. Tel: 43135. *Lathes and all small engineering tools, new and second hand; lists.*

315 Moore, Harold & Son, Ltd., Bailey Works, Bailey Street, Sheffield 1. Tel: 27311. *Offcuts of nylon rod and block suitable for school machining in £5 parcels.*

316 Myford Ltd., Beeston, Nottingham, NG9 1ER. Tel: 254222. *Machine tools and lathes.*

317 Norman Rose (Elec) Ltd., Norman House, 8 St. Chads Place, London, WC1X 9HS. *Chassis boxes.*

318 Nylonic Engineering Co. Ltd., Woodcock Hill Industrial Estate, Hanefield Road, Rickmansworth, Herts, WD3 1PN. Tel: 76261. *Industrial machining plastics – technical and machining manuals.*

319 Prima Plastics Ltd., Platt's Eyot, Lower Sunbury Road, Hampton-on-Thames, Middlesex, TW12 2HF. Tel 01-979 1904. *Glassfibre, resin etc.*

320 Proops Bros. Ltd., 52 Tottenham Court Road, London, W.1. *Cheap and surplus lines.*

321 R.S. Components Ltd., (formerly Radiospares), P.O. Box 427, 13–17 Epworth Street, London, E.C.2. Tel: 01-253 1222. *Electronics components; 24 hour delivery.*

322 Rubert & Co. Ltd., Acru Works, Demmings Road, Cheadle, Cheshire, SK8 2PG. Tel: 061-428 5855. *Acrulite casting resin which can be machined.*

323 Smith, G. W. & Co. Ltd., 11–12 Paddington, London, W.2. *Small electric motors.*

324 J. M. Steel & Co. Ltd., Kingsway House, 18–24 Paradise Street, Richmond, Surrey. *Temperature indicating crayons.*

325 Strainstall Ltd., Hareleco House, Denmark Road, Cowes, Isle of Wight, PO31 7TB. *Strain gauges.*

326 Telcon Plastics Ltd., Farnborough Works, Green Street Green, Orpington, Kent, BR6 6BH. Tel: Farnborough 55685/8. *Low density polythene powder for plastic coating metal.*

327 Telegraph Construction and Maintenance Co. Ltd., Metal Division, Telcon Works, Manor Royal, Crawley, Surrey. *Bi metal strip.*

328 Transducers (C.E.L.) Ltd., 3 Trafford Road, Reading, Berks. *Strain gauges.*

329 Trylon Ltd., Thrift Street, Wollaston, Northants, NN9 7QJ. Tel: 275. *Polyester materials, glassfibre, and technical leaflets.*

330 Tyzacks Ltd., 341–345 Old Street, London, E.C.1. Tel: 01-739 8301. *Comprehensive stock of tools and machines for all trades.*

331 Vinatex Ltd., New Lane, Havant, Hants, PO9 2NQ. Tel: 6350. *Vinamold – remeltable rubber to make molds for plaster casting etc.*

332 Visijar Laboraties Ltd., Pegasus Road, Croydon Airport, Croydon, Surrey, CR0 4PR. *Perspex; rods, sheet, and blocks.*

333 West Hyde Developments Ltd., Ryefield Crescent, Northwood Hills, Northwood, Middlesex, HA6 1NN. *Instrument cases, diecast boxes, range of plug-in printed circuits; components.*

334 Whiston, K. R. Ltd., New Mills, Stockport, Cheshire. *Chassis punches, wide range of tools, materials, and surplus stock. Hammerite paint.*

335 Young, E. C., Homerton Bridge, London, E.9. *Timber at special prices for schools – ask for educational catalogue.*

336 Z & I Aero Services Ltd., 44a Westbourne Grove, London, W.2. *Huge range of valves, cathode ray tubes, and transistors.*

GREENHOUSES AND EQUIPMENT

337 Access Frames Ltd., Yelverton Road, Crick, Rugby, Warwick. *Low priced overhead mist and watering lines. Automatic watering controls. Garden frames. Propagators.*

338 Aerovap Ltd., Shernfold Park, Frant, Tunbridge Wells, Kent. *Automatic pest control.*

339 Alitex Ltd., St. John's Works, Station Road, Alton, Hants. *Well designed aluminium framed glasshouses. 7 ft 5 in. to 33 ft span, and multispan; reasonably priced; erection service.*

340 Alton Glasshouses Ltd., Bewdley, Worcs., DY12 2UJ. *6 ft to 10 ft span cedarwood houses — sliding door with no sill, particularly suitable for wheelchairs in special schools.*

341 Bayliss Precision Components Ltd., Compton, Ashbourne, Derbyshire, DE6 1DA. *Non-electric automatic ventilation.*

342 Bast Glazing Clips Ltd., Cambridge Road, Comberton, Cambridge. *Minibright greenhouse.*

343 Boil Home Garden Dept., Wright Rain Ltd., Crow Arch Road, Ringwood, Hants, BH24 1PA. Tel: 2251. *Automatic non-electric watering systems; trickle irrigation, capillary bench, and mist propagation equipment.*

344 Cambridge Glasshouse Co. Ltd., Comberton, Cambridge, CB3 7BY. *10 ft to 85 ft span 'Alumabright' houses; erection service.*

345 Cameron Irrigation Co. Ltd., Harwood Industrial Estate, Littlehampton, Sussex. Tel: 3985. *Automatic watering and 'Cameron' dilutor for continuous fertiliser application.*

346 Cedarworth Ltd., Donnington, Telford, Shropshire, TF2 7NF. *Ready glazed, cedarwood houses. 6 ft to 10 ft span.*

347 Crittal — Hope Ltd., Horticultural Department, Baintree, Essex. *Aluminium or glavanised steel glass houses 6–10 ft span. Frames.*

348 Edenlite (Greenhouses) Ltd., 29 Station Road, Stationhouse, Witney, Oxfordshire. *Aluminium houses up to 8 ft span.*

349 Ekco Heating and Appliances Ltd., Drury Lane, Hastings, Sussex. Tel: 429141. *Greenhouse tubular electrical heaters.*

350 Electricity Council, Trafalgar Buildings, 1 Charring Cross, London, S.W.1. Tel: 01-930 6757. *Free 88 page book — 'Electric Growing'.*

351 Evenproducts Ltd., Evesham, Worcs. *Evenshower overhead watering system.*

352 Findlay-Irvine Ltd., Bog Road, Penicuik, Edinburgh. Tel: 72111. *Fan heaters.*

353 Halls Cedarwood Ltd., Garden Buildings, Paddock Wood, Kent. *Self erection cedarwood houses; aviaries.*

354 Hartley (Clearspan) Ltd., Greenfield, Oldham, Lancs. *Superbly engineered aluminium glasshouses.*

355 Humex Ltd., 5 High Road, Byfleet, Weybridge, Surrey, KT14 7QF. *Automatic heating, ventilating, watering, propagating and shading equipment.*

356 K.D.G. Instruments Ltd., Fleming Way, Crawley, Sussex, BH10 2QE. Tel: 25151. *Thermostats.*

357 Lewden Metal Products Ltd., 5 Argall Avenue, London, E.10. Tel: 01-539 0233. *Electrical outlets.*

358 Loheat Ltd., Everland Road, Hungerford, Berks. *Low voltage soil warming equipment.*

359 Nethergreen Products, P.O. Box 3, Alderley Edge, Cheshire, SK9 7JJ. *Capillary and trickle systems.*

360 Robinsons of Winchester Ltd., Robinson Close, Winnall Manor Close, Winnall, Winchester, Hants. Tel: 61777. *Very low cost commercial houses in aluminium and timber.*

361 Simplex Ltd., Dairy Equipment Co. Ltd., Horticultural Division, Sawston, Cambridge, CB2 4LJ. *All types of automatic greenhouse equipment. Aspirated thermostats, soil sterilisers, and mercury-fluorescent lights.*

362 Solplan Ltd., Balena Close, Creekmor, Poole, Dorset, BH17 7DX. *High quality automatic control equipment for commercial size houses.*

363 Thermoforce Ltd., Wakefield Road, Cockermouth, Cumbria, CA13 0HS. *Non-electric automatic ventilation for the small greenhouse.*

364 Venner Ltd., Kingston By-Pass, New Maldon, Surrey. Tel: 01-992 2442. *Time clocks.*

365 George Ward (Morley) Ltd., Darlaston, Staffs. *Plastic plant pots and boxes of high quality.*

366 Woodman, E. J. & Sons Ltd., 19–25 High Street, Pinner, Middlesex. *Most gardening equipment.*

VISUAL AIDS AND MATERIALS

Models

367 Catalin Ltd., 54 Farm Hill Road, Waltham Abbey, Essex. *Chemical models.*

368 Crystal Structures, Ltd., Swaffam Road, Bottisham, Cambridge, CB5 9EA. *Molecular models; orbital models.*

369 Educational and Scientific Plastics Ltd., 76 Holmethorpe Avenue, Holmethorpe, Redhill, Surrey, RH1 2PR. *Anatomical models.*

370 General Dental Council, 37 Wimpole Street, London, W.1. Tel: 01-486 2171. *Model jaws, Flannelgraph of teeth, and film strips.*

371 Invicta Plastics Ltd., Oadby, Leicester, LE2 4LB. Tel: 0533 717211. *Viz-aid models.*

372 Medical Art and Science Co. Ltd., 13–15 Lancing Road, West Croydon, Surrey. *Anatomical models.*

373 Scientific Teaching Apparatus Ltd., Colquhoun House, 27–37 Broadwick Street, London, W.1. Tel: 01-437 9641. *Molecular models, wall charts.*

374 St. John's Ambulance Brigade, St. John's Gate, Clerkenwell, London, E.C.1. *Coloured anatomy charts.*

375 Taskamster Ltd., Morris Road, Leics. *Spheres for making molecular models.*

Wallcharts and pictures

376 Adam, Rouilly & Co. Ltd., 10 Winchester Road, Swiss Cottage, London, N.W.3. Tel: 01-636 2703. *Wide range of wall charts.*

377 Arnold, E. J. & Sons, Butterly Street, Leeds, LS10 1AX. *Wallcharts.*

378 British Museum (Natural History), Cromwell Road, South Kensington, London, S.W.7. Tel: 01-589 6323. *List of postcards available.*

379 British Safety Council, 62 Chancellors Road, London, W.6. *Safety charts.*

380 Geological Museum, Exhibition Road, South Kensington, London, S.W.7. Tel: 01-589 3444. *List of postcards available.*

381 Hulton Educational Productions Ltd., 55–59 Saffron Hill, London, E.C.1. *Wallcharts.*

382 Moore & Wright Ltd., Handsworth Road, Sheffield, Yorks, S13 9BR. *Micrometer wallcharts.*

383 Mullard Education Service, Mullard House, Torrington Place, London, W.C.1. Tel: 01-580 6633. *Wallcharts.*

384 Murray, John Ltd., 50 Albermarle Street, London, W.1. Tel: 01-493 4361. *Wallcharts.*

385 Philip, George & Son Ltd., 12–14 Long Acre, London, WC2E 9LP. *Large selection of charts.*

386 Philip & Tacey Ltd., North Way, Andover, Hants. Tel: Andover 61171. *Charts.*

387 Pictorial Charts Educational Trust, 132 Uxbridge Road, London, W13 8QU. *Wide range; subscriber system for schools.*

388 Royal Botanic Gardens, Kew, Surrey. *Plant photographs.*

389 Royal Society for the Prevention of Accidents, 6 Buckingham Place, London, S.W.1. *Safety charts.*

390 Home Safety Division, Royal Oak Centre, Purley, CR2 2UR.

391 Stamfords Ltd., 12 Long Acre, London, W.C.2. *Wall maps, land classification.*

392 Times Publishing Co. Ltd., 200 Grays Inn Road, London, W.C.1. Tel: 01-837 1234. *Charts.*

393 Warne, Frederick & Co. Ltd., 1–4 Bedford Court, London, WC2E 9JB. *Natural history charts.*

394 Zoological Society, Regents Park, London, N.W.1. Tel: 01-722 3333. *Fish wallcharts.*

Films, loops, and filmstrips

395 Aerofilms Ltd., 4 Albemarle Street, London, W.1. Tel: 01-493 5211. *Catalogue of aerial photographs and slides.*

396 Animal Wildlife Slides, The Slide Centre, Portman House, 17 Broderick Road, London, SW17 7DZ. *Over 1000 wildlife slides.*

397 Association of Agriculture, 78 Buckingham Gate, London, S.W.1. Tel: 01-222 6115. *List of sources of agricultural films.*

398 Banta Educational Supplies Ltd., 58 Hopton Road, London, S.W.16. Tel: 01-769 2203. *35 mm slides.*

399 Boulton-Hawker Films Ltd., Hadleigh, Ipswich, Suffolk, 1P7 5BG. *Biological films.*

400 British Film Institute, Scientific Film Library, 81 Dean Street, London, W1V 6AA. *Film catalogue. Wide range of science films.*

401 British Transport Films, Melbury House, Melbury Terrace, London, N.W.1. Tel: 01-262 3232. *Natural history films.*

402 Camera Talks Ltd., 31 North Row, London, W.1. Tel: 01-493 2761. *Filmstrip publishers.*

403 Central Council for Health Education, Tavistock House, North Tavistock Square, London, W.C.1. *Health films and filmstrips.*

404 Central Electricity Generating Board Film Library, Sudbury House, 15 Newgate Street, London, E.C.1. Tel: 01-248 1202. *Scientific and conservation films.*

405 Central Film Library, Government Buildings, Bromyard Avenue, London, W.3. Tel: 01-743 5555. *Catalogue of films.*

406 Central Film Library of Wales, 42 Park Place, Cardiff. *Catalogue of films.*

407 Central Office of Information, Hercules Road, Westminster Bridge Road, London, S.E.1. Tel: 01-928 2345. *Catalogue of films.*

408 Common Ground Ltd., 44 Fulham Road, London, S.W.3. *Film strip publishers.*

409 Disney, Walt, Productions Ltd., 16 mm Division, 68 Pall Mall, London, S.W.1. *Films on science, natural history, geology.*

410 Ealing Scientific Films Ltd., Greycaine Road, Bushey Mill Lane, Watford, Herts, WD2 4PW. *Large selection of 8 mm cassette films.*

411 Educational Film Bureau, 7 Walton Terrace, Aylesbury, Bucks. *Catalogue of films.*

412 Educational Foundation for Visual Aids, 33 Queen Anne Street, London, W.1. Tel: 01-636 5742. *Subject catalogue of films and filmstrips.*

413 Educational Products Ltd., East Ardsley, Wakefield, Yorks. *Comprehensive catalogue: film strip catalogue.*

414 Eothen Films Ltd., 103 Wardour Street, London, W.1. *Film loops in cassettes.*

415 Esso Petroleum Ltd., Victoria Street, London, S.W.1. Tel: 01-834 6677. *Nuffield project films on teaching technique.*

416 Educational Library, Foundation Film Library, Brooklands House, Weybridge, Surrey.

417 Gateway Educational Films, 470 Green Lanes, London, N.13. Tel: 01-882 0177. *Educational films, loop films, 8 mm films, and filmstrips.*

418 Gaumont British Film Library, 1 Aintree Road, Perivale, Middlesex. *Catalogue of hire films.*

419 Halas & Bachelor Cartoon Films Ltd., 3–7 Kean Street, London, W.C.2. *Animated loopfilms on biological topics.*

420 Imperial Chemical Industries Ltd., Film Library, Thames House North, Millbank, London, S.W.1. *Film loan.*

421 International Scientific Film Library, 31 Rue Vartier, Brussels 4, Belgium. *World scientific films.*

422 Kodak Lecture Service, Kodak Ltd., P.O. Box 114, London, N.20. *Filmstrips on photography.*

423 London Natural History Society, 40 Frinton Road, Kirby Cross, Frinton-on-Sea, Essex. *Ornithology films.*

424 McGraw-Hill Publications, Shoppenhangers Road, Maidenhead, Berks, SL6 5BR. *Filmstrips, including excellent animal taxonomy series.*

425 Macmillan Loops, Globe Books, Brunel Road, Houndmills, Basingstoke, Hants. *Loopfilms.*

426 Mullard Film Library, 85–129 Oundle Road, Peterborough, PE2 9PY. *Films on electronics and physics.*

427 National Audio-Visual Aids Library, 2 Paxton Place, Gipsy Road, London, S.E.27. Tel: 01-670 4247. *Films and filmstrips for hire and sale.*

428 National Society for Clean Air, 134–137 North Street, Brighton, BN1 1RG. *Pollution films.*

429 Petroleum Films Bureau, 4 Brook Street, London, W1Y 2AY. *Wide range of films and filmstrips.*

430 Pfiser Ltd., Film Library, 108 West Terry Road, London, E.14. *Films on biology; mould.*

431 Pitman, Sir Isaac Ltd., Pitman House, 39 Parker Street, London, W.C.2. *Filmstrips.*

432 Plymouth Films Ltd., 4 Borrington Villas, Plympton, Plymouth, Devon. *Natural history films.*

433 Random Film Library, 25 The Burroughs, Hendon, London, N.W.4. Tel: 01-202 4177. *Free catalogue – distributes for various organisations.*

434 Rank Film Library, P.O. Box 70, Great West Road, Brentford, Middlesex, TW8 9HR. *Distribute wide range of films for various sponsors.*

435 Marion Ray Ltd., 36 Villiers Avenue, Surbiton, Surrey. *Filmstrip and slide producer.*

436 Ronan Picture Library, Ballards Place, Cowlings, Newmarket, Suffolk. *Scientific and technological pictures.*

437 Royal Institute of Chemistry, 30 Russell Square, London, W.C.1. Tel: 01-580 3482. *Chemical filmstrips; list of sources of chemistry films.*

438 Royal Microscopical Society, Film Library, Imperial Cancer Research Laboratories, Buttonhole Lane, London, N.W.7. *Microscopy and metallurgy.*

439 Royal Society for the Protection of Birds, The Lodge, Sandy, Beds. *Films on most aspects of bird life.*

440 Sound Services Ltd., 27 Charles Street, Cardiff. *Distribute films for numerous sponsors.*

441 Taylor, E. H. Ltd., Beehive Works, Welwyn Garden City, Herts. *Film on beekeeping.*

442 Town & Country Productions Ltd., 21 Cheyne Road, London, S.W.3. Tel: 01-352 7950. *Film hire.*

443 Unilever Film Library, Unilever House, Blackfriars, London, E.C.4. Tel: 01-353 7474. *Science films.*

444 Visual Publications Ltd., 197 Kensington High Street London, W.8. Tel: 01-937 1568. *Filmstrips – free catalogue.*

445 Warner-Pathe Distributors Ltd., 135 Wardour Street, London, W.1. *Films for hire.*

446 Wellcome Film Library, The Wellcome Building, 183 Euston Road, London, W.1. *Research films.*

447 Diana Wyllie Ltd., 3 Park Road, Baker Street, London, N.W.1. *Filmstrips slides and slide storage systems.*

Materials and equipment

448 Block & Anderson Ltd., Banda House, Cambridge Grove, Hammersmith, London, W.6. Tel: 01-748 4121. *Banda duplicators.*

449 Dymo Ltd., Pier Road, North Feltham Trading Estate, Feltham, Middlesex. Tel: 01-890 0961. *Tapewriters.*

450 E.M.I. Ltd., Electronics Ltd., Broadcast Equipment Division, Central Publicity Department, 135 Blythe Road, Hayes, Middlesex. *Closed circuit television.*

451 Hurst, A. & Co. Ltd., Kirkham House, 54A Tottenham Court Road, London, W1P 0DN. *Transparent adhesive film for covering charts etc.*

452 Kodak Ltd., Education Service, Victoria Road, Ruislip, Middlesex.

453 Letraset Ltd., St. Georges House, 195–203 Waterloo Road, London, S.E.1. *Instant rub-down lettering and shading.*

454 Mapograph Co. Ltd., 440 Chiswick High Street, London, W.4. Tel: 01-994 5635. *Rubber rollers to print maps and science outlines.*

455 Matthews Drew & Shellborn, Visual Aid Centre, Holborn, London, E.C.1. Tel: 01-242 6631. *Comprehensive list of materials for the production of visual aids.*

456 Offrex Ltd., Offrex House, Stephen Street, London, W.1. *Overhead projectors, Fordigraph duplicator systems, and loopfilm projectors.*

457 Pye HDT Ltd., St. Andrews Road, Cambridge. *Close circuit T.V.*

458 Rank Audio-Visual, Ltd., Great West Road, Brentford. Tel: 01-568 9222. *All types of projectors.*

459 Sundeala Board Co. Ltd., Columbia House, 3 Aldwych, London, W.C.2. Tel: 01-240 3022. *Pin boarding.*

460 S.W. Optical Instruments Ltd., Hooper's Pool, Southwick, Trowbridge, Wilts., BA14 9NQ. Tel: Trowbridge 3562 and 3194. *Photomicrograph prints.*

461 Transart Visual Aids Ltd., East Chadley Lane, Godmanchester, Hunts., PE18 8AU. *Fliptran O.H.P. transparency system.*

EDUCATIONAL PUBLISHERS

462 Addison-Wesley Publishing Co. Ltd., West End House, 11 Hills Place, London, W1R 2LR. Tel: 01-734 8817.

463 Allen (George) & Unwin Ltd., Park Lane, Hemel Hempstead, Herts, HP2 4PE. Tel: Hemel Hempstead 3244. *And* 40 Museum Street, London, W.C.1. Tel: 01-405 8577.

464 Allen (W. H.) & Co. Ltd., 43 Essex Street, London, WC2R 3JG. Tel: 01-5800127.

465 Angus & Robertson (UK) Ltd., 2 Fisher Street, London, WC1R 4QA. Tel: 01-405 9547.

466 Arnold (Edward) (Publishers) Ltd., 25 Hill Street, London, W.1.

467 Batsford (B. T.) Ltd., 4 Fitzhardinge Street, London, W1H 0AH. Tel: 01-935 0537.

468 Arnold (E. J.) & Son Ltd., Butterley Street, Hunslet Lane, Leeds, LS10 1AX. Tel: 0532-35541.

469 Bell (G.) & Sons Ltd., York House, Portugal Street, London, WC2A 2HL. Tel: 01-405 0805.

470 Benn (Ernest) Ltd., 25 New Street Square, London, EC4A 3JA. Tel: 01-353 3212.

471 Black (A & C) Ltd., 4, 5, & 6 Soho Square, London, W1V 6AD. Tel: 01-734 0845.

472 Blackie & Son Ltd., Bishopbriggs, Glasgow, G64 2NZ. Tel: 041-772 1046.

473 Blackwell (Basil) & Mott Ltd., 5 Alfred Street, Oxford, OX1 4HB. Tel: 0865 22164.

474 Blandford Press (The), 167 High Holborn, London, WC1V 6PH. Tel: 01-836 9551.

475 Blond Educational Ltd., Iliffe House, Iliffe Avenue, Oadby, Leicester, LE2 4ZB. Tel: 053722-4408. *And* 120 Golden Lane, London, EC1Y 0TU. Tel: 01-253 0855.

476 Books for Schools Ltd., Longman House, Burnt Mill, Harlow, Essex. Tel: Harlow 26721.

477 British Museum, Great Russel Street, Bloomsbury, London, WC1B 3DG. Tel: 01-636 1555.

478 Brockhampton Press (The) Ltd., Arlen House, Salisbury Road, Leicester, LE1 7QS. Tel: 0533-21091.

479 Burke Publishing Co. Ltd., 14 John Street, London, WC1N 2EJ. Tel: 01-242 6724/5.

480 Cambridge University Press, Bentley House, 200 Euston Road, London, N.W.1. Tel: 01-387 5030.

481 Cape (Jonathan) Ltd., 30 Bedford Square, London, WC1B 3EL. Tel: 01-636 5764.

482 Cassell & Co. Ltd., 35 Red Lion Square, London, WC1R 4SJ. Tel: 01-242 6281.

483 Chamber (W. & R.) Ltd., 11 Thistle Street, Edinburgh, EH2 1DG. Tel: 031-225 4463. *And* 6 Dean Street, London, W1V 6LD. Tel: 01-437 1709.

484 Chapman (Geoffrey) Ltd., Educational Division, C.C.M. Ltd., Blue Star House, Highgate Hill, London, N.19. Tel: 01-272 7531.

485 Charles & Son Ltd., 55 West Regent Street, Glasgow, G2 2DU. Tel: 041-332 7161.

486 Chatto and Windus Educational Ltd., Frogmore, St. Albans, Hertfordshire, AL2 2NF. Tel: St. Albans 59101.

487 Church Information Office, Church House, Dean's Yard, London, SW1P 3NZ. Tel: 01-222 9011.

488 Collet's Holdings Ltd., Denington Estate, Wellingborough, Northamptonshire, NN8 2QT. Tel: 09-333 4351.

489 Collins (William), Sons & Co. Ltd., 144 Cathedral Street, Glasgow, G4 0ND. Tel: 041-552 4488. *And* 14 St. James's Place, London, S.W.1. Tel: 01-493 5321.

490 Concordia Publishing House Ltd., 117–123 Golden Lane, London, EC1Y 0TL. Tel: 01-253 8884.

491 Crowell Collier Macmillan Publishers Ltd., Blue Star House, Highgate Hill, London, N.10. Tel: 01-272 7531.

492 Darton Longman & Todd Ltd., 85 Gloucester Road, London, SW7 4SU. Tel: 01-370 5031.

493 Davis & Moughton Ltd., Ludgate House, 23 Waterloo Place, Leamington Spa, Warwickshire. Tel: 0926 24003.

494 Dent (J. M.) & Sons Ltd., 10/13 Bedford Street, London, WC2E 9HG. Tel: 01-836 6311.

495 Dobson Books Ltd., 80 Kensington Church Street, London, W8 4BZ. Tel: 01-229 6022.

496 Dryad Ltd., Northgates, Leicester, LE1 4QR. Tel: Leicester 50405.

497 Gerald Duckworth & Co. Ltd., 43 Gloucester Crescent, London, NW1 7DY. Tel: 01-485 3484.

498 Educational Productions Ltd., 26–29 Maunsel Street, London, SW1P 2QS. Tel: 01-834 1067. ·

499 Educational Explorers Ltd., 40 Silver Street, Reading, Berks, RG1 2SU. Tel: 0734 83103.

500 Educational Supply Association Ltd., ESA-Creative Learning Ltd., Pinnacles, P.O. Box 22, Harlow, Essex, CM19 5AY. Tel: Harlow 21131.

501 Encyclopaedia Britannica International Ltd., Mappin House, 156–163 Oxford Street, London, W1N 9DL. Tel: 01-637 3371.

502 English Universities Press Ltd., Saint Paul's House, 8 Warwick Lane, London, EC4P 4AH. Tel: 01-248 5797.

503 European Schoolbooks Ltd., 100 Great Russell Street, London, WC1B 3LE. Tel: 01-636 4724.

504 Evans Brothers Ltd., Montague House, Russell Square, London, WC1B 5BX. Tel: 01-636 4724.

505 Faber & Faber Ltd., 3 Queen Square, London, WC1N 3AU. Tel: 01-278 6881.

506 Field Enterprises Educational Corporation, Canterbury House, Sydenham Road, Croydon, Surrey, CR9 2LR. Tel: 01-686 6421.

507 Forbes Publications Ltd., Hartree House, Queensway, London, W2 4SH. Tel: 01-229 9322.

508 Foulsham (W.) & Co. Ltd., Yeovil Road, Slough, SL1 4JH. Tel: 01-75 26769 & 30956.

509 Ginn & Co. Ltd., 18 Bedford Row, London, WC1R 4EJ. Tel: 01-405 8823.

510 Glasgow (Mary) Publications Ltd., 140–142 Kensington Church Street, London, W.8. Tel: 01-229 9531.

511 Hachette, 4 Regents Place, Regent Street, London, W1R 6BH. Tel: 01-734 5259.

512 Hamlyn Group Ltd., School Library Service, Hamlyn House (5th Floor), 42 The Centre, Feltham, Middlesex, Tel: 01-751 8400.

513 Harrap (George G.) & Co. Ltd., P.O. Box 70, 182–184 High Holborn, London, WC1V 7AX. Tel: 01-405 9935.

514 Hart-Davis (Rupert) Educational Books Ltd., 3 Upper James Street, Golden Square, London, W1R 4BP. Tel: 01-734 8080.

515 Heath (D. C.) (Europe) Ltd., 1 Westmead, Farnborough, Hampshire, Tel: 0252 41196.

516 Heinemann Educational Books Ltd., 48 Charles Street, London, W1X 8AH. Tel: 01-493 9103.

517 Holmes McDougall Ltd., 30 Royal Terrace, Edinburgh, EH7 5FL. Tel: 031-556 1431/3.

518 Hope (Thomas) & Hudson (Sankey) Ltd., Ashtons Mill, Chapleton Street, Manchester, M1 2NH.

519 Hulton Educational Publications Ltd., Raans Road, Amersham, Bucks. Tel: 02403 4196.

520 Hutchinson Educational Ltd., 3 Fitzroy Square, London, W1P 6JD. Tel: 01-387 2888.

521 Initial Teaching Publishing Co. Ltd., 9 Southampton Place, London, WC1A 2EA. Tel: 01-405 9791.

522 International Textbook Co. Ltd., 158 Buckingham Palace Road, London, S.W.1. Tel: 01-730 7216.

523 Jackdaw Publications Ltd., 24 Tottenham Court Road, London, W.1.

524 Longman Group Ltd., Longman House, Burnt Mill, Harlow, Essex. Tel: 0279 26721. *And* Bentinck Street, London, W1X 0AS. Tel: 01-499 7911.

525 Logos Publications Ltd., 54–58 Caledonian Road, London, N1 9RN. Tel: 01-278 6552.

526 Lund Humphries (Percy) & Co. Ltd., 12 Bedford Square, London, WC1B 3JB. Tel: 01-636 7676.

527 Lutterworth Press, Luke House, Farnham Road, Guildford, Surrey. Tel: 0483 77536 (5 lines).

528 Macdonald & Co. (Publishers) Ltd., St. Giles House, 49–50 Poland Street, London, W1A 2L9. Tel: 01-437 0686.

529 McGraw-Hill Publishing Co. Ltd., McGraw-Hill House, Shoppenhangers Road, Maidenhead, Berkshire, SL6 2QL. Tel: 0628 23432.

530 Macmillan Education Ltd., Houndmills, Basingstoke, Hants. Tel: 0256 29242. *And* 4 Little Essex Street, London, WC2R 3LF. Tel: 01-836 6633.

531 Methuen Educational Ltd., 11 New Fetter Lane, London, EC4 P4EE. Tel: 01-583 9855.

532 Mills & Boon Ltd., 17–19 Foley Street, London, W1A 1DR. Tel: 01-580 9074.

533 Muller (Frederick) Ltd., Ludgate House, 107 Fleet Street, London, E.C.4. Tel: 01-353 1040.

534 Murray (John) (Publishers) Ltd., 50 Albermarle Street, London, W1X 4BD. Tel: 01-493 4361.

535 National Christian Education Council and Denholm House Press, Robert Denholm House, Nutfield, Redhill, Surrey, RH1 4HW. Tel: Nutfield Ridge 2411.

536 National Magazine Co. Ltd., Chestergate House, Vauxhall Bridge Road, London, SW1V 1HF. Tel: 01-834 2331.

537 Nelson (Thomas) & Sons Ltd., 36 Park Street, London, W1Y 4DE. Tel: 01-493 8351.

538 Nisbet (James) & Co. Ltd., Digswell Place, Welwyn, Herts. Tel: 07073 25491.

539 Novello & Co. Ltd., Borough Green, Sevenoaks, Kent. Tel: Borough Green 3261.

540 Oliver & Boyd Ltd., Tweeddale Court, 14 High Street, Edinburgh, EH1 1YL. Tel: 031-556 4622.

541 Oxford University Press, 37 Dover Street, London, W1X 8AH. Tel: 01-629 8494.

542 Penguin Books Ltd., Bath Road, Harmondsworth, Middlesex. Tel: 01-759 1984.

543 Pergamon Press Ltd., Headington Hill Hall, Oxford, OX3 0BW. Tel: 0865 64881.

544 Philip (George) & Son Ltd., 12/14 Long Acre, London, WC2E 9LP. Tel: 01-836 7863.

545 Philograph Publications Ltd., North Way, Andover, Hants. Tel: 0264 61171.

546 Pitman (Sir Isaac) & Sons Ltd., 39 Parker Street, London, WC2B 5PB. Tel: 242 1655.

547 Routledge & Kegan Paul Ltd., 68–74 Carter Lane, London, EC4V 5EL. Tel: 01-248 4821.

548 Schofield & Sims Ltd., 35 St. John's Road, Huddersfield, HD1 5DT. Tel: 0484 30684.

549 Science Research Associates Ltd., Reading Road, Henley-on-Thames, Oxfordshire, RG9 1EW. Tel: Henley 2435.

550 The Society for Promoting Christian Knowledge, Holy Trinity Church, Marylebone Road, London, NW1 4DU. Tel: 01-387 5282.

551 Stillit Books Ltd., 72 New Bond Street, London, W1Y 0QY. Tel: 01-493 1177.

552 Transworld Publishers Ltd., Cavendish House, 57/59 Uxbridge Road, Ealing, London, W5 5SA. Tel: 01-579 2652.

553 University of London Press Ltd., Saint Paul's House, 8 Warwick Lane, London, EC4P 4AH. Tel: 01-253 6992.

554 Vallentine Mitchell & Co. Ltd., 67 Great Russell Street, London, WC1B 3BT. Tel: 253 1816 (trade counter).

555 Van Nostrand Reinhold Co. Ltd., 25/28 Buckingham Gate London, SW1E 6LQ. Tel: 01-834 3726.

556 Ward Lock Educational Co. Ltd., Warwick House, 116 Baker Street, London, W.1. Tel: 01-486 3271.

557 Warne (Frederick) & Co. Ltd., 40 Bedford Square, London, WC1B 3HE. Tel: 01-580 9622.

558 Watts (Franklin) Ltd., 18 Grosvenor Street, London, W1X 9FD. Tel: 01-499 7322.

559 Wills & Hepworth Ltd., Derby Square, Loughborough, OE11 0AL. Tel: 05093 4786.

Appendix – Organic Nomenclature

Every attempt has been made in the preparation of this manual to comply with the recommendations of the Working Party on Chemical Nomenclature of the Education (Research) Committee of the Association of Science Education. The Working Party attempted a rationalisation of the Nomenclature Rules of the Commissions of the International Union of Pure and Applied Chemistry ('I.U.P.A.C. rules') and made recommendations on how these rules should be interpreted and applied at school level. Among the difficulties mentioned in the report, which forced the inevitable compromise, are:

1. '... the I.U.P.A.C. rules are, for the most part, drawn up to enable research workers to name new and complicated compounds systematically, and so contain many refinements whose scope and purpose are irrelevant at school level.'
2. 'different contexts may require the use of different names ...'
3. 'well established practice cannot be changed overnight, and an acquaintance with traditional trivial names is still necessary.'

In addition, some other practical considerations which arise in the context of this laboratory manual are:

1. The permanent labels which exist on many bottles in laboratories.
2. Chemicals and solutions may well be handled by unqualified or inexperienced staff or helpers who are not completely familiar with the systematic names.
3. The names which will be used when reordering supplies of chemicals, and hence the names used in manufacturers' current catalogues, will not be systematic names.
4. Where the chemistry of the material is not under discussion it may well be preferable to use a trivial name for relatively complex materials. For example, titles such as aniline dye, and citrate buffer may with advantage be retained. The wide range of complex materials used in biochemical solutions in particular need short titles, and here there is every reason why the trivial name should be retained.

Bearing all these points in mind, organic nomenclature in this manual has been guided by the following principles:

1. The names used should be sufficiently short and simple for satisfactory general use in industrial, commercial, and school laboratories, especially for common substances, while still maintaining as much information of the structure as possible.
2. The older name should wherever possible be used alongside the systematic name to avoid possible confusion. Occasionally the trivial name is considerably shorter or is the name used by all suppliers, and so the trivial name is retained.

The following tables list the common name and the systematic name of some frequently encountered organic substances. Much fuller lists, together with a discussion of the whole question of chemical nomenclature at school level, are contained in the booklet *Chemical Nomenclature, Symbols and Terminology* (1972) available from The Association for Science Education, College Lane, Hatfield, Herts.

Appendix – Organic Nomenclature

Common current name	Recommended (systematic) name
acetaldehyde	ethanal
acetamide	ethanamide
acetates	ethanoates
acetic acid	ethanoic acid
acetic anhydride	ethanoic anhydride
acetone	propanone
acetyl chloride	ethanoyl chloride
acetylene	ethyne
acrylic acid	propenoic acid
alanine	2-aminopropanoic acid
alcohol (ethyl)	ethanol
allyl alcohol	prop-2-en-1-ol
n-amyl	pentyl
aniline	phenylamine
benzaldehyde	benzenecarbaldehyde
benzoic acid	benzenecarboxylic acid
benzylamine	(phenylmethyl)amine
bromoform	tribromomethane
n-butyl alcohol	butan-1-ol
carbon tetrachloride	tetrachloromethane
catechol	benzene-1,2-diol
chloral hydrate	2,2,2-trichloroethanediol
chloroform	trichloromethane
citric acid	2-hydroxypropane-1,2,3-tricarboxylic acid
o-cresol	2-methylphenol
diethyl ether	ethoxyethane
ethyl alcohol	ethanol
ethylene	ethene
ethylene glycol	ethane-1,2-diol
formaldehyde	methanal
formic acid	methanoic acid
gallic acid	3,4,5-trihydroxybenzenecarboxylic acid
glycerol	propane-1,2,3-triol
glycine	aminoethanoic acid
lactic acid	2-hydroxypropanoic acid
malic acid	2-hydroxybutanedioic acid
methyl alcohol	methanol
dimethyl ether	methoxymethane
methyl ethyl ketone	butanone
oxalic acid	ethanedioic acid
phosgene	carbonyl chloride
phthalic acid	benzene-1,2-dicarboxylic acid
picric acid	2,4,6-trinitrophenol
polyvinyl chloride	poly(chloroethane)
pyruvic acid	2-oxopropanoic acid
resorcinol	benzene-1,3-diol
salicylic acid	2-hydroxybenzenecarboxylic acid
succinic acid	butanedioic acid
sulphanilic acid	4-aminobenzenesulphonic acid
l-tartaric acid	(-)2,3-dihydroxybutanedioic acid
toluene	methylbenzene
o-toluidine	2-methylphenylamine
vinyl chloride	chloroethene
o-xylene	1,2-dimethylbenzene

Recommended (systematic) name	Common current name
4-aminobenzenesulphonic acid	sulphanilic acid
aminoethanoic acid	glycine
2-aminopropanoic acid	alanine
benzenecarbaldehyde	benzaldehyde
benzenecarboxylic acid	benzoic acid
benzene-1,2-dicarboxylic acid	phthalic acid
benzene-1,2-diol	catechol
benzene-1,3-diol	resorcinol
tribromomethane	bromoform
butanedioic acid	succinic acid
butan-1-ol	n-butyl alcohol
butanone	methyl ethyl ketone
carbonyl chloride	phosgene
2,2,2-trichloroethanediol	chloral hydrate
chloroethene	vinyl chloride
trichloromethane	chloroform
tetrachloromethane	carbon tetrachloride
ethanal	acetaldehyde
ethanamide	acetamide
ethanedioic acid	oxalic acid
ethane-1,2-diol	ethylene glycol
ethanoates	acetates
ethanoic acid	acetic acid
ethanoic anhydride	acetic anhydride
ethanol	ethyl alcohol
ethanoyl chloride	acetyl chloride
ethene	ethylene
ethoxyethane	diethyl ether
ethyne	acetylene
2-hydroxybenzenecarboxylic acid	salicylic acid
3,4,5-trihydroxybenzenecarboxylic acid	gallic acid
(-)2,3-dihydroxybutanedioic acid	l-tartaric acid
2-hydroxybutanedioic acid	malic acid
2-hydroxypropane-1,2,3-tricarboxylic acid	citric acid
2-hydroxypropanoic acid	lactic acid
methanal	formaldehyde
methanoic acid	formic acid
methanol	methyl alcohol
methoxymethane	dimethyl ether
methylbenzene	toluene
1,2-dimethylbenzene	o-xylene
2-methylphenol	o-cresol
2-methylphenylamine	o-toluidine
2,4,6-trinitrophenol	picric acid
2-oxopropanoic acid	pyruvic acid
pentyl-	n-amyl-
phenylamine	aniline
(phenylmethyl)amine	benzylamine
poly(chloroethene)	polyvinyl chloride
propane-1,2,3-triol	glycerol
propanone	acetone
propenoic acid	acrylic acid
prop-2-en-1-ol	allyl alcohol

Appendix – Mathematical Symbols

$\gg$	much greater than	$+$	positive, plus or add
$>$	greater than	$\mp$	negative or positive, minus or plus
$\equiv$	identical with		
$\simeq$	approximately equal	$\pm$	positive or negative, plus or minus
$\neq$	not equal to		
Σ	sum of all the terms specified	$\div$ or :	divided
		$=$ or $::$	equal to
∞	infinity	antilog	antilogarithm
$\ll$	much less than	$\sin^{-1}$ or	inverse sine (and sim-
$<$	less than	arc sin	ilarly for other functions)
$\not<$	not less than		
$\|n\|$	absolute value of n	e	base of natural log- arithm
$\perp$	perpendicular		
$\angle$	angle	$\because$	because
$\propto$	varies as, or propor- tional to	$\geqslant$	equal to or less than
		$\leqslant$	equal to or greater than
$\log x$	logarithm of x to base 10	$\sqrt{\ }$	square root
		$\sqrt[n]{\ }$	nth root
$\log_e x$	logarithm of x to base e	j	operator 90°
$\therefore$	therefore	h or a	operator 120°
$\triangle$	increment (large or small)	$\int$	integration
		sin	sine
δ	small increment	cos	cosine
$\|\|$	parallel	tan	tangent
$-$	negative, minus, or sub- tract	sec	secant
		cosec	cosecant
$\times$ or $\cdot$	multiplied by	cot	cotangent
f(x) or F(x)	function of x	sinh	hyperbolic sine
		cosh	hyperbolic cosine
		tanh	hyperbolic tangent

General Index

Attention is drawn to the separate Index of Suppliers

Index of Suppliers

The list of suppliers' addresses appears on pages 233–45; the numbers given in this index refer to those given to each supplier in the list.